shire

Measurement and Instrumentation Systems

Books

Measurement and Instrumentation Systems

W. Bolton

An Imprint of Butterworth-Heinemann

Newnes
An imprint of Butterworth-Heinemann
Linacre House, Jordan Hill, Oxford OX2 8DP
A division of Reed Educational and Professional Publishing Ltd

 A member of the Reed Elsevier plc group

OXFORD BOSTON JOHANNESBURG
MELBOURNE NEW DELHI SINGAPORE

First published 1996

British Library Cataloguing in Publication Data
A catalogue record for this book is available from the British Library

ISBN 0 7506 3114 7

Printed and bound in Great Britain by
Hartnolls Limited, Bodmin, Cornwall

Contents

Preface

This book is aimed at all those students who require a basic, but comprehensive and accessible, introductory text on measurement and instrumentation systems. It is seen as being of relevance to first-year undergraduates and HNC/HND students in mechanical engineering, electrical/electronic engineering, chemical engineering, instrumentation and control and applied physics.

The text is in three main sections:

I   Chapters 1 to 4 cover the basic concepts and terms associated with measurement and instrumentation systems in general, including the determination of the accuracy of measurements.

II   Chapters 5 to 9 cover the basic elements of instrumentation systems: sensors, signal conditioning and processing, data acquisition and processing systems and data presentation.

III Chapters 10 to 13 discuss the systems suitable for common measurements: force, torque, pressure, strain, displacement, velocity, acceleration, flow and temperature.

In order to give some mathematical backup to the text, Appendices are included which deal with distributions, differential equations, the Laplace transform, the Fourier series and sampled data systems. Throughout the text, references are included for more in-depth study and an Appendix gives some general references. Problems are included with each chapter and answers to all are given.

W. Bolton

1 Measurement systems

This chapter is a consideration of the basic principles associated with measurement and measurement systems. *Measurement* is the operation of determining the value of a quantity. A *measurement/instrumentation system* is the means used to carry out a measurement. Chapters 2, 3 and 4 expand on these basic principles to give, in these four chapters, a consideration of the principles associated with measurement systems in general.

1.1 Measurement systems

The purpose of a *measurement system* is to give the user a numerical value corresponding to the variable being measured. Thus a thermometer may be used to give a numerical value for the temperature of a liquid. We must, however, recognise that, for a variety of reasons, this numerical value may not actually be the true value of the variable. Thus, in the case of the thermometer there may be errors due to the limited accuracy in the scale calibration or reading errors due to the reading falling between two scale markings. We thus consider a measurement system to be a system which has an input of the true value of the variable being measured and an output of the measured value of that variable (Figure 1.1).

Figure 1.1 *A measurement system*

1.1.1 Functional elements of measurement systems

In general, a measurement system can be considered to consist of several elements, each of which is used to carry out a particular function. These functional elements are:

1   *Sensor*
    This is the element of the system which is effectively in contact with the process for which a variable is being measured and gives an output which depends in some way on the value of the variable. Sensors take information about the variable being measured and change it into a form which enables the rest of the measurement system to give a value to it. For example, a thermocouple is a sensor which has an input of temperature and an output of a small e.m.f. (Figure 1.2). Another

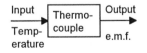

Figure 1.2 *The thermo-couple sensor*

example is a resistance thermometer which has an input of temperature and an output of a resistance change.

2   *Signal conditioner*
The *signal conditioner* puts the output from the sensor into a suitable condition for processing so that it can be displayed or handled by a control system. In the case of the thermocouple the signal conditioner may be an amplifier to make the e.m.f. big enough, with processing then being used to transform the voltage into a temperature reading on a meter (Figure 1.3(a)). In the case of the resistance thermometer there might be a Wheatstone bridge which transforms the resistance change into a voltage change and then an amplifier to make the voltage big enough for display (Figure 1.3(b)), with processing then being used to covert the voltage into a reading on a meter of temperature. Examples of signal conditioners are Wheatstone bridges which convert resistance changes into voltage changes, amplifiers which are used to make signals bigger, and oscillators which convert an impedance change into a variable frequency.

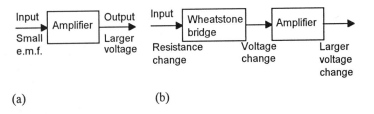

(a)   (b)

Figure 1.3   *Examples of signal conditioning*

3   *Signal processor*
The *signal processor* processes the signal so that it is suitable for display or onward transmission in some control system. This may mean taking into account non-linearity before determining the result or combining signals from more than one sensor or a sensor over a period of time. Examples of signal processing elements are analogue-to-digital converters which put analogue signals into digital ones for perhaps handling by a computer, and filters which eliminate noise from a signal.

4   *Data presentation*
This presents the measured value in a form which enables an observer to recognise it (Figure 1.4). This may be via a display, e.g. a pointer moving across the scale of a meter or perhaps information on a visual display unit (VDU). Alternatively, or additionally, the signal may be recorded, e.g. on the paper of a chart recorder or perhaps on magnetic disc. Alternatively, or additionally, the output from signal processing may be transmitted into other elements in a control system so that it can be compared with a required value and action initiated to control the variable (see Section 1.2).

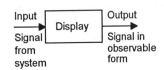

Figure 1.4 *An example of a data presentation element*

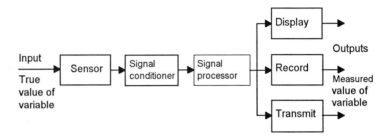

Figure 1.5 *Measurement system*

Figure 1.5 shows how these basic functional elements form a measurement system.

As an illustration of how a measurement system can be broken down into its functional elements, consider a load cell that is used to determine the weight of a container. The load cell is a cylinder on which strain gauges are mounted, the weight that is being measured compressing the cylinder and so changing the resistance of the strain gauges. The strain gauges are connected into a Wheatstone bridge and so the resistance change is transformed into a voltage change, the out-of-balance voltage of the bridge. This voltage will be small, a few millivolts, and so it is amplified. Since the output is to be displayed on a visual display unit screen, the amplified voltage is converted into a digital signal by an analogue-to-digital converter, perhaps also being processed by a microprocessor in order to correct the output for non-linearity or to carry out some calculations on the data before display. Figure 1.6 shows the system and how the signal changes as it progresses through the system.

As another example, consider a thermocouple used for the measurement of temperature: the sensor is the thermocouple. The output from the thermocouple is conditioned by a circuit to give compensation for the cold junction not being at 0°C and an amplifier to make the signal bigger. The signal is then transmitted to a 0 to 10 V meter for display. Figure 1.7 shows the system.

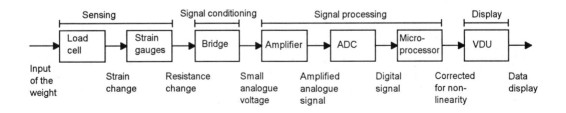

Figure 1.6 *Weight measurement system*

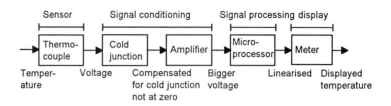

Figure 1.7 *Temperature measurement system*

The term *transducer* is commonly used for an element which transforms input signals from some form into an equivalent electrical form, though some define it as being from just from one form to another. This term may thus be used for many forms of sensor but also signal conditioning elements.

1.1.2 Information transporting systems

Measurement systems can also be regarded as information transporting systems (Figure 1.8), taking information about some variable, processing it and then distributing it to where it is required and in the form required. The system can be considered to have three basic elements, namely:

1 *Data acquisition*
   Information is acquired from the object about some variable and converted into data. This involves the sensor and signal conditioning.

2 *Data processing*
   This is concerned with processing the signal carrying the data into a suitable form and condition to make the information available. This could involve filtering to reduce noise, amplification to make the signal bigger, combining signals from more than one sensor, etc.

3 *Data distribution*
   This involves the distribution of the information to where it is required. This can involve display, recoding and/or transmission.

Figure 1.8 *Information transport system*

1.2 Signals   Information in measurement and control systems is conveyed by means of signals. This might be by a change in a signal from on to off, or by a sequence of on-off signals, or perhaps by the size of the signal or the frequency of the signal. Figure 1.9 shows some possible signal forms.

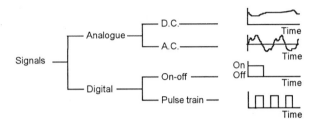

Figure 1.9 *Forms of signals*

Signals can be classified as analogue or digital. An *analogue signal* contains information which shows as a continuous variation of the signal in time in relation to the way the signal changes. Analogue electrical signals may be d.c. or a.c. With d.c. signals, the size of the signal is a measure of the input information. For example, a thermocouple sensor may give a d.c. voltage of 10 mV for a temperature of 20°C and when the temperature changes to 30°C the voltage becomes 15 mV. The size of the voltage is related to the size of the temperature. With a.c. signals, the information may be conveyed in the amplitude of the wave form and/or its frequency. An example of such a signal is the output from a microphone when sensing some sound.

A *digital signal* contains information of on to off, or high level to low level signal changes. Digital signals are not continuous but consist of discrete pulses to represent the information. Digital signals may be just on-off signals or a train of pulses. An on-off signal may arise from a simple switch input or a logic switch. The information conveyed is that there is or there is not an input. Pulse trains consist of a series of what might be termed on-off signals. The information is conveyed by the number of such pulses occurring, the rate at which they occur or the times at which they occur. For example, an encoder might be used to give, say, 30 pulses for each complete revolution of a shaft. By counting the number of pulses in the resulting pulse train the number of revolutions of the shaft can be determined.

1.3 Control systems   A control system can be considered to be a system which for some particular input or inputs is used to control its output to some particular value (Figure 1.10(a)) or give a particular sequence of events or event if certain conditions are met (Figure 1.10(b)). An example of a control system giving the required value of some variable is the domestic central heating system where the variable is the temperature in the house and the control system is designed to operate a furnace so that the required temperature is obtained. An example of a control system giving the required sequence of events is the domestic washing machine where the controller determines the sequence of the pumping water into the machine, heating it, spinning, emptying, etc. The washing machine controller controls the sequence of operations determining when particular events in the sequence are to occur as a result of certain conditions being met, e.g. switching off the heater for the water when the right temperature is attained and then switching on the next event.

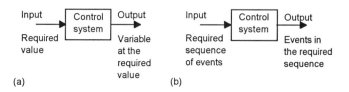

Figure 1.10 *Control systems*

1.3.1 Closed-loop control systems

When there is a requirement to control some variable to a particular value then measurements are required to determine when that value is attained and the measurement system is included in a *closed-loop control system*. Figure 1.11 shows the basic functional elements of a closed-loop control system. To illustrate these functions, consider the domestic central heating system. With such a system the variable being controlled is measured by some measurement system and its value is continuously fed back to be compared with the value required. On the basis of that comparison the controller initiates action to rectify any discrepancy between the actual and the required values. Thus the domestic central heating system has a temperature measurement system which gives a response related to the actual temperature in the house. This is then compared with the value which has been set for the temperature and if the actual temperature is less than the set temperature, the controller causes the central heating furnace to be switched on and pump hot water through the radiators.

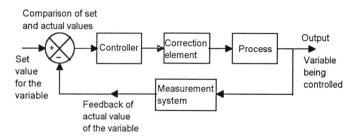

Figure 1.11 *Functional elements of a closed-loop control system*

The functions of the elements in a closed-loop control system are:

1 *Comparison element*

This element compares the required value of the variable with the actual measured value of that variable. The general symbol for the summation point for signals is a circle divided into four quarters with + or – signs to indicate how the signal fed to a particular quarter is to be processed. Thus for the symbol used in Figure 1.11, the signal to the – quarter is to be subtracted from that to the + quarter. The device used might be an operational amplifier with the set value and the measured value as two inputs and an output as the difference between the signals.

2 *Controller*

This determines what action is to be taken when the comparison element indicates a difference between the set value and the measured value. This could be a microprocessor system.

3 *Correction element*

This produces an action which results in a change in the value of the variable being controlled. This might, for example, be a switch which is operated to switch on a heater. The term *actuator* is used for the element of a correction unit that provides the power to carry out the control process, i.e. the bit that moves, grips or applies forces to an object. Thus a valve might be operated by a controller and used to vary the flow of liquid along a pipe and consequently make a piston move in a cylinder and so produce the required linear motion. The piston/cylinder arrangement is the actuator.

4 *Process*

This is the process by which the variable is being controlled. With the domestic central heating system it can be considered to be the house in which the temperature is being controlled.

5 *Measurement system*

This produces a signal related to the variable being controlled and is used to feed back the actual state of the variable so it can be compared with the required value.

As an example of such a control system consider a simple system used to control the speed of rotation of a shaft (Figure 1.12). The required shaft speed is selected by setting the position of the movable contact of the potentiometer. This determines the voltage supplied to one of the inputs of the differential amplifier. The other voltage input to the differential amplifier is obtained from the measurement system, a tachogenerator coupled by gearing to the motor shaft, and is a voltage related to the actual speed of the shaft. The difference between these two signals is then fed to the motor and so the speed of rotation of the shaft is adjusted until there is no difference between the set voltage and the feedback voltage.

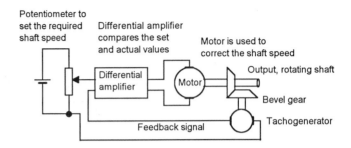

Figure 1.12 *Shaft speed controller*

1.3.2 Discontinuous control systems

The above discussion of closed-loop control systems was concerned with a continuous feedback signal, the variable continuously being monitored by the measurement system. The advent of microprocessors has meant that, because they are basically logic systems, control systems are frequently now designed to operate with on/off signals rather than continuous feedback. The issue then often becomes one that can be phrased as: give an output when this input and that output are both on or perhaps start this output when another output finishes. Such a method of control is particularly useful for controlling the sequence of events. Sequence control can involve such situations as controlling:

1 Events so that they occur in a prescribed sequence, e.g. event 2 is switched on only when event 1 has finished, event 3 switched on when event 2 has finished.

2 Events so that they occur at specific times and so give a timed sequence. Thus event 1 might be switched on 30 s after the start, event 2 at 240 s after the start, etc.

3 Events so that they only occur when certain conditions are realised. Thus event 2 might occur when event 1 is complete and a sensor gives a particular output.

As an example of a microprocessor-controlled system, consider a simplified version of the control system for a domestic washing machine (Figure 1.13). The microprocessor, together with input and output circuits, memory, an analogue-to-digital converter (ADC) and the power supply, constitute the programmable controller. The program to be adopted, i.e. the set values to be used and the sequence of operations to be adopted, is determined by the program which is set by pressing keys or selecting using dials on the fascia of the machine.

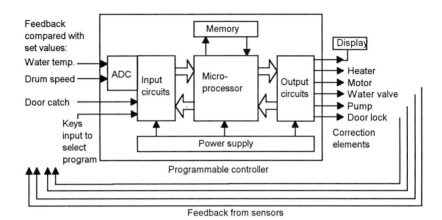

Figure 1.13 *Washing machine control system*

1.4 Automated measurements

Microprocessor controlled systems are increasingly being used for the control of the measurement process with instruments. The microprocessor assumes control of the input of data to the instrument and its processing, perhaps controlling the sequencing of measurements made, as well as their processing. The form of the system might thus be as shown in Figure 1.14. The input to the microprocessor has to be a digital signal and so an analogue-to-digital converter (ADC) has to be used. If the display is to be analogue then the output from the microprocessor has to pass through a digital-to-analogue converter (DAC).

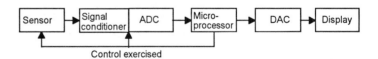

Figure 1.14 *Microprocessor-controlled measurement system*

When there are many sensors, while it would be possible for each sensor to have its own analogue-to-digital converter, its own microprocessor and its own digital-to-analogue converter, a more cost-effective system is to sample each sensor in turn (Figure 1.15). The sample is then processed by the microprocessor and displayed. The device used to sample the outputs from the sensors is termed a *multiplexer*.

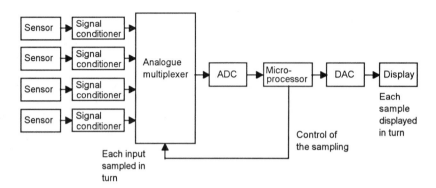

Figure 1.15 *Multiple-input system*

1.4.1 Computerised measurements

A computer can be used to automate the collection, processing and presentation of data. The computer takes control of the sequencing and processing of the measurements. This might involve repeated measurements being made of the same quantity and then processing the results to obtain a mean value with its standard error or perhaps taking measurements of a number of different quantities and combining them in some way to provide the required data.

The hardware required is likely to be a computer into which data acquisition cards or other interface boards have been plugged (Figure 1.16). These boards provide the interface circuitry between sensors and the computer to enable the incoming signals to be conditioned and processed in such a way that the software program run by the computer can make sense of the signals and give an appropriate screen display. In addition to using plug-in cards, some instruments are designed so that they can plug directly into the serial port of a computer.

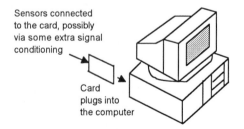

Figure 1.16 *Computer with plug-in board*

With modern software the display on the screen can often look like the display of an instrument with the results of the processing thus displayed in a graphical manner. With software such as LabView the user is able to build a virtual instrument on the screen, with knobs, sliders, switches, graph displays, etc. by selecting the elements from menus and so effectively building their own instrument.

Such automation of measurements is particular useful where there are repetitive operations such as the testing or calibration of many components, complex operations where there are many inputs and outputs of data, remote or hazardous operations, or high-speed operations. Data acquisition and processing systems are discussed in more detail in Chapter 8.

1.5 Specification terms

The following are some of the terms used to specify measurement/ instrumentation systems and their behaviour:

1 *The application for which it can be used*

   This specifies the quantity for which the system can be used to obtain values and the range over which measurements can be made. The *range* is the limits between which measurements can be made. For example, an instrument may be specified as a digital temperature measurement system with a range of 0 to +150°C. The term *dead band* is used if there is a range of input values for which there is no output. For example, bearing friction in a flow meter using a rotor might result in no output until a particular flow rate threshold has been reached.

2 *Conditions required*

   This might be a specification of power requirements and environmental conditions under which it can be used. Thus it might be specified as requiring a power supply of 110 V or 240 V, the selection to be by the setting of an internal switch, and to be operated when the relative humidity is less than 90%. Some might specify a temperature range within which the system can be used.

3 *Accuracy*

   The accuracy of a system is the extent to which the reading it gives might be wrong. For example, a thermometer might have an accuracy specified as ±0.1°C. This means that if the value given by the instrument is, say, 45.5°C, the actual temperature is between 45.4 and 45.6°C.

   Accuracy is often specified as a percentage of the full range output or full-scale deflection (f.s.d.). For example, a system may have an accuracy specified as ±1% of f.s.d. With a full-scale deflection of, say, 10 A, the accuracy is thus ±0.1 A. The accuracy specified in this way implies a constant accuracy over the range of the instrument.

   The accuracy is a summation of all the possible errors that are likely to occur, as well as the accuracy to which the system has been calibrated. The term *error* is used for the difference between the result of a measurement and the true value of the quantity being measured. Where the accuracy is constant over the range of the instrument, we can represent it by drawing a constant error band on a graph showing the relationship between the actual reading and the true value (Figure 1.17). The value of the reading quoted has to have a ± value added to it to indicate that there is an error band extending either side of it.

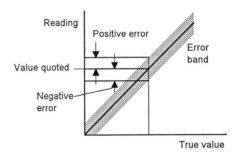

Figure 1.17 *Error band*

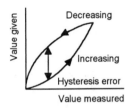

Figure 1.18 *Hysteresis error*

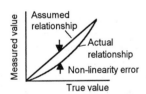

Figure 1.19 *Non-linearity error*

The term *hysteresis error* (Figure 1.18) is used for the difference in the outputs given for the same value of the quantity being measured according to whether that value has been reached by a continuously increasing change or a continuously decreasing change. The hysteresis error is often expressed in terms of the maximum hysteresis error as a percentage of the full-scale deflection. A gear system might show hysteresis in that, as a result of play in the gears, the angular rotation for a particular input depends on whether the input is increasing or decreasing.

The term *non-linearity error* is used for the error that occurs as a result of assuming a linear relationship between the input and output over the working range (Figure 1.19). Few systems have a truly linear relationship and thus errors occur as a result of assuming linearity. For example, with a metal resistance element it might be assumed that the resistance varies linearly with the temperature. This is, however, only an approximation. With modern instrument systems, non-linearity is readily corrected if the system involves a computer or microprocessor for signal processing since it can 'look-up' in a data table the relationship between the output and the true value of the input.

See Chapter 2 for a more detailed discussion of errors and their possible causes.

4 *Resolution*

The resolution of a system is the smallest change in input which can be processed by it. For example, a digital voltmeter might have a four-digit decimal display with a fixed decimal point so that the maximum reading that can be obtained is 999.9. The resolution is thus 0.1, since no change smaller than this can change the reading given by the instrument.

5 *Sensitivity*
   The sensitivity specifies how much output you get per unit input, i.e. it is the ratio output/input. For example, a thermocouple might have a sensitivity of 20 µV/°C. The term is also frequently used to indicate the sensitivity to inputs other that those being measured. Thus the sensitivity might be specified for changes in temperature or perhaps fluctuations in the mains power supply.

6 *Stability and drift*
   Stability is the ability of a system to give the same output when used to measure a constant input over a period of time. The term *drift* is used to describe the change in output that occurs over time. The drift may be expressed as a percentage of the full range output. The term *zero drift* is used for the changes that occur in output when there is zero input.

8 *Repeatability*
   Suppose you use a particular steel rule to measure the length of a constant length rod and repeat the measurement over a number of days. The results obtained might be:

   20.1 mm, 20.2 mm, 20.1 mm, 20.0 mm, 20.1 mm, etc.

   The results of the measurement give values scattered about some value. The term *repeatability* is used for the ability of a measurement system to give the same value for repeated measurements of the same value of a variable. The most common cause of lack of repeatability are random fluctuations in the environment, e.g. changes in temperature and humidity.

9 *Dynamic characteristics*
   The term *static characteristics* is used for the values given when steady-state conditions occur, i.e. when the system has settled down after having received some input. The term *dynamic characteristics* is used for the behaviour of the system between the time that the input value changes and the time that the value has settled down to its steady-state value. For example, when the current is switched on to a moving coil ammeter, the meter pointer moves from zero to generally overshoot and oscillate for a short time before settling down to give the steady-state reading. As another example we might have an input to a measurement system that is changing with time. With the measurement system taking some time to react to the changing input, the reading given will lag behind the true value.
   Terms commonly used to specify dynamic characteristics include rise time, response time and settling time (Figure 1.20). The *rise time* is the time taken for the output to rise to some specified percentage of the steady-state value, e.g. when the input abruptly changes from zero to some constant value, this might be the time to reach 95% of that value. Sometimes the rise time refers to the time taken for the output to rise from 10% of the steady-state value to 90 or 95% of the steady-state

value. The *response time* is generally defined as being the time taken for the output which just rises to its steady-state value without oscillation to reach 63% of its steady-state value or, in the case of one that shows oscillations, the time taken to reach its first peak of the oscillation. The *settling time* is the time taken for the output to settle to within some percentage, e.g. 2.5%, of the steady-state value.

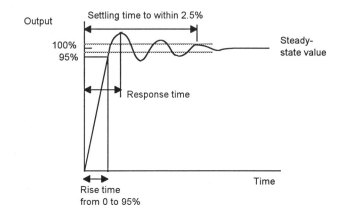

Figure 1.20 *Rise, response and settling times*

10 *Calibration*

In order to deliver the required accuracy, a measurement system must have been calibrated to give that accuracy. *Calibration* is the process of comparing the output of a measurement system against standards of known accuracy. The standards may be other measurement systems which are kept specially for calibration duties or some means of defining standard values. In many companies some instruments and items, such as standard resistors and cells, are kept in a company standards department and used solely for calibration purposes. Thus a certificate of calibration may be issued with a system, this indicating the organisation carrying out the calibration, the test used and the results with their errors. See Section 1.6 for further discussion.

1.6 Quality

The British Standard BS5750 (European Standard EN 29001 and International Standard ISO 9001) is the standard which lays down the standard for a quality system. The term *quality* is used to mean that a product is one which is fit for its purpose or meets requirements, a quality system being one which makes sure that a company delivers quality products. The standard definition for quality is that it is the totality of features and characteristics of a product or service that bear on its ability to meet stated or implied needs. In everyday language the term quality tends to be used to indicate the best available. For example, a Rolls Royce might be considered to be a quality car but a small hatchback not. But in the way the term quality is used in engineering, both cars can be quality cars if they both meet the needs of those buying them, i.e. both are fit for the purpose for which they were bought. If either of the cars breaks down regularly or the paint work blisters or some other defects occur which means that the purchaser does not consider his or her needs are being met, then they are not considered quality goods.

In order to have a quality system it is necessary for a company to exercise control over its measurement systems. For example, it is not possible for a company to state that a product meets, say, a particular length specification if the measurement system used to measure the lengths does not meet the accuracy requirements of that specification. Thus a company in following the standard is expected to provide, control, calibrate and maintain inspection, measuring and test equipment suitable to demonstrate the performance of the product to the specified requirements.

The standard lays down procedures that have to be followed when selecting, using, calibrating, controlling and maintaining measurement standards and measuring equipment. These include:

1   The company has to establish and maintain an effective system for the control and calibration of measurement standards and measuring equipment. This might involve in-company calibration or the use of a suitable calibration service.

2   All the personnel involved in the calibrating should have adequate training.

3   The calibration system used must be periodically and systematically reviewed to ensure that it continues to be effective.

4   All measurements, whether for calibration purposes or measurements of products, must take into account all the errors and uncertainties involved in the measurement process.

5   The procedures used for calibration need to be documented.

6   A separate calibration record should be kept for each measurement instrument. This record is likely to contain a description of the instrument and its reference number, the calibration date, the calibration results, how frequently the instrument is to be calibrated and probably details of the calibration procedure to be used, details of any repairs or modifications made to the instrument, and any limitations on its use.

7   The calibration should be carried out using equipment which can be traceable back to national standards.

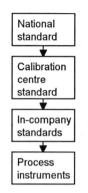

Figure 1.21 *Traceability chain*

1.6.1 Traceable standards

The equipment used in the calibration of an instrument in everyday company use is likely to be *traceable* back to national standards in the following way (Figure 1.21):

1 National standards are used to calibrate standards for calibration centres.

2 Calibration centre standards are used to calibrate standards for instrument manufacturers.

3 Standardised instruments from instrument manufacturers are used to provide in-company standards.

4 In-company standards are used to calibrate process instruments.

There is a simple traceability chain from the instrument used in a process back to national standards.

1.6.2 National standards

The *national standards* are defined by international agreement and are maintained by national establishments, e.g. the National Physical Laboratory in Great Britain and the National Bureau of Standards in the United States. There are seven such *primary standards/base units of measurement*, and two supplementary ones. The seven are:

1 *Mass*
   The mass standard, the kilogram, is defined as being the mass of an alloy cylinder (90% platinum–10% iridium) of equal height and diameter, held at the International Bureau of Weights and Measures at Sèvres in France. Duplicates of this standard are held in other countries.

2 *Length*
   The length standard, the metre, is defined as the length of the path travelled by light in a vacuum during a time interval of duration 1/299 792 458 of a second.

3 *Time*
   The time standard, the second, is defined as a time duration of 9 192 631 770 periods of oscillation of the radiation emitted by the caesium-133 atom under precisely defined conditions of resonance.

4 *Current*
   The current standard, the ampere, is defined as that constant current which, if maintained in two straight parallel conductors of infinite length, of negligible circular cross-section, and placed one metre apart

in a vacuum, would produce between these conductors a force equal to 2×10^{-7} N per metre of length.

5 *Temperature*
The kelvin (K) is defined so that the temperature at which liquid water, water vapour and ice are in equilibrium (known as the triple point) is 273.16 K.

6 *Luminous intensity*
The candela is defined as the luminous intensity, in a given direction, of a specified source that emits monochromatic radiation of frequency 540×10^{12} Hz and that has a radiant intensity of 1/683 watt per unit steradian (a unit solid angle, see below).

7 *Amount of substance*
The mole is defined as the amount of a substance which contains as many elementary entities as there are atoms in 0.012 kg of the carbon-12 isotope.

The two supplementary standards are:

1 *Plane angle*
The radian is the plane angle between two radii of a circle which cuts off on the circumference an arc with a length equal to the radius (Figure 1.22).

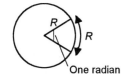

Figure 1.22 *Radian*

2 *Solid angle*
The steradian is the solid angle of a cone which, having its vertex in the centre of the sphere, cuts off an area of the surface of the sphere equal to the square of the radius (Figure 1.23).

Figure 1.23 *The steradian*

Primary standards are used to define national standards, not only in the primary quantities but also in other quantities which can be derived from them. For example, a resistance standard of a coil of manganin wire is defined in terms of the primary quantities of length, mass, time and current. Typically these national standards in turn are used to define reference standards which can be used by national bodies for the calibration of standards which are held in calibration centres.

1.6.2 Terms

The following are terms used in referring to standards:

1 *Fundamental/absolute standard*
This is the standard of which the values have been given in terms of the relevant base units without recourse to another standard of the same quantity.

2 *Primary standard*
This is the standard of a particular quantity which has the highest quality.

3 *Secondary standard*
This is a standard with a value determined by direct or indirect comparison with a primary standard.

4 *Reference standard*
This is the standard used at a location which is the one from which measurements are made.

5 *Working standard*
This is a standard which is intended to verify measuring instruments of lower accuracy at some location and is not specifically reserved as a reference standard.

6 *National standard*
This is the standard for a quantity which is recognised by a national decision as the basis for fixing the value in a country of all the standards of that quantity. In general, it is also the primary standard.

7 *International standard*
This is the standard for a quantity which is recognised by international agreement as the basis for fixing the values of all other standards of that quantity.

1.7 Methods of measurement

Measurement is concerned with the determination of the value of a quantity. Some of the most commonly used methods of measurement are, though the titles are not exclusive and a particular method may appear under more than one heading:

1 *Direct method*
The value of the quantity is obtained directly without the need for calculation from the values of other measured quantities. The measurement of a length by means of a ruler is an example of such a measurement.

2 *Indirect method*
The value of the quantity is obtained from measurements of other quantities which are linked to the required quantity by a known relationship. An example of this is the determination of the density of an object from measurements of its mass and volume.

3 *Direct comparison method*
This method is based on the comparison of the value of a quantity to be measured with the known value of the same quantity. An example of

this is the determination of a mass by means of a balance which compares the mass with that of known weights.

4 *Substitution method*
With this method, the value of a quantity is determined by replacing it by a known value of the same quantity which has been chosen in such a way that the reading by an indicating device is the same for both. This method can be used to determine the value of an electrical resistance by substituting for it known resistances until the same current or balance condition for a bridge is obtained.

5 *Null method*
This method of measurement involves bringing to zero a quantity being used as an indicator. An example of this type of method is the Wheatstone bridge for the measurement of electrical resistance where the current through a galvanometer is brought to zero to give the balance condition for the bridge and hence a specified relationship between the resistance being measured and known resistances.

6 *Deflection method*
This method of measurement involves the value of the quantity being measured being indicated by the deflection of a moving element or indicating device. An example of this is the measurement of current by a moving coil meter.

Problems

1 Explain the functions of the sensor, signal conditioner, signal processing and display elements of a measurement system.

2 Explain the functions of the data acquisition, data processing and data presentation elements of a measurement system.

3 Explain what is meant by closed-loop control system and state the functional elements of such a system.

4 A voltmeter is specified as having ranges of 0 to 2 V and 0 to 20 V, with an accuracy of ±1% of the full-scale deflection. Give the readings with their accuracies when the instrument gives voltage readings of (a) 1.0 V on the 0 to 2 V scale and (b) 4.0 V on the 0 to 20 V scale.

5 A galvanometer is specified as having a d.c. sensitivity of 4.00 mA per centimetre of scale. What will be the change in reading of such an instrument when the current changes by 12 mA?

6 A copper–constantan thermocouple, according to tables, gives an e.m.f. of 0 mV at 0°C, 4.277 mV at 100°C and 9.286 mV at 200°C. It is to be used to measure temperatures in the range 0 to 200°C. What will be the non-linearity error, as a percentage of the full-scale output, at 100°C if a linear relationship is assumed over the full range?

7   An iron–constantan thermocouple, according to tables, gives an e.m.f. of 0 mV at 0°C, 5.628 mV at 100°C and 21.846 mV at 400°C. It is to be used to measure temperatures in the range 0 to 400°C. What will be the non-linearity error, as a percentage of the full-scale output, at 100°C if a linear relationship is assumed over the full range?

8   A pressure measurement system is specified as having a non-linearity error of ±0.15% of full range and a hysteresis error of ±0.05% of full range. What will be value of a pressure and its total error due to non-linearity and hysteresis when the system gives a reading of 1000 kPa on its 0 to 1400 kPa range?

9   A force measurement sensor is specified as having a temperature sensitivity of ±1% of full range. What will be the error arising from an uncompensated temperature change of 5°C when the sensor is used on its 0 to 10 kN range?

10 A thermometer when plunged into a liquid gives the following readings. What is the 95% rise time?

Time (s)	0	30	60	90	120	150	180
Temp. (°C)	20	28	34	39	43	46	49

Time (s)	210	240	270	300	330	360	390
Temp. (°C)	51	53	54	55	55	55	55

11 A load cell has a total error due to non-linearity, hysteresis and non-repeatability of ±0.1% of the indicated value. What will be the error when it is used to determine a load of 50 kN?

12 A rotary variable differential transformer is specified as having a sensitivity of 1.1 (mV/V input)/degree. What will be its output when there is an input voltage of 3 V and a rotation of 10 degrees?

13 A pyrometer is specified as having a repeatability of ±0.4% of full-scale reading when used on its 500 to 3000°C range. What will be its value and error due to repeatability when the reading is 1200°C?

14 Explain what is meant by traceability chain when applied to a measurement system used for calibration by a company.

2 Measurement errors

This chapter is an introduction to measurement errors with a consideration of the importance of stating them when any measurement result is quoted, the sources and types of potential errors and how they can be combined when a measured value involves values from more than one measurement. The least squares method is introduced as a method of fitting experimental data to an equation.

2.1 Why estimate errors?

When a physical quantity is measured, the value obtained should not be expected to be exactly equal to the true value. For example, in a measurement of the acceleration due to gravity, the result calculated from a single measurement of the time taken for a ball to fall through a measured height might be 9.90 m/s². It is not likely that this will be the true value, indeed if the experiment is repeated it is extremely likely that a different value for the acceleration would be obtained. This is because there will be some inaccuracy inherent in the measurement of the time and some in the measurement of the height. Thus with a measurement result we need to give an estimate of the error associated with it. *Error* is defined as being the difference between the result of a measurement of a quantity and its true value:

error = measured value − true value

Thus the result of the measurement of the acceleration due to gravity g might be quoted as:

$g = 9.90 \pm 0.15$ m/s²

This means that we expect the acceleration due to gravity to be somewhere in the range $9.90 - 0.15 = 9.75$ m/s² to $9.90 + 0.15 = 10.05$ m/s² (Figure 2.1). The more accurate the measurement the smaller will be the error range. We cannot be certain that the value lies within this range but there is a certain probability that it will.

The quoting of the errors associated with a measurement are important if any meaningful conclusions are to be drawn from the result. For example, an earlier measurement of the acceleration due to gravity might have given a value of:

$g = 9.80 \pm 0.05$ m/s²

Does the later result indicate a change in the acceleration? Without quoting the errors the values of 9.80 and 9.90 m/s² might seem to indicate a change. However, when we consider the errors we have the later measurement

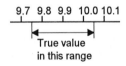

Figure 2.1 *Accuracy of measured value*

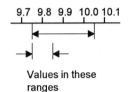

Values in these
ranges

Figure 2.2 *Accuracy of measured values*

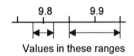

Values in these ranges

Figure 2.3 *Change in value*

giving a value expected to be between 9.75 and 10.05 m/s² and the earlier measurement a value between 9.75 and 9.85 m/s². Since these error ranges overlap (Figure 2.2), the true value might not have changed and we have no evidence to suggest that it did.

However, suppose we had the two results:

Earlier measurement: $g = 9.80 \pm 0.02$ m/s²
Later measurement: $g = 9.90 \pm 0.05$ m/s²

The earlier measurement indicates a value in the range 9.78 to 9.82 m/s² while the later measurement indicates a value in the range 9.85 to 9.95 m/s² (Figure 2.3). The ranges do not overlap and it thus seems reasonable to conclude that the acceleration due to gravity might have changed.

Without quoting the error with an experimental result, it is not possible to judge the significance of a result and the result is really useless. Errors must be quoted with all experimental results.

2.1.1 Fractional and percentage errors

Sometimes the error is specified in the form of the *fractional error*, where:

$$\text{fractional error} = \frac{\text{error in quantity}}{\text{size of quantity}}$$

More frequently, however, the *percentage error* is quoted:

$$\text{percentage error} = \frac{\text{error in quantity}}{\text{size of quantity}} \times 10$$

For example, a speed quoted as 2.0 ± 0.2 m/s might have its error quoted as ±10%.

In the case of some instruments, the error is quoted as a percentage of the full-scale reading possible with an instrument. For example, an ammeter used to measure a current using its 0 to 5 A range and having an error quoted as ±1% will give for all its readings on that range an error of ±0.05 A.

2.2 Sources of error

Common sources of error with measurements are:

1 *Instrument construction errors*
   These result from such causes as tolerances on the dimensions of mechanical components and the values of electrical components used in instruments and are inherent in the manufacture of an instrument. For example, a cheap ruler might have divisions that are not equally spaced. In addition, there can be errors due to the accuracy with which the instrument has been calibrated.

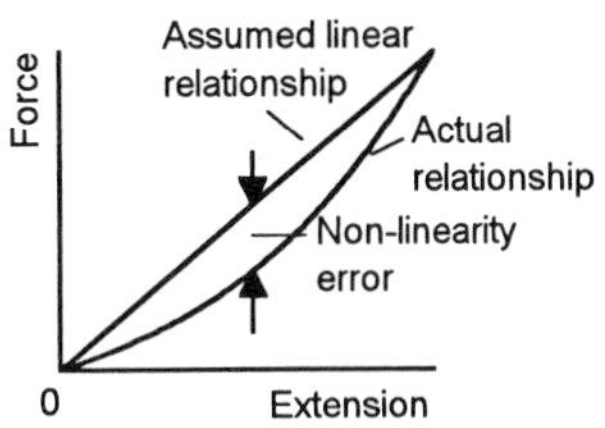

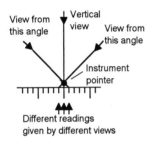

Figure 2.4 *Non-linearity error*

Figure 2.5 *Parallax error*

2 *Non-linearity errors*

In the design of many instruments a linear relationship between two quantities is often assumed, e.g. with a spring balance a linear relationship between force and extension. A linear relationship may be an approximation or may be restricted to a narrow range of values. Thus an instrument may have errors due to a component not having a perfectly linear relationship (Figure 2.4).

3 *Operating errors*

These can occur for a variety of reasons and are often referred to as *human errors*. They can also include errors due to carelessness in reading a scale, perhaps reading the wrong scale. They may occur in reading the position of a pointer on a scale. If the scale and the pointer are not in the same plane then the reading obtained depends on the angle at which the pointer is viewed against the scale (Figure 2.5). These are called *parallax errors*. To reduce the chance of such errors occurring, some instruments incorporate a mirror alongside the scale. Positioning the eye so that the pointer and its image are in line guarantees that the pointer is being viewed at the right angle. Digital instruments, where the reading is displayed as a series of numbers, avoids this problem of parallax. Errors may also occur due to the limited resolution of an instrument and the ability to read a scale. Such errors are termed *reading errors* (see Section 2.2.1). Operating errors can also arise when an instrument has to be brought into contact with an object being measured, e.g. a micrometer, as a result of slightly different contact forces occurring.

4 *Environmental errors*

Errors can arise as a result of environmental effects. For example, when making measurements with a steel rule, the temperature occurring when the measurement is made might not be the same as that for which the rule was calibrated. Another example might be the presence of draughts affecting the readings given by a balance.

5 *Insertion errors*

In some measurements the insertion of the instrument into the position to measure a quantity can affect its value. For example, inserting an ammeter into a circuit to measure the current can affect the value of the current due to the ammeter's own resistance (see Section 2.2.2). Similarly, putting a cold thermometer into a hot liquid can cool the liquid and so change the temperature being measured.

6 *Unrepresentative samples*

There are many situations where a sample of a material is taken and measurements made on it with the assumption being made that these are the properties of the material as a whole. For example, an engineer might carry out a tensile test of a sample of steel in order to forecast the properties of that steel when it is used in constructing a bridge. The sample must be chosen in such a way that its properties are representative, otherwise errors will occur.

2.2.1 Reading errors

One particular form of operating error associated with the scale markings on an instrument is the *reading error*. When the pointer of an instrument falls between two scale markings (Figure 2.6) there is some degree of uncertainty, i.e. possible error, as to what the reading should be quoted as. Thus instrument readings should not be quoted as precise numbers but some indication given of the associated uncertainty, i.e. the possible extent to which the reading could be in error. The worse the reading error could be is that the value indicated by a pointer is somewhere between two successive markings on the scale. In such circumstances the reading error can be stated as a value ± half the scale interval. For example, a rule might have scale markings every 1 mm. Thus when measuring a length using the rule and the reading is nearest to the 23.4 mark, the result might be quoted as 23.4 ± 0.5 mm. However, it is often the case that we can be more certain about the reading and indicate a smaller error.

23.4 reading
in this area

Figure 2.6 *Reading error*

With digital displays there is no uncertainty regarding the value displayed but there is still an error associated with the reading. This is because the reading of the instrument goes up in jumps, a whole digit at a time. We cannot tell where between two successive digits the actual value really is. Thus the degree of uncertainty is ± the smallest digit.

2.2.2 Insertion errors

When the act of making a measurement modifies the variable being measured, the term *loading* is used. As an illustration of such errors, consider a frequently encountered loading problem, that due to a voltmeter being used to measure the voltage across some resistor. Any electrical network may be considered to behave as a circuit containing a single source of e.m.f. in series with a resistance (this assumes we are considering a d.c. circuit) and thus we can represent a network by such an arrangement supplying a current through a load resistor (Figure 2.7). In the absence of the voltmeter the current through the load resistor, resistance R_L, is I and so the potential difference V_L across it is IR_L. If the rest of the circuit has a resistance R and the supplied e.m.f. is E, then:

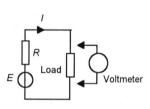

Figure 2.7 *Loading by a voltmeter*

$$E = I(R + R_L)$$

and so the potential difference across the load is:

$$V_L = IR_L = E\left(\frac{R_L}{R + R_L}\right)$$

When the voltmeter, resistance R_m, is connected across the resistance, then we have a resistance R_m in parallel with R_L. For resistances in parallel, the equivalent resistance R_e is given by:

$$\frac{1}{R_e} = \frac{1}{R_L} + \frac{1}{R_m} = \frac{R_m + R_L}{R_L R_m}$$

The total resistance of the circuit has changed and the circuit current has changed, becoming $E/(R + R_e)$. Thus the voltage V_m indicated by the voltmeter is:

$$V_m = E\left(\frac{R_e}{R + R_e}\right)$$

With V_m not equal to V_L there is an error. This is because the voltmeter does not have an infinite resistance. The error is $(V_m - V_L)$ and the fractional error is $(V_m - V_L)/V_L$. Hence:

$$\text{fractional error} = \frac{V_m - V_L}{V_L} = \frac{E\left(\dfrac{R_e}{R + R_e}\right) - E\left(\dfrac{R_L}{R + R_L}\right)}{E\left(\dfrac{R_L}{R + R_L}\right)}$$

$$= \frac{\left(\dfrac{R_e(R + R_L) - R_L(R + R_e)}{(R + R_e)(R + R_L)}\right)}{\left(\dfrac{R_L}{R + R_L}\right)}$$

$$= \frac{R(R_e - R_L)}{R_L(R + R_e)}$$

Since R_e will be less than R_L, the fractional error will be negative. The error will only be zero when R_e is equal to R_L and this will only occur when the voltmeter resistance is effectively infinite.

To illustrate the above, consider a voltmeter with a resistance of 10 kΩ which is placed in parallel with a load resistance of 1 kΩ and which is in a circuit of total resistance 2 kΩ. What will be the fractional error in the measured voltage? Without the voltmeter we have a circuit with a resistance of 1 kΩ in series with 1 kΩ. The equivalent resistance when the voltmeter is in parallel with the load is given by

$$\frac{1}{R_e} = \frac{1}{10} + \frac{1}{1}$$

Thus $R_e = 0.91$ kΩ. Hence, using the equation developed above, the fractional error in the voltage reading is:

$$\text{fractional error} = \frac{1(0.91 - 1)}{1(1 + 0.91)} = -0.047$$

2.3 Random and systematic errors

All errors, whatever their source, can be described as being either random or systematic. *Random errors* are ones which can vary in a random manner between successive readings of the same quantity. These may be due to personal errors by the person making the measurements, e.g. varying reaction times in timing events, applying varying pressures when using a micrometer screw gauge, parallax errors, etc., or perhaps due to random electronic fluctuations (termed noise) in the instruments or circuits used, or perhaps varying frictional effects. *Systematic errors* are errors which do not vary from one reading to another. These may be due to some defect in the instrument such as a wrongly set zero so that it always gives a high or low reading, or perhaps incorrect calibration, or perhaps an instrument is temperature dependent and the measurement is made under conditions which differ from those for which it was calibrated, or there is a loading error. Random errors can be minimised by the use of statistical analysis, systematic errors require the use of a different instrument or measurement technique to establish them.

2.4 The mean value and its error

Random errors mean that sometimes the error will give a reading that is too high, sometimes a reading that is too low. The error can be reduced by repeated readings being taken and calculating the mean (average) value. The *mean* or *average* $\bar{x}$ of a set of readings is given by:

$$\bar{x} = \frac{x_1 + x_2 + \dots x_n}{n}$$

where x_1 is the first reading, x_2 the second reading, ... x_n the nth reading. The more readings we take the more likely it will be that we can cancel out the random variations that occur between readings. The *true value* might thus be regarded as the value given by the mean of a very large number of readings.

2.4.1 Error of a reading

If we consider a single result, what is its likely error from the mean value? Some indication as to the degree of uncertainty is given from considering the spread of results obtained when repeated measurements are made. Consider the two following sets of readings:

20.1, 20.0, 20.2, 20.1, 20.1 and 19.5, 20.5, 19.7, 20.6, 20.2 s

Both sets of readings have the same average of 20.1, but the second set of readings is more spread out than the first and thus shows more random fluctuations.

The *deviation* of any one reading from the mean is the difference between its value and the mean value. Table 2.1 shows the deviations for the two sets of results.

Table 2.1 *Deviations from the mean*

1st set of readings		2nd set of readings	
Reading (s)	Deviation (s)	Reading (s)	Deviation (s)
20.1	0.0	19.5	−0.6
20.0	−0.1	20.5	+0.4
20.2	+0.1	19.7	−0.4
20.1	0.0	20.6	+0.5
20.1	0.0	20.2	+0.1

The second set of readings has greater deviations than the first set. It seems likely that if we had only considered one reading of the less spread out set of readings it would have had a greater chance of being closer to the mean value than any one reading in the more spread out set. The spread of the readings is thus taken as a measure of the certainty we can attach to any one reading being close to the mean value, the bigger the spread the greater the uncertainty. The spread of the readings is specified by a quantity termed the *standard deviation*. The standard deviation, symbol σ, is given by:

$$\text{standard deviation} = \sqrt{\frac{\left(d_1^2 + d_2^2 + \ldots d_n^2\right)}{n-1}}$$

where d_1 is the deviation of the first result from its average, d_2 the deviation of the second reading, ... d_n the deviation of the nth reading from the average. See Appendix A for further discussion.

Note that sometimes the above equation is written with just n instead of $(n-1)$ on the bottom line. With just n it is assumed that the deviations are all from the true value, i.e. the mean when there are very large numbers of readings. With smaller numbers of readings, the deviations are taken from the mean value without assuming that it is necessarily the true value. To allow for this, $(n-1)$ is used. In fact, with more than a very few readings, the results using n and $(n-1)$ are the same, to the accuracy with which the standard deviations are usually quoted. The square of the standard deviation is termed the *variance*.

Consider calculations of the standard deviations for the two sets of data used in Table 2.1. Table 2.2 shows the results of squaring the deviations with the consequential calculations of the standard deviations. The first set of readings has a standard deviation of 0.07 and the second set 0.48. The second set of readings has thus a much greater standard deviation than the first set, indicating the greater spread of those results.

Table 2.2 *Calculation of standard deviations*

1st set of readings

Reading (s)	Deviation (s)	(Deviation)^2 s^2
20.1	0.0	0.00
20.0	−0.1	0.01
20.2	+0.1	0.01
20.1	0.0	0.00
20.1	0.0	0.00

Sum of (deviation)^2 = 0.02, hence standard deviation = $\sqrt{\dfrac{0.02}{5-1}}$ = 0.07.

2nd set of readings

Reading (s)	Deviation (s)	(Deviation)^2 s^2
19.5	−0.6	0.36
20.5	+0.4	0.16
19.7	−0.4	0.16
20.6	+0.5	0.25
20.2	+0.1	0.01

Sum of (deviation)^2 = 0.94, hence standard deviation = $\sqrt{\dfrac{0.94}{5-1}}$ = 0.48.

It is important to realise that the standard deviation is a measure of the spread of the results. A consideration of the statistics involved shows that we can reasonably expect about 68.3% of the readings will lie within plus or minus one standard deviation of the mean, 95.45% within plus or minus two standard deviations and 99.7% within plus or minus three standard deviations. This assumes that the scatter of the readings about the mean is completely random and can be represented by the form of distribution termed a *normal* or *Gaussian distribution* (see Appendix A for further discussion). This type of distribution is very commonly met in engineering and scientific measurements. Thus the standard deviation lets us know how far from the mean we can expect any one reading to be.

With a normal distribution, 95% of the readings will fall within plus or minus 1.96 standard deviations of the mean. Thus if we obtained a value which differed by more than 1.96 standard deviations from the mean, it could be said to be significantly different from most of the readings values since it would not be one of the 95% and so expected to occur often. The reading is said to be *significant at the 5% level*. Since 95% of the readings in a normal distribution lie within plus or minus 1.96 standard deviations of the mean, we can be 95% confident that a reading taken at random will occur within these limits. These limits thus define what is termed a *95% confidence interval*. For a reading to be *significant at the 1% level*, its deviation from the mean must be at least 2.576 standard deviations. A *99% confidence interval* would thus be defined by readings being within plus or minus 2.576 standard deviations of the mean.

2.4.2 Error of a mean

If we take a set of readings and obtain a mean, how far from the true value might we expect the mean to be? Essentially what we do is consider the mean value of our set of results to be one of the many mean values which can be obtained from the very large number of results and calculate its standard deviation from that true value mean. To avoid confusion with the standard deviation of a single result from a mean of a set of results, we use the term *standard error* for the standard deviation of the mean from the true value. Thus, the extent to which we might expect a mean of a set of readings to depart from the true value is given by the *standard error of the mean*, this being given by:

$$\text{standard error} = \frac{\text{standard deviation of the set of results}}{\sqrt{n}}$$

We can reasonably expect that there is a 68.3% chance that a particular mean value lies within plus or minus one standard error of the true value, a 95.45% chance within plus or minus two standard errors and a 99.7% chance within plus or minus three standard errors. With a normal distribution, 95% of the readings will fall within plus or minus 1.96 standard deviations of the mean. Thus if we obtained a mean value which differed by more than 1.96 standard deviations from the true value, it is said to be *significant at the 5% level*. For a mean value to be *significant at the 1% level*, its deviation from the true value must be at least 2.576 standard deviations. Since 95% of the means in a normal distribution lie within plus or minus 1.96 standard deviations of the true value, we can be 95% confident that a mean taken at random will occur within these limits. These limits thus define what is termed a *95% confidence interval*. A *99% confidence interval* would be defined by means being within plus or minus 2.576 standard deviations of the mean. It is quite common in engineering and scientific measurements for the error quoted to be the 95% confidence interval. See Appendix A for further discussion.

The greater the number of measurements made, the smaller will be the standard error. Because the factor is $\sqrt{n}$, increasing the number of readings taken by 100 only reduces the error by a tenth. Note that such a reduction in error is only a reduction in the random error, there may still be an unaffected systematic error.

Thus, for the data in Table 2.2, the first set of measurements had a standard deviation of 0.07 s and so a standard error of $0.07/\sqrt{5} = 0.03$ s. Thus, with the mean value of 20.1 s, the chance of the true value being within ±0.03 s of 20.1 is about 68%. The chance of the true value being within ±0.06 s is about 95%. The 95% confidence interval is thus 20.1 ± 0.06 s. For the second set of measurements, the standard deviation was 0.48 s and so there is a standard error of $0.48/\sqrt{5} = 0.21$ s. The mean is 20.1 s and the chance of the true value being within ±0.21 s is 68%, within ±0.42 s about 95%. The 95% confidence interval us thus 20.1 ± 0.42 s. Thus, since the 95% level is usually the one used, the measurements are likely to be just written as: 20.1 ± 0.06 s and 20.1 ± 0.42 s.

As a further example, consider measurements of the electrical resistance of a resistor which gave the following results: 53, 48, 45, 49, 46, 48, 51, 57, 55, 55, 47, 49 Ω. The mean is given by:

$$\text{mean} = \frac{53 + 48 + 45 + 49 + 46 + 48 + 51 + 57 + 55 + 55 + 47 + 49}{12}$$

and is 50.25 Ω. Table 2.3 shows the derivation of deviations for the data.

Table 2.3 *Derivation of the deviations*

Resistance (Ω)	Deviation (Ω)	(Deviation)^2 Ω^2
53	+2.75	7.5625
48	−2.25	5.0625
45	−5.25	27.5625
49	−1.25	1.5625
46	−4.25	18.0625
48	−2.25	5.0625
51	+0.75	0.5625
57	+6.75	45.5625
55	+4.75	22.5625
55	+4.75	22.5625
47	−2.25	5.0625
49	−1.25	1.5625

The sum of (deviation)^2 = 168.25 and hence the standard deviation is $\sqrt{(168.25/11)} = 3.91$ Ω. The standard error is thus $3.91/\sqrt{12} = 1.13$ Ω and we can write our estimate of the resistance at the 68% confidence level as 50.25 ± 1.13 Ω and at the 95% confidence level as 50.25 ± 2.26 Ω. It is usually this 95% level result that is the one quoted.

Note that in the above discussion of the percentage of results that can be expected to lie within a certain number of standard deviations of the mean or the percentage of mean values that can be expected to lie within a certain number of standard errors of the mean, it has been assumed that large numbers of readings were involved and so we could use the normal distribution. With small numbers of readings a different distribution should really be used, the *Student t-distribution*. Thus for just five readings, the 95% interval is given by plus or minus 2.77 standard deviations, with ten readings 2.26, with 20 readings 2.09 and only with large numbers of readings does it become 1.96. See Appendix A for further discussion.

2.5 Combining errors

A measurement of a quantity might be indirect and require several quantities to be measured and their values inserted into an equation. For example, in a determination of the density ρ of a solid, measurements might be made of the mass m of the body and its volume V. Then the values are inserted into the equation:

$$\rho = \frac{m}{V}$$

in order to obtain the value for the density. The mass and volume measurements will each have errors associated with them. How then do we determine the consequential error in the density? This type of problem is very common. The following illustrates how we can determine the error in such situations.

2.5.1 The worst possible error

Consider the calculation of the quantity Z from two measured quantities A and B where $Z = A + B$. If the measured quantity A has an error $\pm\Delta A$ and the quantity B an error $\pm\Delta B$ then the worst possible error we could have in Z is if the quantities are at the extremes of their error bands and the two errors are both positive or both negative. Then we have:

$$Z + \Delta Z = (A + \Delta A) + (B + \Delta B)$$

$$Z - \Delta Z = (A - \Delta A) + (B - \Delta B)$$

Subtracting one equation from the other gives the worst possible error in Z as:

$$\Delta Z = \Delta A + \Delta B$$

When we add two measured quantities the worst possible error in the calculated quantity is the sum of the errors in the measured quantities.

If we have the calculated quantity Z as the difference between two measured quantities, i.e. $Z = A - B$, then, in a similar way, we can show that the worst possible error is given by

$$Z + \Delta Z = (A + \Delta A) - (B + \Delta B)$$

$$Z - \Delta Z = (A - \Delta A) - (B - \Delta B)$$

and so subtracting the two equations gives the worst possible error as:

$$\Delta Z = \Delta A + \Delta B$$

When we subtract two measured quantities the worst possible error in the calculated quantity is the sum of the errors in the measured quantities.

If we have the calculated quantity Z as the product of two measured quantities A and B, i.e. $Z = AB$, then we can calculate the worst error in Z as being when the quantities are both at the extremes of their error bands and the errors in A and B are both positive or both negative:

$$Z + \Delta Z = (A + \Delta A)(B + \Delta B) = AB + B\Delta A + A\Delta B + \Delta A\Delta B$$

The errors in A and B are small in comparison with the values of A and B so we can neglect the quantity $\Delta A\Delta B$ as being insignificant. Then:

$$\Delta Z = B\Delta A + A\Delta B$$

Dividing through by Z gives:

$$\frac{\Delta Z}{Z} = \frac{B\Delta A + A\Delta B}{Z} = \frac{B\Delta A + A\Delta B}{AB} = \frac{\Delta A}{A} + \frac{\Delta B}{B}$$

Thus, when we have the product of measured quantities, the worst possible fractional error in the calculated quantity is the sum of the fractional errors in the measured quantities. If we multiply the above equation by 100 then we can state it as the percentage error in Z is equal to the sum of the percentage errors in the measured quantities. If we have the square of a measured quantity, then all we have is the quantity multiplied by itself and so the error in the squared quantity is just twice that in the measured quantity. If the quantity is cubed then the area is three times that in the measured quantity.

If the calculated quantity is obtained by dividing one measured quantity by another, i.e. $Z = A/B$, then the worst possible error is given when we have the quantities at the extremes of their error bands and the error in A positive and the error in B negative, or vice versa. Then:

$$Z + \Delta Z = \frac{A + \Delta A}{B - \Delta B} = \frac{A\left(1 + \frac{\Delta A}{A}\right)}{B\left(1 - \frac{\Delta B}{B}\right)}$$

Using the binomial series we can write this as:

$$Z + \Delta Z = \frac{A}{B}\left(1 + \frac{\Delta A}{A}\right)\left(1 + \frac{\Delta B}{B} - \ldots\right)$$

Neglecting products of ΔA and ΔB and writing A/B as Z, gives:

$$Z + \Delta Z = Z\left(1 + \frac{\Delta A}{A} + \frac{\Delta B}{B}\right)$$

Hence:

$$\frac{\Delta Z}{Z} = \frac{\Delta A}{A} + \frac{\Delta B}{B}$$

The worst possible fractional error in the calculated quantity is the sum of the fractional errors in the measured quantities or, if expressed in percentages, the percentage error in the calculated quantity is equal to the sum of the percentage errors in the measured quantities.

To sum up:

1   When measurements are added or subtracted, the resulting worst error is the sum of the errors.

2   When measurements are multiplied or divided, the resulting worst percentage error is the sum of the percentage errors.

The following examples illustrate the use of the above to determine the worst errors. The resistance R of a resistor is determined from measurements of the potential difference V across it and the current I through it. The resistance is given by V/I. The potential difference has been measured as 2.1 ± 0.2 V and the current measured as 0.25 ± 0.01 A. The percentage error in the voltage reading is $(0.2/2.1) \times 100\% = 9.5\%$ and in the current reading is $(0.01/0.25) \times 100\% = 4.0\%$. Thus the percentage error in the resistance is $9.5 + 4.0 = 13.5\%$. Since we have $V/I = 8.4$ Ω and 13.5% of 8.4 is 1.1, then the resistance is 8.4 ± 1.1 Ω.

The cross-sectional area A of a wire is to be determined from a measurement of the diameter d, being given by $A = \pi d^2/4$. The diameter is measured as 2.5 ± 0.1 mm. The percentage error in d^2 will be twice the percentage error in d. Since the percentage error in d is $\pm 4\%$ then the percentage error in d^2, and hence A since the others are pure numbers, is $\pm 8\%$. Since $\pi d^2/4 = 4.9$ mm^2 and 8% of this value is 0.4 mm^2, the result can be quoted as 4.9 ± 0.4 mm^2.

The acceleration due to gravity g is determined from a measurement of the length L of a simple pendulum and the periodic time T, using:

$$T = 2\pi \sqrt{\frac{L}{g}}$$

If we have $L = 1.000 \pm 0.005$ m and $T = 2.0 \pm 0.1$ s, what is the value of the acceleration due to gravity and its worst possible error? Squaring the equation and rearranging it gives:

$$g = \frac{4\pi^2 L}{T^2}$$

To give the percentage error in g, we need to add the percentage errors in L and T^2. The percentage error in L is 0.5% and that in T is 5.0%. Thus the total percentage error in g is $0.5 + 2 \times 5.0 = 10.5\%$. Putting the values of L and T in the equation gives a value of 9.87 m/s^2 and since 10.5% of this is 1.04 m/s^2 we can quote the result of the measurement, to two significant figures, as 9.9 ± 1.0 m/s^2.

The coefficient of viscosity η of a liquid can be determined by measuring the rate of flow Q through a tube of radius r and length L when there is a pressure difference p between the ends:

$$\eta = \frac{\pi p r^4}{8LQ}$$

We need to add the fractional errors in the quantities in order to give the fractional error in the viscosity. Thus:

$$\text{fractional error in } \eta = \frac{\Delta p}{p} + 4\frac{\Delta r}{r} + \frac{\Delta L}{L} + \frac{\Delta Q}{Q}$$

2.5.2 The standard error

In the above discussion, the worst possible errors were determined. Thus, when we added two quantities, the errors were assumed to be both at the extremes of their error bands and both positive or both negative and so give the maximum possible error. In practice it is likely that this will give an overestimate of the error. For example, consider measurements taken of two lengths, say 10 measurements of each, to give mean values with standard errors of 25 ± 1 mm for one and 40 ± 2 mm for the other (Figure 2.8 illustrates this with just three measurements). We can combine any one of the measurements of the first length with any one of the measurements of the second length to obtain the sum of the lengths. With 10 measurements of each, this will give 100 values for the sum. The worst possible error in the sum of the two measurements is when we use just the extreme measurements of each length, i.e. $25 + 1$ mm and $40 + 2$ mm to give $65 + 3$ mm, and $25 - 1$ mm and $40 - 2$ mm to give $65 - 3$ mm and so 65 ± 3 mm. This is only the error we get in the worst situation; generally the sum of a pair of the measurements will have a smaller error. We can obtain a more realistic error by considering all the 100 possible sums of the two quantities.

Consider the error in a value of Z when we have $Z = A + B$ and there are errors $\pm\Delta A$ and $\pm\Delta B$ in the two quantities added. The error in a result is the difference between the calculated value Z and its true value $\bar{Z}$, i.e. $\Delta Z = Z - \bar{Z}$. We can write this as:

$$(\Delta Z)^2 = (Z - \bar{Z})^2$$

But $Z = A + B$ and $\bar{Z} = \bar{A} + \bar{B}$ Thus

$$(\Delta Z)^2 = [(A + B) - (\bar{A} + \bar{B})]$$

This can be rewritten as:

$$(\Delta Z)^2 = (A - \bar{A})^2 + (B - \bar{B})^2 + 2(A - \bar{A})(B - \bar{B})$$

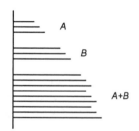

Figure 2.8 *The possible sums from 3 measurements of A and B*

The above equation gives the error for one combination of a measurement of A with one measurement of B and assumes that the errors are both the same sides of their true values, i.e. both errors are positive. However, the errors may be above or below their mean values and may be small or large. If we now consider all the possible combinations of A and B from each set of results, then we will have an equation of the above form for each set of errors. If we consider all the possible cases, add them together and then divide by the number of cases used we can obtain the standard error of Z. We would expect to find the $2(A - \bar{A})(B - \bar{B})$ term having the average value of 0. Thus on average we would expect:

$$(\Delta Z)^2 = (A - \bar{A})^2 + (B - \bar{B})$$

and so:

$$(\Delta Z)^2 = (\Delta A)^2 + (\Delta B)^2$$

The same expression is obtained when $Z = A - B$.

The following is another way of arriving at the same result. The worst error in Z, when both errors are positive and each quantity is at the extreme end of its error band, is given by (see previous section):

$$\Delta Z = \Delta A + \Delta B$$

Consider that each quantity, A and B, has been arrived at from a set of measurements of each and the mean and standard error obtained. If we consider all the possible sums of the quantities A and B, one or both of the errors could be negative or not at the extremes of the error band. Writing the equation in the form:

$$(\Delta Z)^2 = (\Delta A + \Delta B)^2$$

$$= (\Delta A)^2 + (\Delta B)^2 + 2\Delta A \Delta B$$

then adding together all the possible values the errors could take and dividing by the number considered, we would expect the $2\Delta A \Delta B$ term to have the value of zero since there will be as many situations with it having a negative value as having a positive value, and we obtain:

$$(\Delta Z)^2 = (\Delta A)^2 + (\Delta B)^2$$

The same expression is obtained when $Z = A - B$.

Now consider the standard error when we have $Z = AB$ with errors $\pm\Delta A$ and $\pm\Delta B$. The worst error in Z is when the errors in A and B are both positive, or both negative. Then (see previous section):

$$\frac{\Delta Z}{Z} = \frac{\Delta A}{A} + \frac{\Delta B}{B}$$

Squaring both sides of the equation gives:

$$\left(\frac{\Delta Z}{Z}\right)^2 = \left(\frac{\Delta A}{A} + \frac{\Delta B}{B}\right)^2$$

$$= \left(\frac{\Delta A}{A}\right)^2 + \left(\frac{\Delta B}{B}\right)^2 + 2\left(\frac{\Delta A}{A}\right)\left(\frac{\Delta B}{B}\right)$$

But one or both of the errors could be negative or not at the extremes of the error band. Adding all the possible values the errors could take and dividing by the number considered, we would expect the $2(\Delta A/A)(\Delta B/B)$ term to have the value of zero since there will be as many situations with it having a negative value as having a positive value. Thus

$$\left(\frac{\Delta Z}{Z}\right)^2 = \left(\frac{\Delta A}{A}\right)^2 + \left(\frac{\Delta B}{B}\right)^2$$

The above equation involves the fractional errors; it could equally well be in terms of the percentage errors. The same equation is obtained when we have $Z = A/B$.

To sum up:

1   When measurements are added or subtracted, the resulting square of the standard error is the sum of the squares of the errors.

2   When measurements are multiplied or divided, the resulting square of the standard percentage error is the sum of the squares of the percentage errors.

For an equation involving a power, e.g. $Z = AB^n$, then the error is given by:

$$\left(\frac{\Delta Z}{Z}\right)^2 = \left(\frac{\Delta A}{A}\right)^2 + n\left(\frac{\Delta B}{B}\right)^2$$

To illustrate the above, consider the examples used in the previous section for the determination of the worst errors. The resistance R of a resistor is determined from measurements of the potential difference V across it and the current I through it. The potential difference has been measured as 2.1 ± 0.2 V and the current measured as 0.25 ± 0.01 A. The percentage error in the voltage reading is $(0.2/2.1) \times 100\% = 9.5\%$ and in the current reading is $(0.01/0.25) \times 100\% = 4.0\%$. Thus the percentage error in the resistance V/I is given by:

(% error in $R)^2 = 9.5^2 + 4.0^2$

Hence the percentage error is 10.3%. Since $V/I = 8.4\ \Omega$ and 10.3% of 8.4 is 0.87, then the resistance is quoted as $8.4 \pm 0.9\ \Omega$.

The cross-sectional area A of a wire is determined from a measurement of the diameter d. The diameter is measured as 2.5 ± 0.1 mm. Since the percentage error in d is $\pm 4\%$ then the percentage error in the square of the diameter, i.e. d multiplied by d, is given by:

$$(\% \text{ error in } d^2)^2 = 4^2 + 4^2$$

and so the percentage error is 5.7%. Since area $= \pi d^2/4 = 4.9$ mm^2 and 5.7% of this value is 0.3 mm^2, then the area is 4.9 ± 0.3 mm^2.

The acceleration due to gravity g can be determined from a measurement of the length L of a simple pendulum and the periodic time T. The relationship is:

$$g = \frac{4\pi^2 L}{T^2}$$

If we have $L = 1.000 \pm 0.005$ m and $T = 2.0 \pm 0.1$ s, the percentage error in L is 0.5% and that in T is 5.0%. Thus the total percentage error in g is given by:

$$(\% \text{ error in } g)^2 = 0.5^2 + 2(5.0)^2$$

Hence the error is 7.1%. Putting the values of L and T in the equation gives a value of 9.87 m/s^2 and since 7.1% of this is 0.70 m/s^2 we can quote the result to two significant figures as 9.9 ± 0.7 m/s^2.

The coefficient of viscosity η of a liquid can be determined by measuring the rate of flow Q through a tube of radius r and length L when there is a pressure difference p between the ends, using the equation:

$$\eta = \frac{\pi p r^4}{8LQ}$$

Using the equations developed above for products and quotients of quantities, the fractional standard error in the coefficient of viscosity is related to the other errors by:

$$\left(\frac{\Delta\eta}{\eta}\right)^2 = \left(\frac{\Delta p}{p}\right)^2 + 4\left(\frac{\Delta r}{r}\right)^2 + \left(\frac{\Delta L}{L}\right)^2 + \left(\frac{\Delta Q}{Q}\right)^2$$

2.5.3 Partial differentiation and the worst error

In the above analysis we have considered specific cases where we are required to sum two measurements, or subtract, or multiply, or divide. In general, if we want to find out how a small change in one quantity affects another, we differentiate. For example, if we have $Z = A^2$ then:

$$\frac{\mathrm{d}Z}{\mathrm{d}A} = 2A$$

If we consider finite elements, then we can write:

$$\Delta Z = 2A\Delta A$$

Dividing both sides of the equation by Z gives, since $Z = A^2$:

$$\frac{\Delta Z}{Z} = 2\frac{\Delta A}{A}$$

Thus the fractional error in Z is twice the fractional error in A.

Now consider the situation where Z depends on more than one variable, e.g. $Z = A + B$. As before we can differentiate. However, when we have more than one variable, ordinary differentiation is replaced by *partial differentiation*. This involves differentiating assuming that only one variable is allowed to vary at a time, the others being temporarily constant. Thus, taking A to vary and B to be constant:

$$\frac{\partial Z}{\partial A} = 1$$

Note that partial derivatives are always written using a curly ∂ instead of the straight d used with ordinary derivatives. Taking B to vary and A to be constant:

$$\frac{\partial Z}{\partial B} = 1$$

As before, we can consider finite elements of the quantities. Thus when A varies and B is constant:

$$\Delta Z = \Delta A$$

and when B varies and A is constant:

$$\Delta Z = \Delta B$$

When we have a situation where both A and B can vary, then:

$$\Delta Z = \Delta A + \Delta B$$

This is just the equation we arrived at earlier.

In general, the equation we are using is:

$$\Delta Z = \frac{\partial Z}{\partial A}\Delta A + \frac{\partial Z}{\partial B}\Delta B$$

This equation can be expanded to as many terms as we require.

Consider $Z = A\,e^B$ when there are errors in A and B. Taking B to be constant:

$$\frac{\partial Z}{\partial A} = e^B$$

Taking A to be constant:

$$\frac{\partial Z}{\partial B} = A\,e^B$$

Thus the error in Z is given by:

$$\Delta Z = e^B\,\Delta A + A\,e^B\,\Delta B$$

Consider a measurement of the refractive index n in which the angle of incidence i of a beam of light and its angle of refraction r are measured. We then have:

$$n = \frac{\sin i}{\sin r}$$

Taking r to be constant:

$$\frac{\partial n}{\partial i} = \frac{\cos i}{\sin r}$$

Taking i to be constant:

$$\frac{\partial n}{\partial r} = \frac{-\sin i \cos r}{\sin^2 r}$$

When we are determining the error with both quantities varying, we take just the magnitude of the partial differential, ignoring any minus signs that may occur. This avoids cancellation of one error by another. Thus

$$\Delta Z = \frac{\cos i}{\cos r}\Delta i + \frac{\sin i \cos r}{\sin^2 r}\Delta r$$

The focal length f of a lens can be determined from measurements of the object distance u and the image distance v, with:

$$\frac{1}{f} = \frac{1}{u} + \frac{1}{v}$$

Consider the determination of an equation for the worst possible error in the focal length when there are errors in the object distance and the image distance. The equation can be rearranged to give:

$$f = \frac{uv}{u+v}$$

Thus:

$$\frac{\partial f}{\partial u} = \frac{(u+v)v - uv}{(u+v)^2}$$

and

$$\frac{\partial f}{\partial v} = \frac{(u+v)u - uv}{(u+v)^2}$$

Hence:

$$\Delta f = \left(\frac{(u+v)v - uv}{(u+v)^2} \right) \Delta u + \left(\frac{(u+v)u - uv}{(u+v)^2} \right) \Delta v$$

2.5.4 Partial differentiation and the standard error

The above analysis gives the worst possible errors. To obtain the standard errors we have to square and average the square over a set of measurements. Taking the square of the error gives:

$$(\Delta Z)^2 = \left(\frac{\partial Z}{\partial A} \Delta A + \frac{\partial Z}{\partial B} \Delta B \right)^2$$

and hence:

$$(\Delta Z)^2 = \left(\frac{\partial Z}{\partial A} \right)^2 (\Delta A)^2 + \left(\frac{\partial Z}{\partial B} \right)^2 (\Delta B)^2 + 2\frac{\partial Z}{\partial A} \Delta A \frac{\partial Z}{\partial B} \Delta B$$

We can write such an equation for each measurement. Now consider all the measurements and the average error. The sum of all the last terms will be zero since ΔA and ΔB will have both positive and negative values. Thus, the standard error is given by:

$$(\Delta Z)^2 = \left(\frac{\partial Z}{\partial A} \right)^2 (\Delta A)^2 + \left(\frac{\partial Z}{\partial B} \right)^2 (\Delta B)^2$$

To illustrate the above, consider the determination of an equation for the standard error for the total resistance R of two resistances R_1 and R_2 in parallel due to errors in each:

$$\frac{1}{R} = \frac{1}{R_1} + \frac{1}{R_2}$$

We can write this as:

$$R = \frac{R_1 R_2}{R_1 + R_2}$$

This gives:

$$\frac{\partial R}{\partial R_1} = \frac{R_2(R_1 + R_2) - R_1 R_2}{(R_1 + R_2)^2}$$

$$\frac{\partial R}{\partial R_2} = \frac{R_1(R_1 + R_2) - R_1 R_2}{(R_1 + R_2)^2}$$

Hence:

$$(\Delta R)^2 = \left[\frac{R_2(R_1 + R_2) - R_1 R_2}{(R_1 + R_2)^2} \right]^2 (\Delta R_1)$$

$$+ \left[\frac{R_1(R_1 + R_2) - R_1 R_2}{(R_1 + R_2)^2} \right]^2 (\Delta R_2)$$

2.6 Overall instrument error
Instruments are frequently specified as having a number of errors, e.g. a pressure sensor with a linearity error ±0.5% of full-scale reading and hysteresis error ±0.1% of full-scale reading. What will be the error when the instrument is used to make a single reading? Each error will combine in some way with other errors to increase the overall error of the measurement. The worst possible error (see Section 2.5.1) would be given by just adding together the error terms. Thus for the pressure sensor the worst possible error would be ±0.6%. However, a more realistic estimation of the error (see Section 2.5.2) would be given by using:

(overall error)^2 = ±(sum of the squares of the various errors)

Hence, for the pressure sensor, the overall error is given by:

(overall error)^2 = ±(0.5^2 + 0.1^2)

and hence is ±0.51% of the full-scale reading.

As a further example, consider a displacement measurement system which is specified as having a sensor with a linearity error of ±0.25% of the reading and a repeatability error of ±0.25% of the reading and is to be used with a voltmeter display which has an accuracy of ±0.1% of the reading. The overall error is thus:

(overall error)^2 = ±(0.25^2 + 0.25^2 + 0.1^2)

Thus the overall error is ±0.37% of the reading.

Consider another example. A force measurement system, e.g. a spring balance, is specified as having a non-linearity error of ±0.20%, a repeatability error of ±0.30% and a resolution of 0.5 N. What will be the error in a reading of 200 N? The resolution indicates that there is a reading error of ±0.25 N or ±(0.25/200) × 100 = ±0.125 %. Thus the most likely overall error is given by:

(overall error)^2 = ±(0.20^2 + 0.30^2 + 0.125^2)

The error is thus ±0.38% = ±0.76 N.

2.7 The best straight line

Consider the problem of determining the best straight line through a series of points on a graph, e.g. the points shown in Figure 2.9. We can do this by drawing by eye the best straight line so that those points above the line balance those below the line. To simplify the discussion, we are considering that the errors are entirely in the y measurements. Thus the best line is drawn by considering the deviations of the points in the y directions from the line and trying to achieve a balance. Figure 2.9 illustrates this.

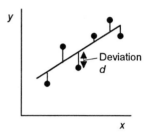

Figure 2.9 *Drawing the best straight line*

The deviation d of any one point from the line is given by:

deviation $d = y - \hat{y}$

where y is the measured value and y the value given by the straight line. The value given by the straight line is:

$\hat{y} = mx + c$

where x is the value of the independent variable, m the gradient of the graph and c its intercept with the y axis. Thus, we can write:

$d = y - (mx + c)$

We can write such an equation for each data point. Thus, for point j we can write for its deviation d_j:

$d_j = y_j - (mx_j + c)$

with y_j and x_j being the pair of data values. We might consider that the best line would be the one that minimises the sum of the deviations. However, a better solution is to find the line that minimises the standard errors of the points in their deviations from the straight line. This means choosing the values of m and c which minimise the sum of the squares of these deviations. This is termed the *least squares estimate* for the best straight line. Hence the best values of m and c are obtained when:

sum of the squares $S = \sum_{j=1}^{n} d_j^2 = \sum_{j=1}^{n} [y_j - (mx_j + c)]^2$

We can consider there to be many values of m and c, i.e. lots of straight lines that can be drawn, and we need to determine the values which result in a minimum value for the sum. If we had a simple equation such as $S = mx^2$ we could find the minimum value of the sum with respect to m by differentiating and then equating the derivative to zero, i.e. finding the value of m that gave $dS/m = 0$. But we have two variables, m and c. Thus, in this case we have to determine the partial derivatives $\partial S/\partial m$ and $\partial S/\partial c$ and equate them to zero. Expanding the equation for the sum gives:

$$S = \sum \left[y_j^2 - 2y_j(mx_j + c) + (mx_j + c)^2 \right]$$

Hence:

$$\frac{\partial S}{\partial m} = -2 \sum x_j(y_j - mx_j - c)$$

and:

$$\frac{\partial S}{\partial c} = -2 \sum(y_j - mx_j - c)$$

Equating both these partial derivatives to zero gives:

$$-2 \sum x_j(y_j - mx_j - c) = 0$$

$$-2 \sum(y_j - mx_j - c) = 0$$

We can rearrange these equations to give:

$$m \sum x_j^2 + c \sum x_j = \sum x_j y_j$$

and

$$m \sum x_j + nc = \sum y_j$$

Thus all we need to do is put the data into the above two equations and solve the pair of simultaneous equations to obtain values for m and c. We then have:

$$m = \frac{n \sum x_j y_j - \sum y_j \sum x_j}{n \sum x_j^2 - (\sum x_j)^2}$$

and:

$$c = \frac{\sum x_j^2 \sum y_j - \sum x_j \sum x_j y_j}{n \sum x_j^2 - \left(\sum x_j\right)^2}$$

We can, however, put these equations in another form. Since the average values of x and y are given by:

$$\bar{x} = \frac{1}{n} \sum x_j$$

$$\bar{y} = \frac{1}{n} \sum y_j$$

we can write m as:

$$m = \frac{\sum x_j(y_j - \bar{y})}{\sum x_j(x_j - \bar{x})}$$

and, since the sum of the deviations of y about the mean value of y must be zero, i.e. $\sum(y_j - \bar{y}) = 0$, then:

$$m = \frac{\sum(x_j - \bar{x})(y_j - \bar{y})}{\sum(x_j - \bar{x})^2}$$

To obtain a value for c, we can substitute this value of m into the equation $m \sum x_j + nc = \sum y_j$ when rewritten in terms of the mean values, namely:

$$c = \bar{y} - m\bar{x}$$

Incidentally, this equation is of the form $\bar{y} = m\bar{x} + c$ and so shows that the least squares line passes through the mean values of the data.

To illustrate the above, consider the following data which was obtained from an experiment in which the resistance R of a coil of metal wire was determined at a number of different temperatures:

Resistance (Ω)	76.4	82.7	87.8	94.0	103.5
Temperature (°C)	20.5	32.5	52.0	73.0	96.0

The relationship between the resistance and temperature θ is envisaged as being of the form:

$$R = m\theta + c$$

We can use the least squares method to determine the values of m and c in the equation.

Table 2.4 shows the calculations of the values of the sum of the temperature data, the sum of the squares of the temperature data, the sum of the resistance data and the sum of the products of the two sets of data.

Table 2.4 *Calculations for the example*

$R\ (\Omega)$	$\theta\ (°C)$	$\theta^2\ (°C)^2$	$R\theta\ (\Omega\ °C)$
76.4	20.5	420.25	1566.20
82.7	32.5	1056.25	1687.75
87.8	52.0	2704.00	4565.60
94.0	73.0	5329.00	6862.00
103.5	96.0	9216.00	9936.00
$\Sigma R = 444.4$	$\Sigma \theta = 274$	$\Sigma \theta^2 = 18\ 725.50$	$\Sigma R\theta = 25\ 617.55$

Using the equation for the gradient, namely:

$$m = \frac{n \Sigma x_j y_j - \Sigma y_j\ \Sigma x_j}{n \Sigma x_j^2 - (\Sigma x_j)^2}$$

and taking the values from the table:

$$m = \frac{5 \times 25\ 617.55 - 444.4 \times 274}{5 \times 18\ 725.50 - 274^2}$$

and hence $m = 0.341$. Using the equation for the intercept, namely:

$$c = \frac{\Sigma x_j^2\ \Sigma y_j - \Sigma x_j\ \Sigma x_j y_j}{n \Sigma x_j^2 - \left(\Sigma x_j\right)^2}$$

and taking the values from the Table:

$$c = \frac{18\ 725.50 \times 444.4 - 274 \times 25\ 617.55}{5 \times 18\ 725.50 - 274^2}$$

Hence $c = 70.19$. The equation is thus:

$$R = 0.341\theta + 70.19$$

Alternatively, to use equations for determining m and c written in terms of the deviations from the means, we need to tabulate the data in a different form. Table 2.5 shows the table. Thus using:

$$m = \frac{\Sigma(x_j - \bar{x})(y_j - \bar{y})}{\Sigma(x_j - \bar{x})^2}$$

gives:

$$m = \frac{1263.59}{3710.30} = 0.341$$

Table 2.5 *Calculations for the example*

$R\,(\Omega)$	$\theta\,(°C)$	$R-\text{mean}\,(\Omega)$	$\theta-\text{mean}\,(°C)$	$(R-\text{mean}) \times (\theta-\text{mean})$	$(\theta-\text{mean})^2$
76.4	20.5	−12.48	−34.3	428.06	1176.49
82.7	32.5	−6.18	−22.3	137.81	497.29
87.8	52.0	−1.08	− 2.8	3.02	7.84
94.0	73.0	5.12	18.2	93.18	331.24
103.5	96.0	14.6	41.2	601.52	1697.44
Mean = 88.88	Mean = 54.8			$\Sigma = 1263.59$	$\Sigma = 3710.30$

Using $c = \bar{y} - m\bar{x}$ gives:

$$c = 88.88 - 0.341 \times 54.8 = 70.19$$

Thus, as before, $R = 0.341\theta + 70.19$.

2.7.1 Errors

In the calculations of the gradient and intercept given above, no estimate was given of the possible error in their values. We used values for y with no assumption of any error in the value used. But we might repeat the measurement of the y value and obtain a different value for the same x value. Indeed we might obtain a distribution of y values for a particular x value.

For example, in an experiment involving a simple pendulum, we might measure the time taken for 20 oscillations when the pendulum length is 0.50 m and repeat the time measurement at that length 5 times. We can thus obtain the error in the time, the y value, at a fixed value of length x. We then obtain another data point by repeating the time measurement for a length of 0.60 m. We will assume that each point has the same standard error in its y values, i.e. the standard error in the time quoted for each of these two lengths is the same. We can thus consider that we have effectively a set of y values at each x value. For each y value at each x value in combination with each of the values at the other x values, we can obtain, by the least squares method, values of the gradient and intercept. Thus we end up with an entire set of straight lines distributed about the mean straight line. What we need to find out is the standard error of the gradients and of the intercepts.

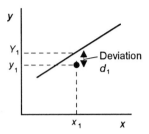

Figure 2.10 *Deviation of a point from a line*

The deviation d_1 of a point at y_1 from the straight line (Figure 2.10), where Y_1 is the value on the line at x_1, is:

$$d_1 = y_1 - Y_1 = y_1 - mx_1 - c$$

We can write similar equations for each point. All the deviations of the measured points about the mean straight line are typical of the deviation about that line that could occur with just one of the points. The standard deviation σ of these points is thus:

$$\sigma^2 = \frac{d_1^2 + d_2^2 + d_3^2 + \dots d_n^2}{n-2} = \frac{1}{n-2} \Sigma d_j^2$$

We are dividing by $(n - 2)$, where n is the number of data points, because there are two restrictions placed on the possible values of y because if we only had two points there would be only one possible line and it would necessarily be the best line. Thus the degree of freedom is reduced by 2 when we consider possible straight lines.

This uncertainty in the value of y will contribute to an uncertainty in the values of m and c. The gradient is given by:

$$m = \frac{\Sigma(x_j - \bar{x})(y_j - \bar{y})}{\Sigma(x_j - \bar{x})^2}$$

If M is the true value of the gradient given by the values of Y, then:

$$M = \frac{\Sigma(x_j - \bar{x})(Y_j - \bar{y})}{\Sigma(x_j - \bar{x})^2}$$

Hence:

$$m - M = \frac{\Sigma(x_j - \bar{x})(y_j - \bar{y})}{\Sigma(x_j - \bar{x})^2} - \frac{\Sigma(x_j - \bar{x})(Y_j - \bar{y})}{\Sigma(x_j - \bar{x})^2}$$

$$= \frac{\Sigma(x_j - \bar{x})(y_j - Y_j)}{\Sigma(x_j - \bar{x})^2}$$

$(m - M)$ is the deviation of the gradient from its true line and $(y_j - Y_j)$ the deviation of the y values from their true values, hence the standard deviation σ_m of the gradient is given by:

$$\sigma_m^2 = \frac{\sigma^2 \Sigma(x_j - \bar{x})^2}{\left[\Sigma(x_j - \bar{x})^2\right]^2} = \frac{\sigma^2}{\Sigma(x_j - \bar{x})^2}$$

Since $\Sigma(x_j - \bar{x})^2 = \Sigma x_j^2 - n\bar{x}^2$, we an write the above equation as:

$$\sigma_m^2 = \frac{\sigma^2}{\Sigma x_j^2 - n\bar{x}^2}$$

and so:

$$\sigma_m^2 = \frac{\Sigma d_j^2}{(n-2)\left(\Sigma x_j^2 - n\bar{x}^2\right)}$$

Similarly the standard deviation σ_c for the intercept values can be derived from the simultaneous equations as:

$$\sigma_c^2 = \frac{\Sigma d_j^2 \Sigma x_j^2}{n(n-2)\left(\Sigma x_j^2 - n\bar{x}^2\right)}$$

Since $\bar{x} = \frac{1}{n}\Sigma x_j$, we can also write the equations as:

$$\sigma_m^2 = \frac{\Sigma d_j^2}{(n-2)\left[\Sigma x_j^2 - \frac{1}{n}\left(\Sigma x_j\right)^2\right]} = \frac{n\Sigma d_j^2}{(n-2)\left[n\Sigma x_j^2 - \left(\Sigma x_j\right)^2\right]}$$

and:

$$\sigma_c^2 = \frac{\Sigma d_j^2 \Sigma x_j^2}{n(n-2)\left[\Sigma x_j^2 - \frac{1}{n}\left(\Sigma x_j\right)^2\right]} = \frac{\Sigma d_j^2 \Sigma x_j^2}{(n-2)\left[n\Sigma x_j^2 - \left(\Sigma x_j\right)^2\right]}$$

There is an alternative way of deriving the standard error for the intercept and that uses the equation:

$$c = \bar{y} - m\bar{x}$$

The error in c arises from errors in two terms, namely $\bar{y}$ and m, the error in x and hence $\bar{x}$ having been assumed to be insignificant. When two quantities with standard errors are added or subtracted, the square of the resulting standard error is equal to the sums of the squares of the two errors. Thus:

$$\sigma_c^2 = \sigma_{\bar{y}}^2 + (\sigma_m \bar{x})^2$$

$\sigma_{\bar{y}}$ is the standard error of the mean of y and thus:

$$\sigma_c^2 = \frac{\Sigma d_j^2}{n(n-2)} + (\sigma_m \bar{x})^2$$

This is, with some manipulation, the same as the equation for σ_c derived from the simultaneous equations.

To illustrate this, consider the standard errors for the example given earlier which gave the straight line $R = 0.341\theta + 70.19$. Taking the data values given in that example we can derive the deviations of the points from the line. Table 2.6 shows the result.

Table 2.6 *Derivation of the deviations*

$R\ (\Omega)$	$\theta\ (°C)$	True $R\ (\Omega)$	Deviation (Ω)	(Deviation)² (Ω^2)
76.4	20.5	77.1805	−0.7805	0.6092
82.7	32.5	81.2725	1.4275	2.0378
87.8	52.0	87.9220	−0.1220	0.0149
94.0	73.0	95.0830	−0.1083	0.0173
103.5	96.0	102.926	0.5740	0.3295
				$\Sigma = 3.0087$

Thus:

$$\sigma_c^2 = \frac{n \Sigma d_j^2}{(n-2)\left[n \Sigma x_j^2 - \left(\Sigma x_j \right)^2 \right]} = \frac{5 \times 3.0087}{3(5 \times 18\,725.50 - 274^2)}$$

and so the standard error in the gradient is 0.016. The standard error in the intercept is given by:

$$\sigma_c^2 = \frac{\Sigma d_j^2 \Sigma x_j^2}{(n-2)\left[n \Sigma x_j^2 - \left(\Sigma x_j \right)^2 \right]} = \frac{3.0087 \times 18\,725.50}{3(5 \times 18\,725.50 - 274^2)}$$

and hence is 1.0. Alternatively, we could calculate σ_c by using:

$$\sigma_c^2 = \frac{\Sigma d_j^2}{n(n-2)} + \left(\sigma_c\,\bar{x}\right)^2 = \frac{3.0087}{5\times3} + 0.016^2 \times 54.8^2$$

and this gives the standard error as 1.0. Thus the equation can be written as:

$$R = (0.34 \pm 0.02)\theta + (70 \pm 1)$$

2.7.2 Least squares regression analysis

The analysis carried out above is the fitting of data points to an equation of the form $y = mx + c$. Such an operation of fitting data to an equation is called *regression* and the technique we have been using is called *least squares*. For more details the reader is referred to texts such as *Introduction to Probability and Statistics for Engineers and Scientists* by S.M. Ross (Wiley 1987) and *Data Reduction and Error Analysis for the Physical Sciences* by P.R. Bevington and D.K. Robinson (McGraw-Hill 1992, 1969).

Problems

1 A thermometer has graduations at intervals of 0.5°C. What is the worst possible reading error?

2 An instrument has a scale with graduations at intervals of 0.1 units. What is the worst possible reading error?

3 Determine the means and the standard deviations for the following sets of results:
(a) The times taken for 10 oscillations of a simple pendulum: 51, 49, 50, 49, 52, 50, 49, 53, 49, 52 s.
(b) The diameter of a wire when measured at a number of points using a micrometer screw gauge: 2.11, 2.05, 2.15, 2.12, 2.16, 2.14, 2.16, 2.17, 2.13, 2.15 mm.
(c) The volume of water passing through a tube per 100 s time interval when measured at a number of times: 52, 49, 54, 48, 49, 49, 53, 48, 50, 53 cm³.

4 Repeated measurements of the forces necessary to break a tensile test specimen gave: 802, 799, 800, 798, 801 kN. Determine (a) the average force, and (b) the standard error of the mean.

5 Repeated measurements of the resistance of a resistor gave: 51.1, 51.3, 51.2, 51.3, 51.7, 51.0, 51.5, 51.3, 51.2, 51.4 Ω. Determine (a) the average resistance, and (b) the standard error of the mean.

6 Repeated measurements of the voltage necessary to cause the breakdown of a dielectric gave: 38.9, 39.3, 38.6, 38.8, 38.8, 39.0, 38.7,

39.4, 39.7, 38.4, 39.0, 39.1, 39.1, 39.2 kV. Determine (a) the average breakdown voltage, and (b) the standard error of the mean.

7 A set of six readings of the resistance of a 1 m length of wire gave the results: 18.6, 20.5, 19.2, 18.4, 20.8, 19.4 Ω. Determine the 95% confidence interval for the resistance, assuming that the distribution is normal.

8 The times for 20 oscillations of a simple pendulum give results with a mean time of 19.80 s and a standard deviation of 1.20 s. Determine the 95% confidence interval.

9 The total resistance of two resistors in series is the sum of their resistances. Determine the worst possible error in the total resistance if the resistors are 50 Ω with 10% accuracy and 100 Ω with 5% accuracy.

10 When two resistors are connected in parallel, the total resistance is given by $R = R_1 R_2/(R_1 + R_2)$. What will be the worst possible error in the total resistance if the resistors are 50 Ω with a 10% accuracy and 100 Ω with a 5% accuracy?

11 The volume of a cube with sides of L is L^3. If the length is measured as 121 ± 2 mm, determine the worst possible error in the volume.

12 The density of a solid is its mass divided by its volume. If the mass is measured as 42.5 ± 0.5 g and the volume as 54 ± 1 cm^3, what will be the worst possible error in the calculated density?

13 Two objects are weighed, giving 100 ± 0.5 g and 50 ± 0.3 g. What will be the average error of the sum of the two weights?

14 The distance s travelled by a car when travelling with a constant speed v is estimated from a reading of the speedometer and a measurement of time t travelled at that speed. If $s = vt$, what will be the distance travelled and the average error if the speed is 60 ± 2 km/h and the time is 1 ± 0.01 h?

15 The volume of a rectangular solid is determined from measurements of its length, height and breadth. If they are 100 ± 1 mm, 50 ± 0.5 mm and 40 ± 0.5 mm, what will be the volume and the average error?

16 The stress σ acting on a rectangular cross-sectional strip of material is determined from measurements of the force F and the cross-sectional area A, where $\sigma = F/A$. What will be the stress and the average error if the force is 20.0 ± 0.5 kN and the area is determined from measurements of the width and breadth of the strip as 5.0 ± 0.5 mm and 10.0 ± 0.5 mm?

17 The area S of a triangle is determined using the equation $S = \frac{1}{2}bc \sin A$. Determine an equation for the worst possible error in the area from measurements of the sides b and c and the angle A.

18 The number of active nuclei N in a sample of radioactive material after a time t is given by $N = N_0\, e^{-\lambda t}$, where N_0 is the number at time $t = 0$. Determine an equation for the worst possible error in N due to errors in N_0 and t.

19 The viscosity η of a liquid can be determined from measurements of the terminal velocity v of a sphere falling through the liquid, with:

$$\eta = \frac{2r^2 g(\rho_s - \rho_l)}{9v}$$

where r is the radius of the sphere, g the acceleration due to gravity, ρ_s the density of the sphere and ρ_l the density of the liquid. Determine an equation for the worst possible error in the viscosity due to errors in r and v, g and the densities being assumed to have insignificant errors.

20 A pressure sensor is specified as having a non-linearity error of $\pm0.15\%$ of the full-scale reading, a hysteresis error of $\pm0.20\%$ of the full-scale reading and a repeatability error of 0.25% of the full-scale reading. What is (a) the overall worst possible error, (b) the overall most likely error?

21 An electronic balance is specified as having, on its 100 g range, a non-linearity error of ±0.0002 g and a repeatability error of ±0.0001 g. What is (a) the overall worst possible error, (b) the overall most likely error?

22 A pressure measurement system is specified as having a non-linearity error of $\pm0.5\%$ of the full-scale reading, a hysteresis error of $\pm0.5\%$ of the full-scale reading and a repeatability error of $\pm0.4\%$ of the full-scale reading. What is (a) the overall worst possible error, (b) the overall most likely error?

23 Determine, by means of the least squares method and assuming that the only significant errors are in the y values, the best straight line relationship to fit the following data:

x	65	68	71	75	77	80	84	87	93	98
y	72	72	80	82	74	78	89	91	96	95

24 The extension e of a spring is measured when different loads W are applied to stretch it and the following data obtained:

W (kg)	1.0	2.0	3.0	4.0	5.0
e (mm)	8	16	27	32	35

If the error is only in the extension values, determine the gradient and the intercept of the best straight line through the points.

25 The mass m of a compound which will dissolve in 100 g of water is measured at different temperatures θ and the following results obtained:

θ (°C)	0	10	20	30	40	50	60	70	80	90	100
m (g)	43.5	49.5	55.2	60.6	65.5	70.2	75.5	80.0	85.0	89.2	94.0

If the error is only in the mass values, determine the gradient and the intercept of the best straight line through the points.

26 The depression d of the end of a cantilever as a result of weight W being added was measured and the following results obtained:

W (kg)	0.25	1.00	2.25	4.00	6.25
d (mm)	2.2	10.0	22.3	39.4	61.7

If the error is only in the depression values, determine the gradient and the intercept of the best straight line through the points.

27 Determine, by means of the least squares method and assuming that the only significant errors are in the y values, the best straight line relationship to fit the following data:

x	1	2	3	4	5	6
y	1	2	2	3	5	5

28 Determine, by means of the least squares method and assuming that the only significant errors are in the y values, the best straight line relationship to fit the following data:

x	1	2	3	4	5	6	7	8	9	10
y	2.3	4.5	5.2	6.0	8.0	9.8	11.1	13.5	14.0	16.6

29 Measurements are made of the effort E to raise load W and the following results obtained:

W (N)	14	42	84	112
E (N)	6.1	14.3	27.0	36.3

Determine, assuming the only significant errors are in E, the equation of the best straight line through the data.

30 The following data is expected to fit a relationship of the form $y = b\,e^{ax}$. By putting the relationship in a form which would give a straight line graph, use the least squares method to determine the values of a and b. The errors are only significant for y.

x	1.00	1.25	1.50	1.75	2.00
y	5.10	5.79	6.53	7.45	8.46

3 Modelling measurement systems

In order to gain an understanding of how measurement systems and their constituent elements react when there are signal inputs to them, we need to develop mathematical models. Then the reaction of those models to different inputs can be studied. If the input signal to a measurement system, or an element of such a system, suddenly changes, the output will not instantaneously change to the new value but some time will elapse before it reaches a steady-state value. The way in which a system reacts to input changes is termed its *dynamic characteristic*. Such characteristics are simplified by using a *transfer function*. This chapter is about dynamic characteristics and the use of the transfer function to describe input-output relationships.

3.1 Zero order elements

Consider a measurement system element such as a potentiometer (Figure 3.1). The input to the element is a displacement of the sliding contact and the output is a voltage which is related to the position of the sliding contact. If y is the input and x the output, then, if the resistance is distributed uniformly along the length of the slide wire, the relationship between the output and input can be described by the equation:

$$x = ky$$

where k is a constant, often termed the *gain* since it indicates how much bigger the output signal is than the input signal. The output is directly proportional to the input.

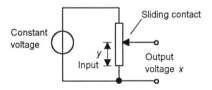

Figure 3.1 *Input and output for a potentiometer*

With the above mathematical model for the potentiometer there is no time term involved in the equation. Thus we should expect that the output reacts instantaneously to any change in input. Thus for a *step input*, i.e. a sudden abrupt change in the input (Figure 3.2(a)), the output is likewise a step (Figure 3.2(b)), since at every instant of time the output has to be directly proportional to the input. Whatever the form of the input, the output is expected to follow it perfectly with no distortion or time lag. An element which has this characteristic is termed a *zero order element*.

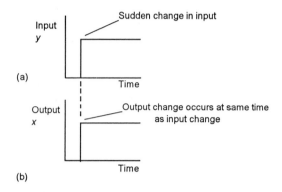

Figure 3.2 *Reaction of a zero order element to a step input*

With a zero order element, if we have a *ramp* change in the input, i.e. an input which increases at a constant rate with time, then the output follows the changing input perfectly. Figure 3.3 shows how the input and output changes might appear.

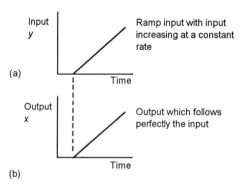

Figure 3.3 *Reaction of a zero order element to a ramp input*

Zero order elements are ideal or perfect elements, no lag occurring between the input and the output. If we have a sinusoidal input then we obtain a sinusoidal output which is perfectly in phase with the input, no lag occurring. In practice, not even the potentiometer is so perfect. There is inevitably some time lag between the input changing and the output changing.

3.2 First order elements

Elements are said to be *first order* when the relationship between the input and the output involves the rate at which the output changes. To illustrate this, consider a thermometer at temperature T_0 inserted into a liquid at a temperature T_1. We can thus think of the thermometer being subject to a step input, i.e. the input abruptly changes from T_0 to T_1. The thermometer will change its temperature until it becomes T_1. Thus we have a measurement system, the thermometer, which has a step input and an output which changes from T_0 to T_1 over some time. The question we are concerned with is how the output, i.e. the reading of the thermometer T, varies with time.

The rate at which energy enters the thermometer from the liquid is proportional to the difference in temperature between the liquid and the thermometer. Thus, at some instant of time when the temperature of the thermometer is T, we can write:

$$\frac{dQ}{dt} = h(T_1 - T)$$

where h is a constant called the *heat transfer coefficient*. If the thermometer has a specific heat capacity c and a mass m, then since the relationship between heat input Q and the consequential temperature change is:

$$Q = mc \text{ (temperature change)}$$

When the rate at which heat enters the thermometer is dQ/dt, we can write for the rate at which the temperature changes:

$$\frac{dQ}{dt} = mc\frac{dT}{dt}$$

Thus:

$$mc\frac{dT}{dt} = h(T_1 - T)$$

We can rewrite this with all the output terms on one side of the equals sign and the input on the other, thus:

$$mc\frac{dT}{dt} + hT = hT_1$$

We no longer have a simple relationship between the input and output but a relationship which involves time. The form of this equation is typical of first order elements.

The solution to the above differential equation indicates how the output T changes with time t when there is the step input (see Appendix B or books such as *Ordinary Differential Equations* by W. Bolton (Longman 1994) or *Introduction to Ordinary Differential Equations* by S.L. Ross (Wiley 1989) for methods for the solution of first order differential equations) and is:

$$T = T_1 + (T_0 - T_1)\, e^{-t/\tau}$$

where $\tau = mc/h$ and is termed the *time constant*. The first term is the *steady-state value*, i.e. the value that will occur after sufficient time has elapsed for all transients to die away, and the second term a transient one which changes with time, eventually becoming zero. Figure 3.4 shows graphically how the temperature T changes with time. After a time equal to one time constant the output has reached about 63% of the way to the steady-state temperature, after a time equal to two time constants the output has reached about 86% of the way, after three time constants about 95% and after about four time constants it is virtual equal to the steady-state value.

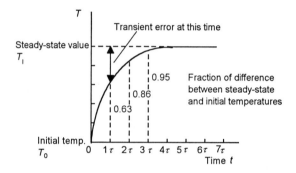

Figure 3.4 *Response of a first order system to a step input*

The error at any instant is the difference between what the thermometer is indicating and what the temperature actually is. Thus:

error $= T - T_1$

and so:

error $= (T_0 - T_1)\, e^{-t/\tau}$

The percentage error is thus:

% error $= e^{-t/\tau} \times 100$

This error changes with time and eventually will become zero. Thus it is a transient error.

To illustrate the above, consider a thermometer which is indicating a temperature of 20°C when it is suddenly immersed in a liquid at a temperature of 60°C. If the thermometer has a time constant of 5 s then the temperature T of the thermometer varies with time according to the equation:

$$T = T_1 + (T_0 - T_1)\, e^{-t/\tau} = 60 - 40\, e^{-t/5}$$

After 5 s the thermometer reading will have reached about 63% of the way to the steady-state value, after 10 s about 86%, after 15 s about 95% and after 20 s it is virtually at the steady-state value. Thus after 5 s the reading is 45.3°C, after 10 s it is 54.6°C, after 15 s it is 58.0°C.

As a further illustration, consider a thermometer which behaves as a first order element with a time constant of 15 s. Initially it reads 20°C. What will be the time taken for the temperature to rise to 90% of the steady-state value when it is immersed in a liquid of temperature 100°C, i.e. a temperature of 92°C? Using the equation:

$$T = T_1 + (T_0 - T_1)\, e^{-t/\tau}$$

This can be rearranged to give:

$$\frac{T - T_1}{T_1 - T_0} = e^{-t/\tau}$$

With $T - T_0$ as 90% of $T_1 - T_0$, then we have $T - T_1$ as 10% of $T_1 - T_0$ and thus:

$$0.10 = e^{-t/15}$$

Taking logarithms gives $-2.30 = -t/15$ and so $t = 34.5$ s.

If a thermometer is required to be fast reacting and quickly attain the temperature being measured, it needs to have a small time constant. Since $\tau = mc/h$, this means a thermometer with a small mass, a small thermal capacity and a large heat transfer coefficient. If we compare a mercury-in-glass thermometer with a thermocouple, then the smaller mass and specific heat capacity of the thermocouple will give it a smaller time constant and hence a faster response to temperature changes.

3.2.1 First order system in general

In general, a first order system has a differential equation which can be written in the form:

$$a_1 \frac{dx}{dt} + a_0 x = b_0 y$$

where x is the input and y the output. Alternatively it can be written as:

$$\tau \frac{dx}{dt} + x = \frac{b_0}{a_0} y$$

where τ is the time constant and is given by (a_1/a_0). The solution of the differential equation for a step input from initial value to final value at time $t = 0$ is of the form:

$$x = \text{steady-state value} + (\text{initial value} - \text{steady-state value})\, e^{-t/\tau}$$

Table 3.1 shows the percentage of the response, i.e. the difference between the initial and steady-state values, that will have been achieved after various multiples of the time constant.

Table 3.1 *First order system response*

Time	% response	% dynamic error
0	0.0	100.0
1τ	63.2	36.8
2τ	86.5	13.5
3τ	95.0	5.0
4τ	98.2	1.8
5τ	99.3	0.7
∞	100.0	0.0

3.2.2 Ramp input

Consider now the response of a thermometer to a ramp input, i.e. a steadily increasing temperature input. Thus if the thermometer is initially at a temperature of T_0 and the temperature T_1 of the liquid in which the thermometer is immersed is given by:

$$T_1 - T_0 = at$$

where a is a constant and t the time, i.e. the temperature is rising at a steady rate from the initial reading, then the differential equation becomes:

$$mc\frac{dT}{dt} + hT = hat$$

and the solution of this differential equation is:

$$T = a(\tau e^{-t/\tau} + t - \tau) + T_0$$

The measurement error at any instant is:

$$error = T - T_1 = T - at - T_0$$

and so:

$$error = a\tau e^{-t/\tau} - a\tau$$

The first term varies with the time t and eventually will die away. It is thus the transient error. The second term is a constant and so is an error that is always going to be present. It is termed the *steady-state error*. The steady-state error means that even when the thermometer reaches the steady-state value it will always indicate a temperature which lags behind the actual temperature. Figure 3.5 illustrates this.

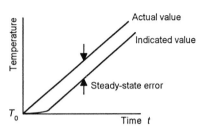

Figure 3.5 *Steady-state error*

3.2.3 Sinusoidal input

Consider another form of input signal, a sinusoidally varying input. Such a periodic form of input is often encountered with measurement system elements. For such an input to a first order system we have, using the symbols given above, a differential equation of the form:

$$mc\frac{dT}{dt} + hT = hA \sin \omega t$$

or, in terms of the time constant,

$$\tau\frac{d\theta}{dt} + T = A \sin \omega t$$

This equation has a solution of:

$$T = C\, e^{-t/\tau} + \frac{A}{\left[1 + (\omega t)^2\right]^{1/2}} \sin(\omega t - \tan^{-1}\omega t)$$

where C is a constant depending on the initial value of T. This equation can be written as:

$$T = C\, e^{-t/\tau} + B \sin\left(\omega t + \phi\right)$$

The first term is the transient response. The second term indicates the steady-state error and shows that the amplitude B of the output will depend on the frequency and that there will be a phase difference ϕ between the input and the output, this phase difference depending on the frequency. The phase difference occurs because we have a changing input and the output fails to keep up with it. Figure 3.6 illustrates these points for the steady state.

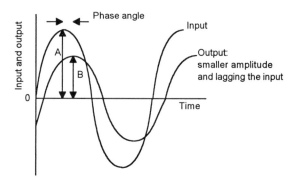

Figure 3.6 *Response of a first order element to a sinusoidal input*

3.3 Second order elements

There are many measurement system elements which behave in a similar manner to a damped mass on a spring when subject to inputs. An obvious example of such a system is a spring balance for the measurement of force. Another example is a diaphragm pressure gauge. Figure 3.7 illustrates the basic features of such systems.

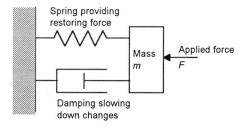

Figure 3.7 *Mass, spring, damper system*

Consider the system when a force F is applied to the mass. The net force applied to the mass is the applied force F minus the force resulting from the compressing, or stretching, of the spring and the force from the damper. The force resulting from compressing the spring is proportional to the change in length x of the spring, i.e. kx with k being a constant termed the spring stiffness. The damper can be thought of as a piston moving in a cylinder, though it can take many other forms. The force arising from the damping is proportional to the rate at which the displacement of the piston is changing, i.e. $c\, dx/dt$ with c being a constant. Thus:

$$\text{net force applied to mass} = F - kx - c\frac{dx}{dt}$$

This net force will cause the mass to accelerate. Acceleration is the rate of change of velocity dv/dt and velocity v is the rate of change of displacement dx/dt. Hence acceleration can be written as d^2x/dt^2. Thus:

$$m\frac{d^2x}{dt^2} = F - kx - c\frac{dx}{dt}$$

We can write this as:

$$m\frac{d^2x}{dt^2} + c\frac{dx}{dt} + kx = F$$

In the absence of damping, a mass m on the end of a spring freely oscillates with a natural angular frequency ω_n given by:

$$\omega_n = \sqrt{\frac{k}{m}}$$

If we define a constant ζ, termed the *damping ratio*, by:

$$\zeta = \frac{c}{2\sqrt{mk}}$$

then we can write the second order differential equation as:

$$\frac{1}{\omega_n^2}\frac{d^2x}{dt^2} + \frac{2\zeta}{\omega_n}\frac{dx}{dt} + x = \frac{F}{k}$$

Consider a step input such that the applied force jumps from zero to F at the time $t = 0$. When we solve this differential equation (see Appendix B or texts such as *Ordinary Differential Equations* by W. Bolton (Longman 1994) or *Introduction to Ordinary Differential Equations* by S.L. Ross (Wiley 1989) for methods for the solution of first order differential equations) we find that there are three possible forms of solution:

1   When we have the damping ratio with a value between 0 and 1:

$$x = C\,e^{-\zeta\omega_n t}\sin\left(\sqrt{1 - \zeta^2}\,\omega_n t + \phi\right) + \frac{F}{k}$$

where C is a constant determined by the damping factor and ϕ a phase difference determined by the damping factor. This describes a sinusoidal oscillation which is damped, the exponential term being the damping factor which gradually reduces the amplitude of the oscillation. Such a motion is said to be *under-damped*. The steady-state response is F/k.

2   When we have the damping ratio with the value 1:

$$x = C_1\,e^{m_1 t} + C_2 t\,e^{m_2 t} + \frac{F}{k}$$

where C_1 and C_2 are constants and m_1 and m_2 are constants. This describes a situation where no oscillations occur but x exponentially changes with time. Such a motion is said to be *critically damped*. The steady-state response is F/k.

3   When we have the damping ratio greater than 1:

$$x = C_1\ e^{m_1 t} + C_2\ e^{m_2 t} + \frac{F}{k}$$

where C_1 and C_2 are constants and m_1 and m_2 are constants. This describes a situation where no oscillations occur but x exponentially changes with time, taking longer to reach the steady-state value than the critically damped motion. Such a motion is said to be *over-damped*. The steady-state response is F/k.

Figure 3.8 shows the form the solution of the second order differential equation takes for different values of the damping ratio. The output is plotted as a multiple of the steady-state value F/k. Instead of just giving the output variation with time t, the axis used is $\omega_n t$. This is because t and ω_n always appear as the product $\omega_n t$ and using this product makes the graph applicable for any value of ω_n.

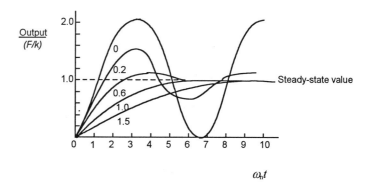

Figure 3.8 *Response of second order system to step input for different damping factors*

3.3.1 Dynamic characteristics

For under-damped systems, the angular frequency ω_d of the damped output is not constant but depends on the damping factor with:

$$\omega_d = \omega_n \sqrt{1 - \zeta^2}$$

Thus the bigger the damping factor the smaller the angular frequency.

The *rise time* is generally taken as the time taken for the output to rise to some particular percentage, e.g. 90 or 95%, of the steady-state output (see Section 1.5). As will be apparent from the graph, the larger the damping ratio the longer the rise time. Thus for an under-damped system, the time taken to reach the steady-state value axis is the time taken for one-quarter of a cycle of the oscillation. Since $\omega_d = 2\pi f = 2\pi/T$, where f is the frequency and T the time for one complete oscillation, then $T/4 = (2\pi/\omega_d)/4$. Hence for a rise time specified for 90% of the steady state output:

$$\text{rise time} = \frac{\pi}{2\omega_d} \times \frac{90}{100}$$

Since the angular frequency depends on the damping factor, the bigger the damping factor the bigger the rise time. Thus to achieve a small rise time the damping factor should be made small. However, the smaller the damping factor the greater the number of oscillations before the system settles down to the steady-state value.

The amplitude of the under-damped oscillations is $C\,e^{-\zeta\omega_n t}$. The *settling time* is defined as the time taken for the oscillations to decay to within some percentage of the steady-state value (see Section 1.5). When $t = 0$ then the amplitude is C. But this equals the steady-state value (see Figure 3.8). If we denote this value by x_s, then the amplitude of the oscillations is $x_{ss}\,e^{-\zeta\omega_n t}$. If we take the settling time t_s as the time taken for this amplitude to drop to, say, $0.2x_{ss}$ then:

$$0.02x_{ss} = x_{ss}\,e^{-\zeta\omega_n t_s}$$

Hence, taking logarithms:

$$\ln 0.02 = -\zeta\omega_n t$$

Hence, since $\ln 0.02 = -3.9$ or approximately -4:

$$t_s = \frac{4}{\zeta\omega_n}$$

Thus the bigger the damping factor the smaller the settling time.

A damping factor of about 0.7 is generally taken as a compromise, giving a reasonably fast rise time without too many oscillations before the system settles down.

To illustrate the above, consider a galvanometric recorder element which has a damping factor of 0.7 and a natural frequency of oscillation of 500 Hz, both being typical values. The angular frequency of the damped oscillation will be:

$$\omega_d = \omega_n\sqrt{1 - \zeta^2} = 2\pi \times 500\sqrt{1 - 0.7^2} = 2244 \text{ rad/s}$$

Hence the 95% rise time will be:

$$\text{rise time} = \frac{\pi}{2\omega_d} \times \frac{90}{100} = \frac{\pi}{2 \times 2244} \times \frac{90}{100} = 0.63 \text{ ms}$$

The 2% settling time will be:

$$t_s = \frac{4}{\zeta\omega_n} = \frac{4}{0.7 \times 2\pi \times 500} = 1.82 \text{ ms}$$

3.3.2 Response of a second order system to a sinusoidal input

In the same way as we derived a differential equation for a second order system when subject to a step input, so we can derive the differential equation when it is subject to a sinusoidal input. When we have an input of the form $A \sin \omega t$, then the solution of the differential equation gives a steady-state output which is a sinusoidal signal with the same angular frequency ω, an amplitude which depends on ω, the natural frequency of the system and the damping factor, and with a phase difference relative to the input (see Section 3.5 for an easier way than solving the differential equation). The ratio of the steady-state output amplitude to the input amplitude is:

$$\text{amplitude ratio} = \frac{1}{\sqrt{\left[1 - (\omega/\omega_n)^2\right]^2 + (2\zeta\omega/\omega_n)^2}}$$

and for the phase difference ϕ between the output and input signals:

$$\phi = \tan^{-1}\left[\frac{2\zeta\omega/\omega_n}{1 - (\omega/\omega_n)^2}\right]$$

Figure 3.9 shows how the amplitude ratio and the phase angle depend on the value of ω/ω_n.

When $\omega = \omega_n$ and the damping ratio is zero, the amplitude ratio becomes infinite and the phase difference zero. When the damping ratio is less than 1, a maximum occurs in the response at a frequency termed the *resonant frequency*. If the equation from the amplitude ratio is differentiated with respect to (ω/ω_n) and set equal to zero, we have the condition:

$$-4\frac{\omega}{\omega_n}\left[1 - \left(\frac{\omega}{\omega_n}\right)^2\right] + 8\zeta^2\left(\frac{\omega}{\omega_n}\right) = 0$$

and hence the angular frequency at which the maximum amplitude occurs is given by:

$$\omega = \omega_n\sqrt{1 - 2\zeta^2}$$

When $2\zeta^2$ is greater than 1 there is no resonance value and the amplitude ratio does not rise above 1. This is a damping factor of 0.71.

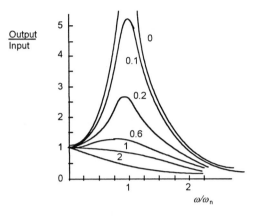

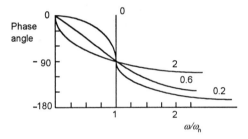

Figure 3.9 *Amplitude ratio and phase angle response for different damping ratios*

3.4 Transfer function

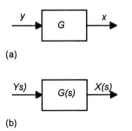

(a)

(b)

Figure 3.10 *(a) Static element, (b) dynamic element in s-domain*

When we deal with the static characteristics of an element, we can refer to its *gain*, this being how much bigger the output is than the input. Figure 3.10(a) shows how we can represent such an element by means of a block diagram, the gain G being the factor by which the input y is multiplied to give the output x. This implies that the output is directly related to the input so that a particular input gives just one output value. However, for many systems the relationship between the output and the input is in the form of a differential equation and such a simple relationship cannot be given. A particular input to a system gives an output which has a value which changes with time. We cannot just divide the output by the input because we have a differential equation rather than an algebraic equation. We can, however, write differential equations in such a way that they become algebraic equations. This method involves using the *Laplace transform*. Signals which were functions of time, i.e. said to be in the time domain, are transformed into signals which are said to be in the s-domain (Figure 3.10(b)). Then we can have a simple relationship between the output and the input, the ratio of the two when transformed by the Laplace transform being called the *transfer function*:

$$\text{transfer function} = \frac{\text{Laplace transform of output}}{\text{Laplace transform of input}}$$

The Laplace transforms are for when all initial conditions are zero, i.e. zero input at time $t = 0$, zero rate of change of output with time at time $t = 0$. To clearly indicate in equations whether we are dealing with signals in the time domain or the s-domain, a signal in the time domain is represented in the form $x(t)$ and the same signal when transformed into the s-domain as $X(s)$, note the use of the capital letter. Thus when we represent such a system by a box with input $Y(s)$ and output $X(s)$, the transfer function $G(s)$ is the factor which multiplies the input to give the output (Figure 3.10(b)).

The procedure adopted to determine the output variation with time for a system where the relationship between the input and output is described by a differential equation is:

1 Transform the differential equation into an equation in the s-domain.

2 Rearrange the transformed equation to give a simple relationship between output and input.

3 To see how the output varies with time, transform the equation back into the time domain.

To obtain the transform of a first derivative we use:

$$\text{transform of } \left\{ \frac{\mathrm{d}}{\mathrm{d}t} f(t) \right\} = sF(s) - f(0)$$

where $f(0)$ is the value of the function of time $f(t)$ when $t = 0$. When we are dealing with a transfer function, the initial conditions are specified as being zero and so $f(0) = 0$. $F(s)$ is the value of the transform for $f(t)$. To obtain the transform of a second derivative we use:

$$\text{transform of } \left\{ \frac{\mathrm{d}^2}{\mathrm{d}t^2} f(t) \right\} = s^2 F(s) - sf(0) - \frac{\mathrm{d}}{\mathrm{d}t} f(0)$$

where $\mathrm{d}f(0)/\mathrm{d}t$ is the value of the rate of change of the function at time $t = 0$. When we are dealing with a transfer function, the initial conditions are specified as being zero and so $sf(0) = 0$ and $\mathrm{d}f(0)/\mathrm{d}t = 0$.

The following are the transforms of some commonly encountered input signals:

1 A unit impulse signal which occurs at a time $t = 0$ has a transform of 1.

2 A unit step signal, i.e. a signal which changes from 0 to a constant value of 1 at time $t = 0$, has a transform of $1/s$.

3   A unit ramp signal, i.e. a signal which starts at time $t = 0$ and increases by 1 every unit time interval and is thus described by the equation $y = 1t$, has a transform of $1/s^2$.

4   A unit amplitude sine wave signal, i.e. a signal described by the equation $y = 1 \sin \omega t$, has a transform of $\omega/(s^2 + \omega^2)$.

5   A unit amplitude cosine wave signal, i.e. a signal described by the equation $y = 1 \cos \omega t$, has a transform of $s/(s^2 + \omega^2)$.

When a function of time is multiplied by a constant then the Laplace transform of that quantity is multiplied by the same constant. For example, if we have a step input of size 10 V then the transform for a unit step is just multiplied by 10 to give $10/s$. If an equation involves the sum of two, or more, separate quantities which are functions of time then the transform of that equation is just the sum of the transforms of the separate terms.

After transformation of a differential equation into the *s*-domain and manipulation of the resulting algebraic equation, it is necessary to transform the equation back into the time domain in order to see how the output changes with time. Tables can be used to do this. The following are some of the commonly encountered items from such a table (see Appendix C for a more detailed table).

$\dfrac{1}{s+a}$ gives e^{-at}

$\dfrac{a}{s(s+a)}$ gives $(1 - e^{-at})$

$\dfrac{b-a}{(s+a)(s+b)}$ gives $e^{-at} - e^{-bt}$

$\dfrac{s}{(s+a)^2}$ gives $(1 - at)\,e^{-at}$

$\dfrac{a}{s^2(s+a)}$ gives $t - \dfrac{1 - e^{-at}}{a}$

$\dfrac{ab}{s(s+a)(s+b)}$ gives $1 - \dfrac{b}{b-a}e^{-at} + \dfrac{a}{b-a}e^{-bt}$

The following illustrate the use of the Laplace transform to solve problems involving inputs to first and second order systems. See Appendix C for an explanation of the basis of the Laplace transform and more information concerning its use. For further information more specialist mathematics texts should be consulted, e.g. *Laplace and z-transforms* by W. Bolton (Longman 1994), *An Introduction to the Laplace Transform and the z Transform* by A.C. Grove (Prentice-Hall 1991).

3.4.1 First order elements

Consider a system where the relationship between the input y and the output x is described by a first order differential equation. Suppose the equation is of the general form:

$$a_1 \frac{dx}{dt} + a_0 x = b_0 y$$

where a_0, a_1 and b_0 are constants. The Laplace transform of this equation, with all the initial conditions zero, is:

$$a_1 s X(s) + a_0 X(s) = b_0 Y(s)$$

We now have an algebraic equation and can gather together all the input terms to give:

$$(a_1 s + a_0) X(s) = b_0 Y(s)$$

Hence the transfer function $G(s)$ is:

$$G(s) = \frac{X(s)}{Y(s)} = \frac{b_0}{a_1 s + a_0}$$

This can be rearranged to give:

$$G(s) = \frac{b_0/a_0}{(a_1/a_0)s + 1} = \frac{G}{\tau s + 1}$$

where G is the steady-state gain of the system (when the input is not changing with time the differential equation becomes just $a_0 x = b_0 y$ and so the steady-state gain is $x/y = b_0/a_0$) and τ the time constant.

Now consider what happens when the first order system is subject to a unit step input. When this happens we have $Y(s) = 1/s$. Since:

$$\text{output transform } X(s) = G(s)Y(s)$$

then:

$$X(s) = \frac{G}{\tau s + 1} \times \frac{1}{s}$$

To find out how the output changes with time, we have to transform back from the s-domain to the time domain. This means finding out what equation in the time domain would fit the above s-domain equation. To help with this, tables of transforms are available. Using the short list given in the previous section (or that in the table in Appendix C), the transform can be made, after some rearrangement, of the form $a/s(s + a)$:

$$X(s) = G \times \frac{1}{s(\tau s + 1)} = G \times \frac{(1/\tau)}{s(s + 1/\tau)}$$

Thus taking a to be $1/\tau$.

$$x = G(1 - e^{-t/\tau})$$

Thus the above equation describes how the output y varies with time t. The dynamic error is the difference between the steady-state value and the value at some instant of time. Thus:

dynamic error $= -G\, e^{-t/\tau}$

The dynamic error declines as the time increases.

As a further illustration of a first order system element, consider a thermocouple which has a steady-state gain (sensitivity is probably the better term here) of 40 μV/°C and a time constant of 10 s. It is described by the differential equation:

$$10\frac{dx}{dt} + x = 40 \times 10^{-6}y$$

where y is the input, a temperature, and x the output, an e.m.f. in volts. Taking the Laplace transform and taking all initial conditions to be zero:

$$10sX(s) + X(s) = 40 \times 10^{-6}Y(s)$$

This gives the transfer function:

$$G(s) = \frac{40 \times 10^{-6}}{10s + 1}$$

Now consider what happens when we have a step input of size 100°C. We then have $Y(s) = 100/s$. Thus:

$$X(s) = G(s)Y(s) = \frac{40 \times 10^{-6}}{10s + 1} \times \frac{100}{s}$$

We need to get this into one of the standard forms for which we know the time domain form. It is like $a/s(s + a)$.

$$X(s) = 40 \times 10^{4}\frac{(1/10)}{s(s + 1/10)}$$

Hence:

$$x = 40 \times 10^{-4}(1 - e^{-t/10})\ \text{V}$$

This equation thus describes how the voltage, in volts, of the thermocouple changes with time.

Now consider what happens when we have a ramp input of 5°C per second to the thermocouple. This ramp input can be described by the equation $y = 5t$. The Laplace transform of the input is $5/s^2$. Hence:

$$X(s) = G(s)Y(s) = \frac{40 \times 10^{-6}}{10s + 1} \times \frac{5}{s^2}$$

We can rearrange this equation to get it into the standard form given earlier, or in the table in Appendix C, of $a/s^2(s + a)$. Thus:

$$X(s) = 200 \times 10^{-6} \frac{1/10}{s^2(s + 1/10)}$$

and so:

$$x = 200 \times 10^{-6} \left(t - \frac{1 - e^{-t/10}}{1/10} \right) \text{ V}$$

3.4.2 Second order elements

For a second order system, the relationship between the input y and the output x is described by a differential equation of the form:

$$\frac{1}{\omega_n^2} \frac{d^2x}{dt^2} + \frac{2\zeta}{\omega_n} \frac{dx}{dt} + x = b_0 y$$

where, b_0 is a constant, ω_n is the natural angular frequency and ζ the damping factor. The Laplace transform of this equation, with all the initial conditions zero, is:

$$\frac{1}{\omega_n^2} s^2 X(s) + \frac{2\zeta}{\omega_n} s X(s) + X(s) = b_0 Y(s)$$

Hence the transfer function $G(s)$ is given by:

$$G(s) = \frac{X(s)}{Y(s)} = \frac{b_0 \omega_n^2}{s^2 + 2\zeta \omega_n s + \omega_n^2}$$

Consider what happens when the second order system is subject to a unit step input. Then $X(s) = 1/s$ and so the output transform is given by:

$$X(s) = G(s)Y(s) = \frac{b_0 \omega_n^2}{s^2 + 2\zeta \omega_n s + \omega_n^2} \times \frac{1}{s}$$

We can rearrange this equation into the form:

$$X(s) = \frac{b_0 \omega_n^2}{s(s + p_1)(s + p_2)}$$

where p_1 and p_2 are the roots of the equation $s^2 + 2\zeta\omega_n + \omega_n^2$. Hence, using the equation for the roots of a quadratic equation, we obtain:

$$p = \frac{-2\zeta\omega_n \pm \sqrt{4\zeta^2\omega_n^2 - 4\omega_n^2}}{2}$$

Thus:

$$p_1 = -\zeta\omega_n + \omega_n\sqrt{\xi^2 - 1}$$

$$p_2 = -\zeta\omega_n - \omega_n\sqrt{\xi^2 - 1}$$

With a damping factor greater than 1 the square root term is real and the system is over-damped. Then the transform is of the form given earlier in this chapter, or in the table in Appendix C, $ab/s(s + a)(s + b)$. Such a transform gives:

$$1 - \frac{b}{b-a}e^{-at} + \frac{a}{b-a}e^{-bt}$$

Hence we have:

$$x = \frac{b_0\omega_n^2}{p_1 p_2}\left[1 - \frac{p_2}{p_2 - p_1}e^{-p_2 t} + \frac{p_1}{p_2 - p_1}e^{-p_1 t}\right]$$

This describes a quantity x which gradually, as t tends to infinity, decays to give the steady state value of $b_0\omega_n^2/p_1 p_2$. The product of the roots when multiplied out has the value ω_n^2 and so the steady-state value is b_0. The dynamic error is the difference between the steady-state value and the value at some instant of time. Thus:

$$\text{dynamic error} = b_0\left[-\frac{p_2}{p_2 - p_1}e^{-p_2 t} + \frac{p_1}{p_2 - p_1}e^{-p_1 t}\right]$$

With a damping factor equal to 1, the square root term for the roots is zero and so $p_1 = p_2 = -\omega_n$. The equation for the transform thus becomes:

$$X(s) = \frac{b_0\omega_n^2}{s(s + \omega_n)^2}$$

This equation can be rearranged, using partial fractions (see Appendix C), to give:

$$X(s) = b_0\left[\frac{1}{s} - \frac{1}{s + \omega_n} - \frac{\omega_n}{(s + \omega_n)^2}\right]$$

Hence, by considering each term individually, we obtain:

$$x = b_0(1 - e^{-\omega_n t} - \omega_n t\, e^{-\omega_n t})$$

This describes a quantity x which, when t becomes infinite, attains the steady-state value of b_0. The dynamic error is the difference between the steady-state value and the value at some instant of time. Thus:

$$\text{dynamic error} = b_0(-e^{-\omega_n t} - \omega_n t \, e^{-\omega_n t})$$

With the damping factor less than 1, the square root term for the roots is imaginary. When this occurs we obtain (using the transform given in the table in Appendix C):

$$x = b_0 \left[1 - \frac{e^{-\zeta \omega_n t}}{\sqrt{1-\zeta^2}} \sin\left(\omega_n \sqrt{1-\zeta^2}\, t + \phi\right) \right]$$

where $\cos\phi = \zeta$. This describes a damped sinusoidal oscillation which eventually dies away to give the steady-state value of b_0. The dynamic error is the difference between the steady-state value and the value at some instant of time. Thus:

$$\text{dynamic error} = b_0 \left[-\frac{e^{-\zeta \omega_n t}}{\sqrt{1-\zeta^2}} \sin\left(\omega_n \sqrt{1-\zeta^2}\, t + \phi\right) \right]$$

As an application of the above, consider a pressure sensor that has been found to have a frequency of 1200 Hz with a damping ratio of 0.5. The angular frequency ω_d is thus $2\pi \times 1200 = 7540$ rad/s. The natural angular frequency ω_n is thus given by:

$$\omega_d = \omega_n \sqrt{1-\zeta^2}$$

$$\omega_n = \frac{\omega_d}{\sqrt{1-\zeta^2}} = \frac{7540}{\sqrt{1-0.5^2}} = 8706 \text{ rad/s}$$

Since the damping factor is less than 1, the system will be under-damped and oscillate with an angular frequency of 7540 rad/s. The dynamic error when the sensor is subject to a unit step input is thus:

$$\text{dynamic error} = b_0 \left[-\frac{e^{-\zeta \omega_n t}}{\sqrt{1-\zeta^2}} \sin\left(\omega_n \sqrt{1-\zeta^2}\, t + \phi\right) \right]$$

$$= b_0 \left[-\frac{e^{-4353t}}{0.87} \sin(7540t + \pi/3) \right]$$

Since b_0 is the steady-state value, the dynamic error as a percentage of the steady-state value is:

$$\% \text{ dynamic error} = \left[-\frac{e^{-4353t}}{0.87} \sin(7540t + \pi/3) \right] \times 100$$

3.5 Measurement systems

A measurement system consists of a number of elements in series, each element having its own transfer function (Figure 3.11). What we are often concerned with is the output from the system as a whole for an input to the system.

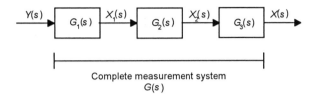

Figure 3.11 *A measurement system and its constituent elements*

The overall transfer function $G(s)$ of the system gives the relationship between the Laplace transforms of the input $Y(s)$ to the system as a whole and its output $X(s)$. Thus:

$$G(s) = \frac{X(s)}{Y(s)}$$

We can, however, write this equation in terms of the transfer functions of the individual elements. Thus:

$$G(s) = \frac{X(s)}{Y(s)} = \frac{X_1(s)}{Y(s)} \times \frac{X_2(s)}{X_1(s)} \times \frac{X(s)}{X_2(s)}$$

where $X_1(s)$ is the output from the first element and the input to the second element, $X_2(s)$ the output from the second element and the input to the third element. Thus if $G_1(s)$ is the transfer function of the first element, $G_2(s)$ that of the second element and $G_3(s)$ that of the third element:

$$G(s) = G_1(s) \times G_2(s) \times G_3(s)$$

The transfer function of the system as a whole is the product of the transfer functions of the series elements.

Thus, if we have a measurement system consisting of a sensor with a transfer function of $2/(s + 4)$ connected to signal conditioning with a transfer function of 10 and a display with transfer function $10/(s + 1)$, then the overall transfer function of the measurement system is:

$$\text{overall transfer function} = \frac{2}{s+4} \times 10 \times \frac{10}{s+1} = \frac{200}{(s+4)(s+1)}$$

Hence we can find out how the output from the display will vary with time when there is an input to the sensor.

3.5.1 Systems with feedback

Consider a simple system with feedback (Figure 3.12). We have an element, or a group of elements combined, with a transfer function $G(s)$ and feedback via a system with a transfer function $H(s)$.

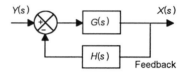

Figure 3.12 *Feedback system*

The input to the element with transfer function $G(s)$, often termed the forward path transfer function for the system, is the input signal to the system $Y(s)$ minus the feedback signal via the element with transfer function $H(s)$. The input to the element $H(s)$ is the system output $X(s)$. Thus the input to the $G(s)$ element is:

input $= Y(s) - H(s)X(s)$

The output from the element is $X(s)$. Hence we can write:

$$G(s) = \frac{X(s)}{Y(s) - H(s)X(s)}$$

This can be rearranged to give:

$$\frac{X(s)}{Y(s)} = \frac{G(s)}{1 + G(s)H(s)}$$

Hence the overall transfer function for the system is:

$$\text{overall transfer function} = \frac{G(s)}{1 + G(s)H(s)}$$

For example, if we had a closed-loop control system with a forward path transfer function of $4/(s + 1)$ and we used a measurement system for feedback with a transfer function $2/(s +5)$, then the overall transfer function of the system is:

$$\text{overall transfer function} = \frac{\dfrac{4}{s+1}}{1 + \dfrac{4}{s+1}\dfrac{2}{s+5}} = \frac{4(s+5)}{(s+1)(s+5)+8}$$

3.6 Frequency response

We can determine the response of an element to a sinusoidal input by using the Laplace transform and the transfer function, the transform of the unit amplitude input signal of sin ωt being $\omega/(s^2 + \omega^2)$. However, if we only want the steady-state sinusoidal response, i.e. the response occurring when the system has settled down after any initial transients due to the initial switching on of the input have subsided, we can do this more simply.

An alternating quantity, say a voltage v, which varies sinsusoidally with time t can be represented by an equation of the form:

$$v = V \sin \omega t$$

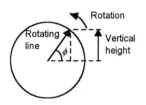

Figure 3.13 *Phasor representation of a sinusoidal signal*

where V is the maximum value of that voltage and $\omega = 2\pi f$ with f being the frequency of the signal. This equation represents the signal in the *time domain*. There is an alternative way of representing the signal and that is by using a phasor. A *phasor* (Figure 3.13) can be considered to be a line of length equal to the amplitude V of the sinusoidal signal and which rotates with a constant angular velocity ω. The vertical height v of the end of the line varies with time and its variation with time is given by $v = V \sin \omega t$. The phasor can be considered to be the snap-shot of the rotating line at time $t = 0$. If at time $t = 0$ the phasor starts with some angle from the reference axis then this angle is termed the phase angle ϕ, the time domain signal then being described by $v = V \sin (\omega t + \phi)$. A phasor is said to be the *frequency domain* representation of an alternating signal.

We can specify a phasor by its length and phase angle. Thus we can write $4\angle 20°$ for a phasor of length 4 units and at a phase angle of $20°$. This type of notation is termed *polar notation*.

Another form of notation that can be used is *complex* or *Cartesian notation*. With this notation we use the Cartesian axes shown in Figure 3.14, the horizontal axis being termed the real axis and the vertical axis the imaginary axis. Thus the line of length r at angle θ is specified by the coordinates of the end of its radial point as $z = a + jb$, where j is the square root of minus 1. The operation of multiplying a quantity by j is equivalent to rotating the line on the graph through $90°$. Since $a = r \cos \theta$ and $b = r \sin \theta$, then $z = r \cos \theta + jr \sin \theta$. Thus we can represent a phasor in the form $2 + j3$, this containing the necessary information to specify the length of the phasor and its phase angle. The length r of the line is given by the Pythagorus theorem as $\sqrt{(a^2 + b^2)}$ and the angle θ is $\tan^{-1} (b/a)$.

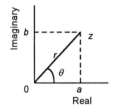

Figure 3.14 *Complex notation*

The differential equation form of describing how a system behaves is the *time domain* description. When we transform this using the Laplace transform then we say we have moved to the *s-domain*. When we describe sinusoidal signals by phasors we say we have moved to the *frequency domain*. Operating in the frequency domain is the simplification that enables the frequency response of systems to be determined when they are subject to sinusoidal inputs.

3.6.1 Frequency response function

If we take a system with a transfer function $G(s)$ and subject it to a sinusoidal input of $a \sin \omega t$, then since such an input has the Laplace transform of $a\omega/(s^2 + \omega^2)$ the output is given by:

$$\text{output(s)} = G(s)\frac{a\omega}{s^2 + \omega^2}$$

The solution of this, in the time domain, is of the form:

output = transient terms + steady-state terms

The steady-state part of the solution is a sinusoidal signal with the same frequency as the input signal, with an amplitude term which depends on the frequency and a phase angle. If we are only concerned with the steady-state terms then we can write the solution in the form:

output = $a|G(j\omega)| \sin (\omega t + \phi)$

Where $|G(j\omega)|$ is the magnitude of the factor by which the input amplitude a has to be multiplied to give the output amplitude. But if we were drawing phasors to represent the input and steady-state output signals then the input phasor would have a length a and the output phasor a length $a|G(j\omega)|$. $G(j\omega)$ is termed the *frequency response function*.

$$G(j\omega) = \frac{\text{output phasor}}{\text{input phasor}}$$

It turns out that if we take the transfer function $G(s)$ and replace s by $j\omega$ then we obtain the *frequency response function*.

For a more detailed discussion of complex numbers and the frequency response function, the reader is referred to *Complex numbers* by W. Bolton (Longman 1995).

3.6.2 Frequency response with a first order system

If we have a first order element with a transfer function of, say:

$$G(s) = \frac{1}{1 + \tau s}$$

then the frequency response function for that element is:

frequency response function $G(j\omega) = \dfrac{1}{1 + j\omega\tau}$

Thus if we have an input to the element of a sinusoidal signal the output phasor is:

output phasor = $G(j\omega)$ × input phasor

$$= \frac{1}{1+j\omega\tau} \times \text{input phasor}$$

To obtain the ratio of the output amplitude to the input amplitude we need to convert this complex number representation to polar notation, i.e. in terms of the length of the phasor and its angle relative to a reference axis. This first requires the complex number to be put in the form $a + jb$. We can do this by multiplying the top and the bottom of the equation by $(1 - j\omega)$. This gives:

$$\text{output phasor} = \frac{1}{1+j\omega\tau} \times \frac{1-j\omega\tau}{1-j\omega\tau} \times \text{output phasor}$$

$$= \frac{1-j\omega\tau}{1+\omega^2\tau^2} \times \text{output phasor}$$

The length of a phasor given by a complex number $(a + jb)$ is $\sqrt{(a^2 + b^2)}$. Thus the ratio of the amplitude of the output phasor to the input phasor is:

$$\text{ratio of amplitudes} = \sqrt{\left(\frac{1}{1+\omega^2\tau^2}\right)^2 + \left(\frac{\omega\tau}{1+\omega^2\tau^2}\right)^2}$$

This simplifies to give the magnitude $|G(j\omega)|$ or ratio of the amplitudes:

$$\text{ratio of amplitudes } r = \frac{1}{\sqrt{1+\omega^2\tau^2}}$$

The phase angle of a phasor described by the complex number $(a + jb)$ is given by $\phi = \tan^{-1}(b/a)$. The phase angle difference between the output phasor and the input phasor is thus:

$$\text{phase angle } \phi = -\tan^{-1}\frac{\omega\tau/(1+\omega^2\tau^2)}{1/(1+\omega^2\tau^2)} = -\tan^{-1}\omega\tau$$

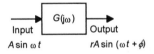

Input   Output
$A\sin\omega t$ $rA\sin(\omega t + \phi)$

Figure 3.15 *The system with its frequency response*

The output is thus a sinusoidal signal with the same frequency as the input but with an amplitude given by the ratio of amplitude equation and with a phase difference from the input given by the phase angle equation (Figure 3.15).

For a measurement system we can take the input to be the true value and the output to be the measured value. The amplitude ratio and phase angle thus indicate how the measured value differs from the true value. The larger the time constant, the larger the phase difference and the smaller the amplitude ratio. A negative phase angle indicates that the measured value lags behind the true value and thus there is a time delay in readings being given. Thus the larger the time constant, the greater the time delay and the amplitude reduction. For there to be zero dynamic error then the phase angle of the output must be zero and the amplitude ratio 1. Then we have an output with zero difference from the input, the true value, at every instant of time.

As an application of the above principles, consider a thermometer which behaves as a first order system with a time constant of 10 s and subject to a sinusoidal temperature input having a frequency of 0.01 Hz. The applied angular frequency is $\omega = 2\pi f = 2\pi \times 0.01 = 0.0628$ rad/s. The phase angle difference between the oscillations of the thermometer reading and the applied oscillations is $-\tan^{-1} \omega\tau = -\tan^{-1}(0.0628 \times 10) = -32.1°$. With one oscillation taking a time of $1/f = 1/0.01 = 100$ s, then a lag of 32.1° means that the reading lags by the fraction 32.1/360 in every cycle, i.e. a time lag of $(32.1/360) \times 100 = 8.91$ s.

3.6.3 Frequency response with a second order system

Consider now a second order element for which we have:

$$G(s) = \frac{\omega_n^2}{s^2 + 2\zeta\omega_n s + \omega_n^2}$$

The frequency response function will be:

$$G(j\omega) = \frac{\omega_n^2}{(j\omega)^2 + 2\zeta\omega_n(j\omega) + \omega_n^2} = \frac{\omega_n^2}{\omega_n^2 - \omega^2 + 2j\zeta\omega_n\omega}$$

$$= \frac{1}{\left[1 - \left(\frac{\omega}{\omega_n}\right)^2\right] + j\left[2\zeta\left(\frac{\omega}{\omega_n}\right)\right]}$$

$$= \frac{\left[1 - \left(\frac{\omega}{\omega_n}\right)^2\right] - j\left[2\zeta\left(\frac{\omega}{\omega_n}\right)\right]}{\left[1 - \left(\frac{\omega}{\omega_n}\right)^2\right]^2 + \left[2\zeta\left(\frac{\omega}{\omega_n}\right)\right]^2}$$

Thus the magnitude $|G(j\omega)|$ or amplitude ratio is:

$$\text{amplitude ratio} = \frac{1}{\sqrt{\left[\left(1 - \frac{\omega^2}{\omega_n^2}\right)^2 + 4\zeta^2\frac{\omega^2}{\omega_n^2}\right]}}$$

and the phase is:

$$\text{phase angle} = -\tan^{-1}\left[\frac{2\zeta\frac{\omega}{\omega_n}}{1 - \frac{\omega^2}{\omega_n^2}}\right]$$

As with the first order system, for there to be zero dynamic error then the phase angle of the output must be zero and the amplitude ratio 1. Then we have an output with zero difference from the input at every instant of time. This is approached when ω is much smaller than ω_n.

In practice some error is generally acceptable and so there is a range of frequencies for which the dynamic error will be below some specified percentage. The term *bandwidth* is often used in specifications for the range of frequencies for which the amplitude ratio is greater than $1/\sqrt{2}$. It thus gives an indication of the range of frequencies for which the dynamic error is small enough to be acceptable.

As an illustration of the above, consider a sensor which is to be used with input frequencies of between 0 and 100 Hz, i.e. $\omega = 2\pi \times 100 = 628$ rad/s, have a damping ratio of 0.7 and give an output which has a dynamic error of $\pm 5\%$ at this frequency. For this dynamic error we must have the amplitude ratio varying between 1.05 and 0.95. We can use these values with the following equation to determine the natural angular frequency ω_n required of the sensor:

$$\text{amplitude ratio} = \frac{1}{\sqrt{\left[\left(1 - \frac{\omega^2}{\omega_n^2}\right)^2 + 4\xi^2 \frac{\omega^2}{\omega_n^2}\right]}}$$

With the amplitude ratio greater than or equal to 0.95 we must have ω/ω_n less than or equal to 0.6 and so a natural angular frequency ω_n which is greater than, or equal to, 1047 rad/s. With the amplitude ratio less than or equal to 1.05 we obtain no real answer for ω_n. The form of the graph of the amplitude ratio against ω/ω_n is shown in Figure 3.16. Thus the dynamic error is within the required range when the angular frequency is between 0 and 1047 rad/s.

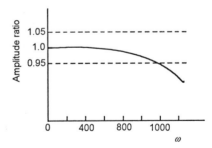

Figure 3.16 *Frequency response of system*

3.6.4 Response to non-sinusoidal signals

The above discussion has concerned sinusoidal inputs. Often, however, in the real world the input is likely to be periodic but not necessarily sinusoidal. The term *periodic* is used for a signal that repeats itself at regular intervals of time. In order to deal with such signals we use the

Fourier series. Jean Baptiste Fourier (1768–1830) showed that a periodic signal, whatever the form of its wave, can be built up from a series of sinusoidal waves of multiples of a basic frequency. Thus we can consider a periodic signal to be represented by:

$$x = A_0 + A_1 \sin(\omega t + \phi_1) + A_2 \sin(2\omega t + \phi_2) + \ldots + A_n \sin(n\omega t + \phi_n)$$

Thus it is possible to consider a system having a number of different input signals, one for each of the terms in the above equation, and then take the output to be the sum of the outputs due to each term.

As an illustration, consider a system having an input of the rectangular waveform shown in Figure 3.16. The Fourier series for that waveform is:

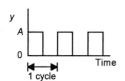

Figure 3.16 *Rectangular waveform*

$$y = A\left(\frac{1}{2} + \frac{2}{\pi} \sin \omega t + \frac{2}{3\pi} \sin 3\omega t + \ldots\right)$$

Sinusoidal components are not found at every frequency and the amplitude of the components decreases as the frequency increases. We can then use the method outlined in the previous section to obtain the amplitude ratio and phase angles for the outputs for each of the above terms when considered separately. The total output is then the waveform resulting from the sum of these output terms. Though they will have the same frequencies as the input terms, the amplitude ratios for each term may differ and so the sum of the output terms gives a waveform which differs from the input waveform.

For further reading on the Fourier series, see Appendix D or more specialist texts such as *Fourier Series* by W. Bolton (Longman), *Lectures on Fourier Series* by L. Solymar (Oxford University Press) or *Fourier Series and Harmonic Analysis* by K.A. Stroud (Stanley Thornes).

Problems

1   A sensor behaves as a capacitance of 2 μF in series with a 1 MΩ resistance. As such the relationship between its input y and output x is given by:

$$2\frac{dx}{dt} + x = y$$

How will the output vary with time when the input is (a) a unit step input at time $t = 0$, (b) a ramp input described by $y = 4t$?

2   A system is specified as being first order with a time constant of 10 s and a steady-state value of 5. How will the output of the system vary with time when subject to a step input?

3   A sensor is first order with a time constant of 1 s. What will be the percentage dynamic error after (a) 1 s, (b) 2 s, from a unit step input signal to the sensor?

4 How long must elapse for the dynamic error of a sensor to drop below 5% if the sensor is first order with a time constant of 4 s?

5 A thermometer originally indicates a temperature of 20°C and is then suddenly inserted into a liquid at 45°C. The thermometer has a time constant of 2 s. (a) Derive a differential equation showing how the thermometer reading is related to the temperature input and (b) give its solution showing how the thermometer reading varies with time.

6 An element has an output x for an input y related by the differential equation:

$$\frac{dx}{dt} + 2x = y$$

Determine how the output varies with time when there is a ramp input of $y = 3t$.

7 A second order system has a natural angular frequency of 2.0 rad/s and a damped angular frequency of 1.8 rad/s. What is the damping factor?

8 What is the 2% settling time for a second order system having a damping factor of 0.5 and a natural angular frequency of 2.0 rad/s?

9 Determine the natural angular frequency and damping factor for a second order system with input y and output x and described by the following differential equation:

$$0.02\frac{d^2x}{dt^2} + 0.20\frac{dx}{dt} + 0.50x = y$$

10 A sensor can be considered to be a mass-damper-spring system with a mass of 10 g and a spring of stiffness 1.0 N/mm. Determine the natural angular frequency and the damping constant required for the damping element if the system is to be critically damped.

11 Determine the transfer functions for systems having inputs and outputs described by the differential equations:

(a) $A\frac{dh}{dt} + \frac{\rho g h}{R} = q$, input q and output h

(b) $LC\frac{d^2v}{dt^2} + RC\frac{dv}{dt} + v = V$, input V and output v

12 Determine whether the second order system described by the following differential equation is under-damped, critically damped or over-damped when subject to a step input y:

$$\frac{d^2x}{dt^2} + 5\frac{dx}{dt} + 6x = y$$

13 Determine the response of the system in problem 5 to a unit step input.

14 Determine whether the systems having the following transfer functions are under-damped, critically damped or over-damped:

(a) $G(s) = \dfrac{10}{s^2 - 6s + 16}$

(b) $G(s) = \dfrac{10}{s^2 + s + 100}$

(c) $G(s) = \dfrac{10}{s^2 + 6x + 9}$

15 Determine the output response of a system with the transfer function:

$$G(s) = \frac{10}{(s+3)(s+4)}$$

when subject to a unit impulse.

16 Determine the frequency response function for a first order system having a transfer function of:

$$G(s) = \frac{5}{1 + 2s}$$

17 Determine the amplitude ratio and the phase angle for the response from a first order system having the transfer function:

$$G(s) = \frac{1}{1 + 5s}$$

when subject to the sinusoidal input sin ωt.

18 Determine the amplitude and phase of the sinusoidal response from a first order system having a transfer function of:

$$G(s) = \frac{4}{s+1}$$

when subject to an input of 2 sin(3t + π/3).

19 A thermocouple has a transfer function of 1/(1 + 10s). Determine the bandwidth of the frequency response.

20 A thermocouple behaves as a first order system with a time constant of 10 s. It is used to measure a temperature which oscillates between 250°C and 290°C in a sinusoidal manner with a frequency of 0.0125 Hz.

Determine the maximum and minimum values the thermocouple will indicate.

21 A thermometer behaves as a first order system with a time constant of 15 s. It is used to measure a temperature which oscillates in a sinusoidal manner with a frequency of 0.010 Hz. By what time does the thermometer reading lag behind the true temperature?

22 A thermocouple with a transfer function of $4 \times 10^{-6}/(1 + 10s)$ is connected to signal conditioning with a transfer function of $100/(1 + s)$ and a display system with transfer function $20/(2 \times 10^{-5}s^2 + 0.01s + 1)$. What is the overall transfer function of the measurement system?

4 Reliability

An important consideration in the purchase of any item, whether it be a measurement system or a car, is: how long can you rely on it performing the required function without breaking down or failing to perform to its specification? We are thus concerned with reliability. The *reliability* of a system or element is defined as the chance that the system or element will operate to a specified level of performance for a specified period of time under specified environmental conditions. It is thus a predictor of future behaviour. Reliability is thus an essential item in quality assurance. This chapter is a consideration of the basic principles associated with reliability. For further reading see texts devoted specifically to the subject, e.g. *Engineering Reliability: New Techniques and Applications* by B.S. Dhillon and C. Singh (Wiley 1981).

4.1 Definitions

Chance, or *probability*, is the frequency in the long run with which an event occurs. If you drop a coin it has two ways it can land, heads uppermost or tails uppermost. If the coin is dropped just once, then we cannot predict which way the coin will land. However, if we drop the coin 100 times then it is likely to land heads uppermost about 50 times. The frequency with which we obtain heads uppermost is about 50/100, i.e. ½. The more times we drop the coin the more the frequency tends to the value ½. The chance, or probability, of the coin landing heads uppermost is thus said to be ½.

The chance is ½ with the coin because there are two possible ways the coin can land, heads or tails uppermost, but just one way it can fall and land heads uppermost. We can thus also define *chance*, or *probability*, as the number of ways a specific event can occur divided by the number of ways all possible events can occur. Thus we have two alternative, but giving identical results, ways of defining chance:

$$\text{Chance} = \frac{\text{number of times, in the long run, an event occurs}}{\text{total number of times considered}}$$

$$= \frac{\text{number of ways an event can occur}}{\text{number of ways all events can occur}}$$

Chance is the frequency in the long run with which an event occurs. In the case of reliability, the event we are considering is that an item meets the required specification. Thus if we start off with N_0 items meeting the required specification and after some time t we have N meeting the required specification, then the chance of an item meeting the specification is N/N_0. This is defined as being the reliability. Thus:

$$\text{reliability} = \frac{N}{N_0}$$

In that time t the number of items failing to meet the specification is $(N - N_0)$. We can define *unreliability* as the chance of an item failing to meet the specification. Thus:

$$\text{unreliability} = \frac{N - N_0}{N_0} = 1 - \frac{N}{N_0} = 1 - \text{reliability}$$

4.1.1 Quantifying failure

Failure is when a system or element fails to perform to the required specification. The failure may be critical in that there is a total loss of function of the system and it can no longer be used or just that the item has gone out of its specification limits but can still be used. The following are some of the terms commonly used in quantifying failure.

For an item which is tested for a time t and repaired each time it fails, then if it fails N_f times the *mean time between failures* (MTBF) is:

$$\text{MTBF} = \frac{t}{N_f}$$

The term *failure rate* λ is used for the average number of failures, per item, per unit time. Thus if over a time t there are N_f failures, then:

$$\text{failure rate } \lambda = \frac{N_f}{t} = \frac{1}{\text{MTBF}}$$

If we consider N items and have a total of N_f failures in a time t then:

$$\text{MTBF} = \frac{Nt}{N_f}$$

and:

$$\text{failure rate } \lambda = \frac{NN_f}{t} = \frac{1}{\text{MTBF}}$$

To illustrate the above, consider a sensor in use in a robot manipulator on a production line. The time interval in days between successive failures of the sensor is recorded as:

12, 20, 15, 26, 32, 17, 16, 31, 22, 19 days

The mean time between failures is thus:

$$\text{MTBF} = \frac{12 + 20 + 15 + 26 + 32 + 17 + 16 + 31 + 22 + 19}{10} = 21 \text{ days}$$

The failure rate is $1/21 = 0.048$/day.

The mean time between failures and the failure rate are terms used for repairable systems and elements. When items are not repairable, a measure of the reliability used is the *mean time to failure* (MTTF). It is the average time before failure occurs. Thus if N items are tested and the times to failure for the individual items are $t_1, t_2, t_3, \ldots t_N$, then:

$$\text{MTTF} = \frac{\sum\limits_{i=1}^{N} t_i}{N}$$

For example, if we have a batch of 10 items and they fail after:

12, 10, 15, 13, 20, 15, 11, 14, 19, 21 days

then the MTTF is:

$$\text{MTTF} = \frac{12 + 10 + 15 + 13 + 20 + 15 + 11 + 14 + 19 + 21}{10} = 15 \text{ days}$$

Typical values of failure rates for measurement elements are: thermocouple 0.2 per year, pressure sensor 0.3 per year, pen recorder 0.3 per year.

4.1.2 Availability

Consider items which can be repaired. Repairs, however, take time and so when an item has failed a time must elapse before it is repaired and back in service. We can define a *mean time to repair* (MTTR) as the average time it takes to repair a particular type of item. Thus if N items are repaired and the times to repair for the individual items are $t_1, t_2, t_3, \ldots t_N$, then:

$$\text{MTTR} = \frac{\sum\limits_{i=1}^{N} t_i}{N}$$

The *availability* of an item is defined as the fraction of the time for which a system or element is operational. An item will, on average, be operational for the MTBF. The item will then not be operational for the MTTR. Thus the availability is:

$$\text{availability} = \frac{\text{MTBF}}{\text{MTBF+MTT}}$$

This can also be written as:

$$\text{availability} = \frac{1}{1 + (\text{MTTR/MTBF})} = \frac{1}{1 + \lambda \times \text{MTT}}$$

where λ is the failure rate.

Thus with an MTTR of 15 days and an MTBF of 21 days, the availability would be 0.58. This means that the item is only operational for 58% of its time.

4.2 Failure patterns

Failure rate λ for an item or system is often not constant over its lifetime, varying with time. Figure 4.1 shows the typical form of the graph showing how the failure rate varies with time. The graph, because of its shape, is frequently referred to as the '*bath tub curve*'. The graph shows three distinct phases: early failure, normal working life and wear-out.

1   *Early failure phase*
    During the early failure phase the failure rate decreases with time and a lot of failures occur. This is a result of manufacturing faults, sub-standard components, material imperfections, fault assembly, bad connections, etc. which quickly result in failure following manufacture of the item. To overcome this, manufacturers may use a 'burn-in' period, i.e. run the items for a period of time so that any such faults manifest themselves before the item is sold.

2   *Normal working life*
    During this phase, all the items with initial manufacturing faults have been eliminated and there is an almost constant failure rate, the failures being due to random causes. This phase represents the useful working life period of use for an item or system.

3   *Wear-out*
    With this phase the failure rate increases due to the wearing out of components.

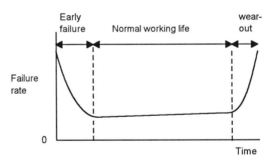

Figure 4.1 *The 'bath tub curve'*

4.3 The exponential law of failure

Consider there to be a constant failure rate λ and a situation where at time $t = 0$ there are N_0 items and no items are repaired but just eliminated from this number when failure occurs. As a result of failures, after a further interval of time δt, the number has decreased by δN. Thus the change in number is $-\delta N$ in time δt and so the failure rate is:

$$\text{failure rate } \lambda = \frac{-\delta N}{N\delta t}$$

In the limit when $\delta t \to 0$, we can write:

$$\lambda = -\frac{1}{N}\frac{dN}{dt}$$

We can solve this differential equation by the separation of the variables method (see Appendix B). Thus if we have N items left at time t, we can write:

$$\int_{N_0}^{N}\frac{1}{N}\,dN = -\int_{0}^{t}\lambda\,dt$$

and so:

$$\ln N - \ln N_0 = -\lambda t$$

This can be written as:

$$N = N_0\,e^{-\lambda t}$$

This exponential equation describes how the number of useable items changes with time due to a constant failure rate.

Since reliability = N/N_0 we can write:

$$\text{reliability} = e^{-\lambda t}$$

The total number of failures in a time t is $(N_0 - N)$. Hence:

$$\text{number of failures} = N_0 - N_0\,e^{-\lambda t}$$

The unreliability is (number of failures/N_0) and so:

$$\text{unreliability} = 1 - e^{-\lambda t}$$

Figure 4.2 shows how the reliability and the unreliability change with time.

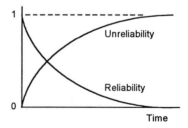

Figure 4.2 *Variation of reliability and unreliability with time*

After a time equal to the MTBF we have $t = 1/\lambda$ and then:

$$\text{reliability} = e^{-1} = 0.37$$

and:

$$\text{unreliability} = 1 - e^{-1} = 0.63$$

The probability of an item failing after it has been operating for a time equal to the MTBF is 63%.

To illustrate the above, consider an instrument which has a mean time between failures of 100 hours. What is the chance of it working successfully after 50 hours? The failure rate, assumed to be constant and not varying with time, is (1/MTBF) and so is 1/100 = 0.01 per hour. Thus:

$$\text{reliability} = e^{-\lambda t} = e^{-0.01 \times 50} = 0.61$$

The chance of it working successfully is 0.61.

4.4 Reliability of a system

Often with systems, when one component of the system fails then the entire system fails. Such a situation is like a number of lamps connected in series, when one lamp fails then all the lamps fail. Consider such a system with each element having its own, independent, reliability (Figure 4.3). What is the chance of the system not failing?

The situation is rather like having three coins and considering the chance of one of them landing heads uppermost when all three are dropped. Consider the ways in which the three coins can land. They are:

Input

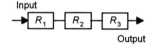

Output

Figure 4.3 *Elements in series*

HTT HHT HHH HTH TTT THT THH TTH

There is just one way in the eight possible ways in which the system has no heads uppermost and so there is a chance of 1/8 that there will be no heads. But 1/8 is just ½ × ½ × ½, i.e. the product of the chances of each coin not landing heads uppermost. Thus if we equate heads with failure, the chance of there being no failure is the product of the chances of each coin 'not failing'. The chances are multiplied.

Thus the reliability of a system having three elements, with independent reliabilities R_1, R_2 and R_3, is:

$$\text{reliability of system} = R_1 \times R_2 \times R_3$$

Since, for a constant failure rate λ, we have reliability = $e^{-\lambda t}$, then:

$$\text{reliability of system} = e^{-\lambda_1 t} \times e^{-\lambda_2 t} \times e^{-\lambda_3 t}$$

where λ_1, λ_2 and λ_3 are the independent failure rates of the three elements. Hence:

$$\text{reliability of system} = e^{-(\lambda_1 + \lambda_2 + \lambda_3)t}$$

Hence the system as a whole has a failure rate given by:

$$\lambda = \lambda_1 + \lambda_2 + \lambda_3$$

The failure rate of a system with a number of elements, for which the failure of one results in the failure of the system as a whole, is the sum of the failure rates of the constituent elements.

To illustrate the above, consider a system which has three integrated circuits, each having a failure rate of 1×10^{-6} per hour. The failure rate of the system as a whole will be:

$$\text{failure rate of system} = 1 \times 10^{-6} + 1 \times 10^{-6} + 1 \times 10^{-6} = 3 \times 10^{-6}/\text{hour}$$

Thus the chance of the system operating for, say, 10 000 hours without failure is:

$$\text{reliability} = e^{-\lambda t} = e^{-3 \times 10^{-6} \times 10\,000} = 0.97$$

Thus there is a 97% chance that the system will not fail in 10 000 hours.

4.4.1 Redundancy

If we have lamps connected in parallel then though one lamp is faulty the others will still remain on. The system as a whole can only fail if all the lamps are faulty. Where vital systems must be monitored and it is important that monitoring can continue, even when one element fails, then the 'parallel' form of connection is used. Such a system gives a higher reliability for the system as a whole, though at the cost of duplication with the parallel elements.

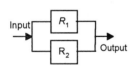

Figure 4.4 *Elements in parallel*

Consider a system involving two elements 'in parallel', i.e. a failure of one will not result in the failure of the system as a whole (Figure 4.4). One has a reliability of R_1 and the other R_2. What is the reliability of the parallel system as a whole? What we are concerned with is the chance of two independent events simultaneously occurring. Consider dropping two coins and the chance of simultaneously falling to give two heads uppermost. The possible ways the coins could land are:

HH HT TH TT

There is just one way in four of obtaining two heads, thus the chance of obtaining two heads is 1/4. If we had three coins, the possible ways the coins could land are:

HHH HTH HHT HTH HTT TTH THT TTT

There are nine ways the coins can land but only one way which gives all heads uppermost. Thus the chance of obtaining three heads is 1/8. Thus, in general, the chance of obtaining all heads is the product of the chances for each coin giving heads. If we take heads to mean failure, then the chances of all independent events failing is the product of their unreliabilities. Thus, since unreliability = 1 − reliability:

Chance of parallel system failing $= (1 − R_1)(1 − R_2)$

Hence the unreliability of a parallel system is the product of the unreliabilities of the individual parallel components. For the two elements in parallel this gives:

reliability of the system $= 1 − (1 − R_1)(1 − R_2) = R_1 + R_2 − R_1R_2$

Consider a system where we have three temperature measurement systems in parallel. If each system has a failure rate of 1.2 per year then, since reliability $= e^{−\lambda t}$, the unreliability of the system as a whole for a year is:

unreliability $= (1 − e^{−1.2}) \times (1 − e^{−1.2}) \times (1 − e^{−1.2})$

$$= 0.70 \times 0.70 \times 0.70 = 0.027$$

Thus the reliability $= 1 − 0.027 = 0.973$. This compares with the reliability of a single element of $e^{−1.2} = 0.30$. Thus the reliability has been increased.

Parallel systems are expensive since there is duplication. To minimise the cost, often only the most unreliable element is in parallel. Thus we might have a system of the form shown in Figure 4.5.

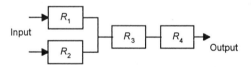

Figure 4.5 *Parallel and series elements*

For the two parallel elements, the combined unreliability is:

unreliability $= (1 − R_1)(1 − R_2)$

and so the reliability is:

reliability $= 1 − (1 − R_1)(1 − R_2) = R_1 + R_2 − R_1R_2$

This element is in series with the other elements and so the reliability of the system as a whole is:

reliability $= (R_1 + R_2 − R_1R_2) \times R_3 \times R_4$

Problems

1   What is the unreliability of an item if it has a reliability of 0.2?

2   A measurement system is found to have failed four times over a period of use of 1000 hours, being repaired each time. What is the mean time between failures and the failure rate?

3   A measurement system has a mean time between failure of 2000 hours and a mean time to repair of 5 hours. What is the availability?

4   100 thermocouples are tested for a year and during that period 8 are found to fail. What is the failure rate?

5   A system has a constant failure rate of 2.0×10^{-6} per hour. What is the chance that the system will fail after (a) 1000 hours, (b) 10 000 hours, (c) 100 000 hours?

6   A system has a mean time between failure of 1000 hours. What is the chance that the system will fail after (a) 1000 hours, (b) 2000 hours?

7   A plant has 100 independent sensors, each having a failure rate per hour of 0.01. How many might be expected to fail after 500 operating hours?

8   A system involves 10 electrical components, each having a constant failure rate of 1×10^{-8} per hour and 15 soldered joints, each having a constant failure rate of 10×10^{-8} per hour. If the failure of any one of the electrical components or any one of the soldered joints would result in the failure of the system as a whole, what is the failure rate for the system as a whole?

9   A measurement system has a sensor with a mean time between failure of 2000 hours, connected to a signal conditioner with a mean time between failure of 8000 hours and a display with a mean time between failure of 5000 hours. If the failure of any one item will result in the failure of the system as a whole, what is the chance of failure occurring after 1000 hours?

10 A measurement system consists of a sensor with a reliability of 0.9 per year, a signal conditioner of reliability 0.85 per year and a display with a reliability of 0.80 per year. What is the chance of a failure occurring after one year?

11 An electronic component includes 40 diodes, each with a failure rate of 0.0002 per 1000 hours; 20 transistors, each with a failure rate of 0.0005 per 1000 hours; 50 resistors, each with a failure rate of 0.000 005 per 1000 hours; and 20 capacitors, each with a failure rate of 0.0001 per 1000 hours. What is the overall failure rate per 1000 hours of the component if the failure of any one of the components results in failure of the component?

12 A critical plant item has two identical power units in parallel. Each has an independent, constant, failure rate of 0.0003 per hour. What is the reliability of the system after 1000 hours if failure of one does not result in failure of the plant, failure only occurring when both fail?

13 A measurement system, for safety, employs two sensors in parallel. If each has a constant failure rate of 0.0001 per hour, what will be the reliability of the parallel sensors after 100 hours if failure of one does not result in failure of the system, failure only occurring when both fail?

14 A measurement system has three pressure sensors, each with a failure rate of 1.2 per year, connected in parallel. A selector, with a failure rate of 0.1 per year, is used to select the output from one of the thermocouples and its output is then fed, via a signal conditioner with a failure rate of 0.1 per year, to a recorder with a failure rate of 0.3 per year. What is the reliability, after a year, of the system as a whole?

5 Sensors

Sensors are devices which take information about a physical stimulus and turn it into a signal which can be measured (see Section 1.1). The term transducer is often used for an element which transforms input signals from some form into an electrically equivalent form or sometimes input signals to electrically equivalent signals or vice versa. Thus, by this definition, sensors might be transducers. Sometimes the term transducer is used for a combination of a sensor and signal conditioning which takes an input and gives as output from the signal conditioner an electrical signal. This chapter is a discussion of the principles of commonly encountered sensors. Chapter 6 extends this discussion to micro sensors, these being sensors that have physical dimensions less than a millimetre and arise from the micro-electronics development of integrated circuits.

5.1 Classification of sensors

There are a number of ways we can classify sensors. The classification adopted in this chapter is in terms of the physical principle involved. The following are some of the commonly used physical principles, each group of sensors being discussed in more detail in later sections in this chapter and as part of measurement systems in Chapters 10 to 13.

1   *Resistive sensors*
    The input being measured is transformed into a resistance change. Such sensors include potentiometers, resistance thermometers, strain gauges and photo conductive cells.

2   *Capacitive sensors*
    The input being measured is transformed into a capacitance change. Such sensors give pressure sensors, displacement sensors and liquid level sensors.

3   *Inductive sensors*
    The input being measured is transformed into a change in inductance. A particularly useful form is the linear variable differential transformer (LVDT).

4   *Electromagnetic sensors*
    These are based on Faraday's laws of electromagnetic induction with the input being measured giving rise to an induced e.m.f. The tachogenerator is an example of such a sensor.

5   *Thermoelectric sensors*
    The input is temperature and the output an e.m.f., the sensor being termed a thermocouple.

6 *Elastic sensors*
The input being measured is transformed into a displacement. Examples of such sensors are springs for the measurement of force and diaphragm pressure gauges.

7 *Piezoelectric sensors*
Forces applied to a crystal displace the atoms in the crystal and result in the crystal acquiring a surface charge. Such sensors are used for the measurement of transient pressures, acceleration and vibration.

8 *Photo voltaic sensors*
These are p-n junctions which produce a change in current when electromagnetic radiation is incident on the junction. They are widely used where the presence of an object is being detected by it breaking a beam of light.

9 *Pyroelectric sensors*
Temperature changes give rise to changes in surface charges. Such sensors are widely used for burglar alarm systems to detect the presence of people by their body heat.

10 *Hall effect sensors*
The action of a magnetic field on a flat plate carrying an electric current generates a potential difference which is a measure of the strength of the field.

Sensors can also be classified as passive or active. *Passive sensors* are ones which require an external power supply; *active sensors* are ones which need no such external power supply. A potentiometer is an example of a passive sensor while a thermocouple is an example of an active one.

5.2 Resistive sensors

With resistive sensors, the input being measured is transformed into a resistance change. Such sensors include potentiometers, resistance thermometers, strain gauges and photo conductive cells. They are used as elements in measurement systems for the measurement of displacement, temperature, strain, force and light intensity.

5.2.1 Potentiometers

A *potentiometer* consists of a resistance element with a sliding contact which can be moved over the length of the element. The resistance element can be in the form of a resistance coil or film of conductive plastic formed in a straight line or a circular shape. Figure 5.1 shows these basic forms, the straight line form being used for monitoring the linear displacement of the sliding contact and the circular one for angular displacement.

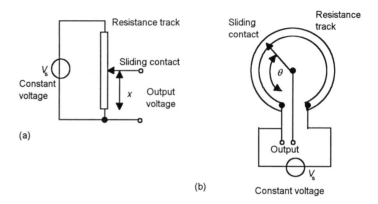

Figure 5.1 *Potentiometers for: (a) linear displacement, (b) angular displacement*

With a constant voltage V_s across the length of the track, the output voltage between the sliding contact and one end of the track is a fraction of the input voltage, being the fraction of the total track resistance between the output terminals. If the track has a constant resistance per unit length then the output voltage is proportional to the linear displacement or the angular displacement of the sliding contact from one end of the track:

$$\text{output voltage} = \frac{x}{L}V_s \text{ or } \frac{\theta}{\theta_t}V_s$$

where L is the total length of the straight track and θ_t the total angle of the circular track. Straight potentiometers are available with lengths up to about 1 m and so displacements up to this amount can be monitored. Helical potentiometers are available with as many as 20 turns and thus angular displacements up to 7200° can be monitored.

The resolution of a wire wound track (Figure 5.2) is limited by the diameter of the wire used and typically ranges from about 0.5 mm for a finely wound one to 1.5 mm for a coarsely wound one. Conductive plastic tracks have no such resolution problem. Errors due to non-linearity of track for wire wound tracks tend to range from less than 0.1% to about 1% and for conductive plastics are about 0.05%. The conductive plastic track is more sensitive to temperature changes which can more markedly affect its accuracy.

An important effect that has to be considered with potentiometers is the effect of the resistance of the load placed across the output (Figure 5.3). This introduces non-linearity, and hence error, into the response of what otherwise might have been a potentiometer with no non-linearity error. The following shows how we can calculate the non-linearity error introduced by loading.

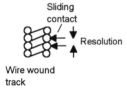

Figure 5.2 *Wire wound track*

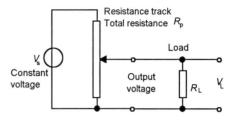

Figure 5.3 *Loading of a potentiometer*

Consider a load of resistance R_L connected across the output from the potentiometer. Suppose the total track length of the potentiometer has a resistance R_p. The resistance of the track between the sliding contact and the end from which measurements are made is the fraction of R_p between those points. If we denote this fraction as f then we have a resistance of fR_p in parallel with R_L. Thus the resistance across the output is:

$$\text{resistance across output} = \frac{fR_pR_L}{fR_p + R_L}$$

The total resistance across the constant source voltage is the above element plus the remainder of the track and so is:

$$\text{total resistance} = \frac{fR_pR_L}{fR_p + R_L} + R_p(1-f)$$

Hence the fraction of the supply voltage V_s which is across the output when loaded is:

$$\frac{V_L}{V_s} = \frac{\dfrac{fR_pR_L}{fR_p + R_L}}{\dfrac{fR_pR_L}{fR_p + R_L} + R_p(1-f)} = \frac{f}{\left(\dfrac{R_p}{R_L}\right)f(1-f)+1}$$

With a load of infinite resistance, i.e. an open circuit, we have $V_L = fV_s$. The above equation thus indicates that the output across the load is no longer directly proportional to f and so we have a non-linearity error. The error introduced by having a load with a finite resistance is thus:

$$\text{error} = fV_s - V_L = fV_s - \frac{f}{\left(\dfrac{R_p}{R_L}\right)f(1-f)+1} = V_s\frac{R_p}{R_L}(f^2 - f^3)$$

For a particular track resistance, the error is a maximum when $d(error)/df$ is 0 and is thus when $2f - 3f^2 = 0$, i.e. when $f = 0.67$. The full-scale output is when $f = 1$, there being no error at this value, and is V_s. The maximum error as a percentage of the full-scale output is thus $14.8(R_p/R_L)\%$.

To illustrate the above, consider a potentiometer with a track of resistance 500 Ω and a voltmeter of resistance 10 kΩ being connected between the sliding contact and one end of the track. What will the meter reading be if the supply voltage is 4 V and the slider is at the mid point of the track? If the voltmeter resistance had been infinite then the output would have been half of the supply voltage, i.e. 2 V. However, taking the effect of the voltmeter resistance into account we have an error of:

$$\text{error} = 4 \times \frac{500}{10\,000}(0.5^2 - 0.5^3) = 0.025 \text{ V}$$

Thus the voltmeter reading would be 2.025 V.

To obtain a small non-linearity error with loading, we need to have a potentiometer with a low track resistance relative to the load resistance. However, a low track resistance means that we must also have a low value of the supply voltage V_s since the power dissipation by the potentiometer is (V_s^2/R_p) and there will be a limit to the power that can be dissipated. A low value of V_s means that the output voltage per unit length of track is low. Thus the consequence of using a lower resistance track is a reduction in sensitivity. A balance has thus to be achieved between minimising the effect of loading and maintaining a reasonable sensitivity.

5.2.2 Resistance temperature detectors

The resistance of most metals increases in a reasonably linear way with temperature over the range −100 to +800°C. In general, the relationship is of the form:

$$R_t = R_0(1 + \alpha t + \beta t^2 + \gamma t^3 + \ldots)$$

where R_t is the resistance at t°C, R_0 the resistance at 0°C and α, β, γ, are temperature coefficients of resistance with their magnitudes having $\alpha > \beta > \gamma$. For an ideal linear relationship we would just have:

$$R_t = R_0(1 + \alpha t)$$

There is thus a non-linearity error due to the β and γ terms. Figure 5.4 shows the general way in which the resistance varies with temperature for three metals.

Platinum has a small non-linearity error, having α with a value of 3.91×10^{-3} K^{-1} and β a value of -5.85×10^{-7} K^{-2}. This gives the linear approximation accurate to within about ±0.3% over the range 0 to 200°C and ±1.2% over the range 200 to 800°C. Platinum is widely used for sensors because it is highly stable, is capable of operating over a wide range

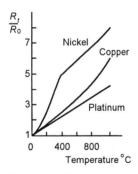

Figure 5.4 *Variation of resistance with temperature*

of temperature, and gives a predictable and reproducible change in resistance with temperature. Platinum temperature sensors are made by winding platinum wire on a glass or ceramic bobbin and sealing with molten glass or by depositing platinum in a thin film on a ceramic substrate and etching it to form the resistance element, then sealing. For coils, the response time can be fairly large, of the order of seconds, by virtue of their bulk. The thin film form has less mass and consequently has a much faster response time, perhaps about 0.1 s.

Another form of resistance temperature detector is the thermistor. *Thermistors* (*therm*ally sensitive re*sistors*) are small pieces of material made from mixtures of metal oxides, such as those of chromium, cobalt, iron, manganese and nickel. These oxides are semiconductors. The material is formed into elements such as rods, discs and beads (Figure 5.5). The resistance of a thermistor is very sensitive to temperature changes, considerably more than metals, and generally decreases with an increase in temperature, though there are some for which the resistance increases with an increase in temperature. The resistance variation with temperature is non-linear, following an exponential variation with temperature of the form:

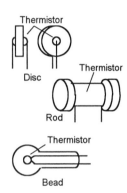

Figure 5.5 *Forms of thermistor*

$$R_T = R_0 \, e^{\beta(1/T - 1/T_0)}$$

where R_T is the resistance at temperature T, R_0 the resistance at temperature T_0 and β a constant which depends on the thermistor concerned. Figure 5.6 shows a resistance–temperature graph for a typical thermistor.

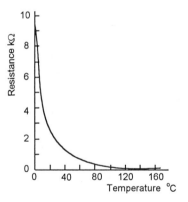

Figure 5.6 *Typical variation of resistance with temperature for a thermistor*

For thermistors that show a decrease in resistance with an increase in temperature, self-heating due to currents can become significant. When this happens, the current through the thermistor increases its temperature. This has the consequential effect of reducing its resistance. The reduction of the resistance in a circuit increases the current and so further increases the temperature and consequentially further reduces the resistance. This effect continues until the heat dissipation of the thermistor equals the power supplied to it.

Thermistors have high sensitivity, are rugged and can be very small, so enabling them to respond rapidly to changes in temperature. Their main disadvantage is their non-linearity and variability. Thermistors are not interchangeable and the same nominal thermistor can have characteristics which differ significantly.

5.2.3 Strain gauges

The electrical resistance strain gauge consists of a resistance element in the form of a flat coil of wire or etched metal foil, or a strip of semiconductor material. The element is wafer-like and has an insulating backing material so that it can be stuck like a postage stamp onto surfaces. Figure 5.7 shows some forms of strain gauges.

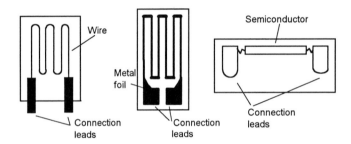

Figure 5.7 *Some forms of strain gauges*

The resistance R of an element of length L and cross-sectional area A is given by:

$$R = \frac{\rho L}{A}$$

where ρ is the resistivity of the material. When the element is stretched its length increases, its cross-sectional area decreases, and there is also a change in its resistivity. The result is that the resistance of the element changes.

The change in resistance ΔR that occurs is the algebraic sum of the changes due to the resistivity, length and area changes. Thus if the resistivity changes by $\Delta \rho$ then, since $dR/d\rho = L/A = R/\rho$, we have $\Delta R/R = \Delta \rho/\rho$. If the length changes by ΔL then, since $dR/dL = \rho/A = R/L$, we have $\Delta R/R = \Delta L/L$. If the area changes by ΔA then, since $dR/dA = -\rho L/A^2 = -R/A$, we have $\Delta R/R = -\Delta A/A$. Thus the resistance change when all three factors change is:

$$\frac{\Delta R}{R} = \frac{\Delta \rho}{\rho} + \frac{\Delta L}{L} - \frac{\Delta A}{A}$$

When an element is stretched it increases in the direction of the stretching and contracts in the transverse direction. The strain ε in the direction of the stretching is the amount by which it stretches divided by the length of the element. Thus $\varepsilon = \Delta L/L$. The transverse strain is related to the strain in the direction of the stretching by:

transverse strain $= -v\varepsilon$

where v is Poisson's ratio. The minus sign is because the cross-section reduces when the length increases. Thus if we consider a circular cross-section element of diameter d, since $A = \frac{1}{4}\pi d^2$ and hence $dA/dd = \frac{1}{2}\pi d = 2A/d$, we have $dA/A = 2dd/d = 2v\varepsilon$. Hence:

$$\frac{\Delta R}{R} = \frac{\Delta \rho}{\rho} + \varepsilon + 2v\varepsilon$$

The fractional change in resistance $\Delta R/R$ divided by the strain ε is called the *gauge factor G*. Thus:

$$G = 1 + 2v + \frac{1}{\varepsilon}\frac{\Delta \rho}{\rho}$$

For most metals Poisson's ratio is about 0.3 and the term involving the resistivity has a value of about 0.4 for metals commonly used for strain gauges. Thus the gauge factor is about 2.0. In practice, the gauge factor is determined from measurement of the change in resistance when known strains are applied. Advance, an alloy of nickel and copper with a low temperature coefficient of resistance and low temperature coefficient of expansion, is a commonly used material used for strain gauges.

We thus use the following equation to determine strain from resistance measurements:

$$\frac{\Delta R}{R} = G\varepsilon$$

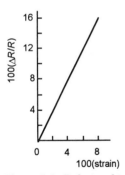

Figure 5.8 *Relationship for Advance*

Figure 5.8 shows how the fractional change in resistance for such a gauge varies with the applied strain, the temperature being constant or its effects eliminated. When the temperature of a strain gauge changes then its resistance changes and also it expands and so suffers strain. Both these effects add a resistance change on top of that due to the gauge being strained. Thus it is an advantage to have a material with a low temperature coefficient of resistance and a low temperature coefficient of expansion.

For semiconductors the resistivity term dominates and very high gauge factors are possible. Silicon p- and n-type gauges are widely used as sensors and have gauge factors of about +100 or more for p-type silicon and −100 or more for n-type silicon. A problem with semiconductor strain gauges is that they have a much greater sensitivity to temperature than metal strain gauges, the resistance changing markedly with temperature and also the gauge factor changing.

5.2.4 Photo conductive sensors

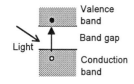

Figure 5.9 *Energy bands for a photoconductive cell*

Photo conductive cells, or *light dependent resistors,* are semiconductors used for their property of changing resistance when electromagnetic radiation is incident on them. When radiation of sufficiently high frequency is incident on such a material, electrons are excited from the valence band to the conduction band with the result that a hole is produced for conduction in the valence band and an electron for conduction in the conduction band (Figure 5.9). The result is an increase in the number of charge carriers and hence a drop in resistance. As the intensity of the radiation falling on the cell increases so the number of charge carriers available for conduction is increased and the resistance decreases.

Cadmium sulphide, generally doped with atoms of other elements, is a commonly used material for photo conductive cells in the visible region. Figure 5.10 shows how the resistance of a typical cadmium sulphide cell changes with the intensity of illumination (note the log axes). Lead sulphide is a commonly used material for photo conductive cells in the near infrared region.

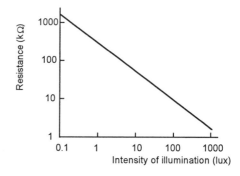

Figure 5.10 *Resistance of a cadmium sulphide cell as a function of the intensity of illumination*

5.3 Capacitive sensors

With capacitive sensors, the input being measured is transformed into a capacitance change. The capacitance C of a parallel plate capacitor (Figure 5.11) depends on the area A of the plates, their separation d and the relative permittivity (dielectric constant) ε_r of the material between them, being given by:

$$C = \frac{\varepsilon_0 \varepsilon_r A}{d}$$

where ε_0 is the permittivity of free space (8.85 pF/m).

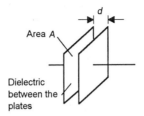

Figure 5.11 *Parallel plate capacitor*

5.3.1 Displacement sensor

Capacitive displacement sensors can be based on the separation between the plates changing (Figure 5.12(a)), the overlap area of the plates changing (Figure 5.12(b)) or the amount of dielectric between the plate changing (Figure 5.12(c)).

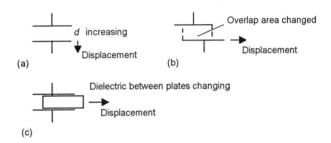

Figure 5.12 *Ways in which displacement can produce a change in capacitance*

For the displacement changing the plate separation (Figure 5.12(a)), if the separation d is increased by a displacement x then the capacitance becomes:

$$C - \Delta C = \frac{\varepsilon_0 \varepsilon_r A}{d + x}$$

Hence the change in capacitance ΔC as a fraction of the initial capacitance is given by:

$$\frac{\Delta C}{C} = -\frac{d}{d + x} - 1 = -\frac{x/d}{1 + (x/d)}$$

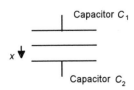

Figure 5.13 *Push-pull sensor*

There is thus a non-linear relationship between the change in capacitance ΔC and the displacement x. This can be overcome by using what is termed a *push-pull displacement sensor* (Figure 5.13). This has three plates with the upper pair forming one capacitor and the lower pair another capacitor. The displacement results in the middle plate moving, so increasing the capacitance of one of the capacitors and decreasing the other. We thus have:

$$C_1 = \frac{\varepsilon_0 \varepsilon_r A}{d+x} \text{ and } C_2 = \frac{\varepsilon_0 \varepsilon_r A}{d-x}$$

When C_1 is in one arm of an a.c. bridge and C_2 in the other, then the resulting out-of-balance voltage is proportional to x.

A particular form of displacement sensor that is often used as a pressure sensor is where one of the plates is a flexible circular diaphragm and the other a fixed plate (Figure 5.14), there being air (relative permittivity 1) between the plates. Changes in the pressure acting on the diaphragm cause it to curve and so change the separation between the capacitor plates by amounts which are not the same over the full plate area. The result is that the average plate separation changes. For a diaphragm of radius a, thickness t, with plates initially with separation d then the change in capacitance ΔC produced by a pressure difference P across the flexible plate is:

$$\frac{\Delta C}{C} = \frac{(1-v^2)a^4}{16Edt^3}P$$

C is the initial capacitance, namely $\varepsilon_0 \pi a^2/d$, E is Young's modulus for the flexible plate material and v is its value of Poisson's ratio.

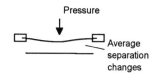

Pressure

Average separation changes

Figure 5.14 *Capacitive pressure sensor*

5.3.2 Liquid level sensor

One particular form of capacitive sensor that involves varying the amount of dielectric between the plates is the liquid level sensor. The capacitor plates are a rod surrounded by a concentric cylinder (Figure 5.15). The capacitor stands in the liquid being monitored. As the liquid level changes, so the amount of the space between the capacitor plates occupied by the liquid changes.

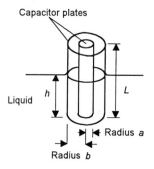

Capacitor plates

Liquid

h

L

Radius a

Radius b

Figure 5.15 *Capacitive liquid level sensor*

The capacitance per unit length of two coaxial cylinders, radii a and b, is $2\pi\varepsilon_0\varepsilon_r/\ln(b/a)$, for its derivation see, for example, *Electric and Magnetic Properties of Materials* by W. Bolton (Longman 1992). We can consider the liquid level sensor to consist of two capacitors in parallel, one having the liquid as the dielectric and being of length h and the other the air above the liquid as the dielectric and of length $(L - h)$. The total capacitance of capacitors in parallel is the sum of their individual capacitances. Hence:

$$\text{capacitance of sensor} = \frac{2\pi\varepsilon_0\varepsilon_r h}{\ln(b/a)} + \frac{2\pi\varepsilon_0(L-h)}{\ln(b/a)}$$

where ε_r is the relative permittivity of the liquid. The equation can be simplified to give:

$$\text{capacitance of sensor} = \frac{2\pi\varepsilon_0}{\ln(b/a)}[L + (\varepsilon_r - 1)h]$$

Thus the capacitance can be used as a measure of the height h of the liquid.

5.4 Inductive sensors

With inductive sensors, the input being measured is transformed into a change in inductance.

5.4.1 Variable reluctance sensors

A basic electrical circuit involves a battery, i.e. a source of e.m.f. (electromotive force) driving a current through electrical resistance (Figure 5.16(a)). We then have:

e.m.f. = current × resistance

In a similar way we can consider a basic magnetic circuit as consisting of a coil of wire wrapped round part of a core of ferromagnetic material (Figure 5.16(b)). A current through the coil of wire generates an m.m.f. (magnetomotive force) which drives magnetic flux through the reluctance of the core. We then have:

m.m.f. = flux × reluctance

For a coil with n turns and carrying a current I, the m.m.f. is nI. The flux in the core produced by this m.m.f. depends on the reluctance of the magnetic circuit. With an electrical circuit, the resistance R of a strip of material depends on its length L, cross-sectional area A and resistivity ρ:

$$R = \frac{\rho L}{A}$$

Likewise for a magnetic element, the reluctance S of a strip of material depends on its length L, cross-sectional area A and relative permeability μ_r:

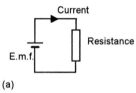

(a)

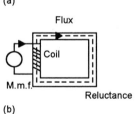

(b)

Figure 5.16 *(a) Electric circuit, (b) magnetic circuit*

$$S = \frac{L}{\mu_r \mu_0 A}$$

where μ_0 is the permeability of free space.

If we have a magnetic circuit of the form shown in Figure 5.17 then we have elements with different reluctances in series. As with electrical circuits where the total resistance of series elements is the sum of their individual resistances, so with magnetic circuits the total reluctance of series elements is the sum of their individual reluctances. By changing the air gap, the reluctance of that element changes and hence the total reluctance of the circuit. Thus if the length of the flux path in the ferromagnetic material is L (the average length is taken as the length down the centre of the core) then its reluctance S_f is:

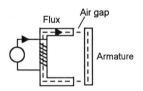

Figure 5.17 *Magnetic circuit*

$$S_f = \frac{L}{\mu_r \mu_0 A}$$

If the width of the air gap is d then the total length of the air gap element is $2d$ and its reluctance S_a is:

$$S_a = \frac{2d}{\mu_0 A}$$

The total reluctance of the magnetic circuit is then:

$$\text{total reluctance} = \frac{L}{\mu_r \mu_0 A} + \frac{2d}{\mu_0 A}$$

If initially the air gap is zero, the armature piece being pushed up to the remainder of the core, then the initial reluctance S_0 is:

$$S_0 = \frac{L}{\mu_r \mu_0 A}$$

Thus, with the air gap:

$$\text{total reluctance} = S_0 + \frac{2d}{\mu_0 A} = S_0 + kd$$

where k is a constant. Thus the reluctance depends on the displacement of the armature.

The flux in the circuit is m.m.f./total reluctance and so is:

$$\text{flux} = \frac{nI}{S_0 + kd}$$

A coil of wire of n turns carrying a current I has a self-inductance given by $n\phi/I$, where ϕ is the flux linked by the coil. Thus:

$$\text{self inductance} = \frac{n^2}{S_0 + kd}$$

The self-inductance can be measured using a bridge circuit and is a measure of the displacement of the armature. The relationship between the self-inductance and the displacement *d* is, however, non-linear. This non-linearity can be overcome by using a *push-pull* form of sensor (Figure 5.18). Movement of the armature increases the self-inductance of one coil and decreases that of the other. By incorporating the coils in different arms of a bridge circuit, an output can be obtained which is a linear function of the armature displacement.

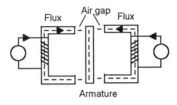

Figure 5.18 *Push-pull variable reluctance sensor*

5.4.2 Linear variable differential transformer

The *linear variable differential transformer* (LVDT) is a transformer with a primary coil and two identical secondary coils wound on the same insulating former (Figure 5.19). There is a ferromagnetic core which can move inside this former, the displacement of this core being the input to the sensor.

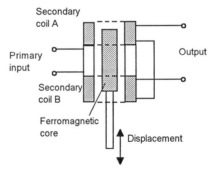

Figure 5.19 *Basic principle of the LVDT*

An alternating current in the primary coil will give rise to an alternating magnetic field. There will thus be an induced e.m.f. produced in the secondary coils, the size of the alternating e.m.f. being dependent on the magnetic flux linking the coils. Since the coils are identical and the same distance from the primary coil, this will depend on the amount of ferromagnetic materials in the core of the coil. The two secondaries are connected in series so that their outputs oppose each other. An alternating voltage input to the primary coil induces alternating e.m.f.s in the secondary coils. Both the secondary coils are identical so that when the ferromagnetic core is central with equal amounts in each secondary coil, the induced e.m.f.s are the same. Since they are connected so that the outputs from the two secondary coils oppose each other, this means that there is zero output when the core has equal amounts in each coil. When the core is moved so that there are different amounts in the two secondary coils, the e.m.f.s induced in each coil will differ and the difference will be a measure of the displacement of the ferromagnetic core.

The e.m.f. induced in a secondary coil by a changing current I in the primary coil is given by:

$$e = M\frac{di}{dt}$$

where M is the mutual inductance, its value depending on the number of turns on the coils and the ferromagnetic core. Thus, for a sinusoidal input current $i = I \sin \omega t$ to the primary coil, the e.m.f.s induced in the two secondary coils A and B can be represented by:

$$v_A = k_A \sin(\omega t - \phi) \text{ and } v_B = k_B \sin(\omega t - \phi)$$

where the values of k_A, k_B and ϕ depend on the degree of coupling between the primary and secondary coils for a particular core position. ϕ is the phase difference between the primary alternating voltage and the secondary alternating voltages. The two outputs are in series so that their difference is the output. Thus:

$$\text{output voltage} = v_A - v_B = (k_A - k_B)\sin(\omega t - \phi)$$

When the core is equally in both coils, k_A equals k_B and so the output voltage is zero. When the core is more in A than in B we have $k_A > k_B$ and when the core is more in B than in A we have $k_A < k_B$. A consequence of k_A being less than k_B is that there is a phase change of 180° in the output when the core moves from more in A to more in B. Thus for $k_A > k_B$ we have:

$$\text{output voltage} = (k_A - k_B)\sin(\omega t - \phi)$$

and for $k_A < k_B$:

$$\text{output voltage} = -(k_A - k_B)\sin(\omega t - \phi) = (k_B - k_A)\sin[\omega t + (\pi - \phi)]$$

Figure 5.20 shows how the size and phase of the output changes with the displacement of the core.

With the form of output given in Figure 5.20, the same amplitude output voltage is produced for two different displacements. To give an output voltage which distinguishes between these two situations, i.e. where the amplitudes are the same but there is a phase difference of 180°, a phase

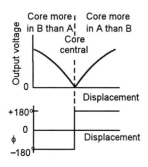

Figure 5.20 *LVDT output*

sensitive demodulator, with a low pass filter, is used to convert the output into a d.c. voltage which gives a unique value for each displacement (Figure 5.21).

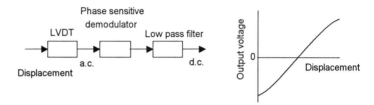

Figure 5.21 *LVDT d.c. output*

The output voltage is not perfectly linear, the non-linearity increasing as the displacement increases from the core central position. LVDT sensors are available to cover displacement ranges from ±0.25 mm to ±250 mm. The frequency of the current supplied to the primary coil can range from 50 Hz to 25 kHz. If the sensor is used to measure transient or alternating displacements, the frequency used for the primary should be ten times greater than the highest frequency component in the displacement signal.

5.5 Electromagnetic sensors

These are based on Faraday's laws of electromagnetic induction. When the flux linked by a conductor changes then the e.m.f. *e* induced in the conductor is proportional to the rate of change of the linked flux Φ, i.e.

$$e = -\frac{d\Phi}{dt}$$

The minus sign is because the direction of the induced e.m.f. is such as to oppose the change producing it. If the conductor is a coil with *n* turns with the flux Φ linked by each turn, then the induced e.m.f. for the coil is:

$$e = -n\frac{d\Phi}{dt}$$

5.5.1 Tachogenerator

The tachogenerator is used for the measurement of angular velocity. One form, termed the *variable reluctance tachogenerator*, consists of a toothed wheel of ferromagnetic material which is attached to the rotating shaft (Figure 5.22). As the wheel rotates, so the flux linked by a pickup coil wound on a permanent magnet changes. The resulting cyclic change in the flux linked produces an alternating e.m.f. in the coil.

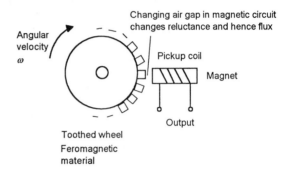

Figure 5.22 *Tachogenerator*

If the wheel contains n teeth and rotates with an angular velocity ω, then the flux change with time for the coil can be considered to be of the form:

$$\Phi = \Phi_0 + \Phi_a \cos n\omega t$$

where Φ_0 is the mean value of the flux and Φ_a the amplitude of the flux variation. The induced e.m.f. e in the N turns of the pickup coil is:

$$e = -N\frac{d\Phi}{dt} = -N\frac{d}{dt}(\Phi_0 + \Phi_a \cos n\omega t) = N\Phi_a n\omega \sin n\omega t$$

This is an equation of the form:

$$e = E_{max} \sin \omega t$$

The maximum value of the induced e.m.f. is thus $N\Phi_a n\omega$ and so is a measure of the angular velocity.

An alternative way of processing the output from the coil is to use a pulse-shaping signal conditioner to transform the output into pulses which can be counted by a counter.

An alternative form of tachogenerator involves using what is essentially an *a.c. generator*. Such a tachogenerator consists of a coil, termed the rotor, which rotates with the rotating shaft in the magnetic field produced by a stationary permanent magnet or electromagnet. An alternating e.m.f. is induced in the rotor and its amplitude and frequency is a measure of the angular velocity of the rotor. The output may be rectified to give a d.c. voltage with a size which is proportional to the angular velocity. Non-linearity for such sensors is typically of the order of ±0.15% of the full range and the sensors typically used for rotations up to about 10 000 revs per minute.

5.6 Thermoelectric sensors

With thermoelectric sensors, the input is temperature and the output an e.m.f., the sensor being termed a *thermocouple*. When two different metals are in contact, there is a potential difference across the junction, the size of the potential difference depending on the metals used and the temperature of the junction. Thus if two different metals are connected as part of a circuit with a voltmeter, as shown in Figure 5.23, then we have two junctions. If the junctions are at the same temperature then the potential differences across each junction are the same size and in such directions as to produce no net voltage across the voltmeter. However, if the two junctions are at different temperatures, there is a net voltage and so the voltmeter gives a reading. The voltmeter reading is related to the temperature difference between the junctions. A thermocouple is thus a closed circuit with two junctions formed by two metals, A and B, at different temperatures, T_1 and T_2, as shown in Figure 5.23.

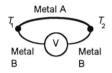

Figure 5.23 *Thermocouple*

The following are the five basic laws governing the behaviour of thermocouples:

1   The e.m.f. of a given thermocouple depends only on the temperatures of the junctions and is independent of the temperatures of the wires connecting the junctions.

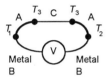

Figure 5.24 *Thermocouple*

2   If a third metal C is introduced into metals A or B, then provided the two new junctions between C and A, or B, are at the same temperature the e.m.f. is unchanged (Figure 5.24). This means that a voltmeter can be introduced into a thermocouple circuit and provided both its terminals are at the same temperature, the voltmeter will not affect the e.m.f.

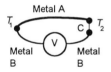

Figure 5.25 *Thermocouple*

3   If a third metal C is introduced between A and B at either junction (Figure 5.25), then provided the two new junctions AC and CB are both at the same temperature, the e.m.f. is unchanged. This means that wires A and B can be soldered or brazed together by a third metal without affecting the e.m.f.

4   If the thermoelectric e.m.f. produced between metals A and C is E_{AC} and that between metals C and B is E_{CB}, then the thermoelectric e.m.f. E_{AB} produced between metals A and B is:

$$E_{AB} = E_{AC} + E_{CB}$$

This is called the *law of intermediate metals*. For example, if we know the thermoelectric e.m.f. for a copper–iron thermocouple with cold junction at 0°C and hot junction at 100°C and for an iron–constantan thermocouple for the same temperature junctions, then the thermoelectric e.m.f for a copper–constantan thermocouple for the same temperature junctions is the sum of the thermoelectric e.m.f.s for the copper–iron and iron–constantan thermocouples.

5   If a thermocouple gives a thermoelectric e.m.f. $E_{1,2}$ when its junctions are at temperatures T_1 and T_2 and e.m.f. $E_{2,3}$ when its junctions are at temperatures T_2 and T_3, then the e.m.f. $E_{1,3}$ it would produce when its junctions are at temperatures T_1 and T_3 is given by:

$$E_{1,3} = E_{1,2} + E_{2,3}$$

This is known as the *law of intermediate temperatures*. Suppose we have a thermocouple, chromel–constantan, with one junction at 200°C and the other at 0°C. The thermoelectric e.m.f. is 13.419 mV. If we have the same thermocouple with one junction at 20°C and the other at 0°C then the e.m.f. is 1.192 mV. Using the law of intermediate temperatures we can determine what the thermocouple e.m.f. would be when its junctions are at 200°C and 20°C.

$$E_{200,0} = E_{200,20} + E_{20,0}$$

Hence:

$$E_{200,20} = E_{200,0} - E_{20,0} = 13.419 - 1.192 = 12.227$$

Table 5.1 lists some commonly encountered thermocouples, their range of use and sensitivity are in the range 0 to 100°C. Reference letters are frequently used for such thermocouples and are indicated in the table.

Table 5.1 *Thermocouples*

Reference letter	Materials	Range °C	Sensitivity $\mu V/°C$
B	Platinum 30% rhodium/platinum 6% rhodium	0 to 1800	3
E	Chromel/constantan	−200 to 1000	63
J	Iron/constantan	−200 to 900	53
K	Chromel/alumel	−200 to 1300	41
N	Nirosil/nisil	−200 to 1300	28
R	Platinum/platinum 13% rhodium	0 to 1400	6
S	Platinum/platinum 10% rhodium	0 to 1400	6
T	Copper/constantan	−200 to 400	43

The value of the e.m.f. produced by a thermocouple depends on the two metals concerned and the temperatures of both junctions. Usually one junction is held at 0°C. The relationship is then of the form:

$$E = a_1T + a_2T^2 + a_3T^3 + \ldots$$

where T is the temperature in °C of the hot junction and a_1, a_2, a_3, ... are constants with $a_1 > a_2 > a_3$. Figure 5.26 shows how the thermoelectric e.m.f. varies with temperature for some commonly used thermocouples.

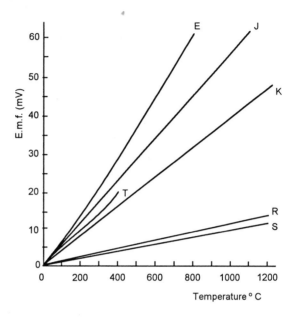

Figure 5.26 *Thermoelectric e.m.f.s of common thermocouples*

Tables are available giving the thermoelectric e.m.f.s for different hot junction temperatures when the cold junction is at 0°C. Table 5.2 gives some values for commonly used thermocouples.

Table 5.2 *Thermoelectric e.m.f.s with cold junction at 0°C*

Temp. °C	E.m.f.s in mV					
	E	J	K	R	S	T
−200	−8.824	−7.890	−5.891			−5.603
−100	−5.237	−4.632	−3.553			−3.378
0	0.000	0.000	0.000	0.000	0.000	0.000
100	6.317	5.268	4.095	0.645	0.645	4.277
200	13.419	10.777	8.137	1.465	1.440	9.286
300	21.033	16.325	12.207	2.395	2.323	14.860
400	28.943	21.846	16.395	3.399	3.260	20.869
500	36.999	27.388	20.640	4.455	4.234	
600	45.085	33.096	24.902	5.563	5.237	
700	53.110	39.130	28.128	6.720	6.720	
800	61.022	45.498	33.277	7.924	7.924	
900	68.783	51.875	37.325	9.175	9.175	
1000	76.357	57.942	41.269	10.471	10.471	
1100		63.777	45.108	12.503	11.817	
1200		69.536	48.828	13.193	13.193	

Table 5.2 is based on the cold junction being kept at 0°C. To obtain a cold junction at 0°C the junction is commonly kept in a mixture of ice and water. If the cold junction is left at room temperature, then a correction has to be made since not only is this temperature not 0°C but can vary. Compensation circuits (Figure 5.27) can be used to generate a potential difference which is the same as the e.m.f. that would be generated by the thermocouple with one junction at 0°C and the other at room temperature, so enabling the law of intermediate temperatures to be used. This correction potential difference can be produced by using a resistance thermometer in a Wheatstone bridge (see Section 7.1.4). The bridge is balanced at 0°C and the out-of-balance potential difference which occurs when the resistance thermometer is at room temperature provides the correction potential difference.

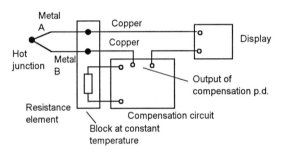

Figure 5.27 *Compensation for cold junction at room temperature*

5.7 Elastic sensors

With elastic sensors, the input being measured is transformed into a displacement, the displacement often being transformed into another signal form by some other sensor. A useful model for elastic sensors is provided by the mass-spring-damper system described in Section 3.3. Elastic sensors are second order systems.

5.7.1 Springs and load cells

The simple *spring balance* is an example of an elastic sensor in that when forces are applied to the free end of the spring, it stretches and so the force is transformed into a displacement (Figure 5.28).

A wide variety of elastic members are used to transform forces into displacements which can then be transformed into electric signals by other sensors (Figure 5.29). The term *load cell* is generally used to describe such systems. Typically load cells consists of elastic members which can be classified as rings, beam types such as cantilevers, and columnar types. Strain gauges are the usual method used to transform the displacements into electrical signals.

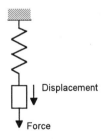

Figure 5.28 *Spring balance*

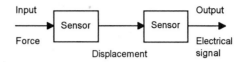

Figure 5.29 *The basic elements of a load cell*

The *proving ring* is an example a ring-type load cell. This is a steel ring which deforms from its circular shape under the action of forces. The amount of deformation is a measure of the forces and can be transformed into an electrical signal by strain gauges attached to the ring (Figure 5.30(a)) or using an LVDT to monitor the change in the diameter of the ring (Figure 5.30(b)). The reduction in the mean diameter d of the ring in the direction of the applied force F is given by:

$$\text{reduction in diameter} = \left(\frac{\pi}{2} - \frac{4}{\pi}\right)\frac{Fd^3}{16EI}$$

where E is Young's modulus for the ring material and I is the second moment of area. For a rectangular cross-section ring I has the value $bt^3/12$, where b is the breadth of the ring and t its thickness. The above equation then becomes:

$$\text{reduction in diameter} = 0.223\frac{Fd^3}{Wwt^3}$$

Proving rings are capable of high accuracy and are typically used for forces in the range 2 kN to 2000 kN.

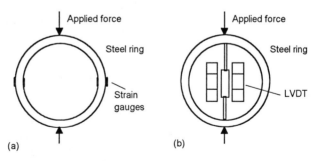

Figure 5.30 *Examples of proving rings*

Figure 5.31 shows examples of beam-type load cells. Both are forms of cantilevers with strain gauges used to monitor the bending. For a cantilever of length L, uniform thickness t, uniform width w, with strain gauges used to monitor the strain a distance x from the clamped end and a force F applied at the free end, then the strain experienced by the strain gauges is given by:

$$\text{strain} = \frac{6(L-x)F}{wt^2E}$$

where E is Young's modulus for the cantilever material.

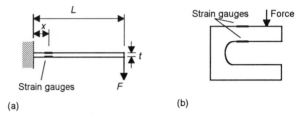

Figure 5.31 *Bending beam load cells*

Figure 5.32 is an example of a columnar-type load cell. These cells often take the form of rectangular or circular cross-sectional solid sections or rectangular or circular cross-sectional hollow sections. Strain gauges are generally mounted so that for each one in the longitudinal direction there is a matching strain gauge in the transverse direction. This enables the effects of temperature on the resistances of the strain gauges to be eliminated when both are mounted in a Wheatstone bridge (see Chapter 7).

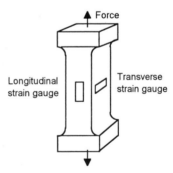

Figure 5.32 *Columnar load cell*

With a columnar load cell having a cross-sectional area A and subject to tensile force F, the longitudinal strain is:

$$\text{longitudinal strain} = \frac{F}{AE}$$

where E is Young's modulus for the material. The transverse strain is then:

$$\text{transverse strain} = -v\frac{F}{AE}$$

where v is Poisson's ratio for the material.

5.7.2 Diaphragms

With a *diaphragm* (Figure 5.33), a pressure difference between its two sides results in a displacement of the centre of the diaphragm by an amount which is related to the pressure difference. The diaphragm may be circular, flat and clamped around its edges. For such a diaphragm, the relationship between the pressure difference P and deflection y of the centre is given by:

$$P = \frac{16Et^3}{3a^4(1-v^2)}\left[y + 0.5\left(\frac{y}{t}\right)^2\right]$$

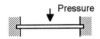

(a) *Flat diaphragm*

(b) *Corrugated diaphragm*

Figure 5.33 *Diaphragms*

where E is Young's modulus for the diaphragm material, v its Poisson's ratio, t the thickness of the diaphragm and a its diameter to the clamped edge. Greater movement is possible with a corrugated diaphragm. The displacement of a diaphragm can be monitored by a range of other sensors. One possibility is to use the diaphragm as one plate of a parallel plate capacitor, another to use a piezoelectric sensor to monitor the displacement (see Section 5.8).

Another form of diaphragm sensor uses strain gauges. A specially designed diaphragm strain gauge is often used. This consists of four strain gauges to form the arms of a Wheatstone bridge. Two of the gauges are near the centre of the diaphragm and designed to measure the strain in the circumferential direction while the other two are round the outer circumference of the diaphragm and are designed to measure radial strain (Figure 5.34). The circumferential strain is a maximum at the centre of a diaphragm and the radial strain a maximum, and tensile if the circumferential strain is compressive, at the circumference of the diaphragm.

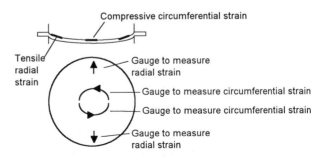

Figure 5.34 *Strains measured by special diaphragm strain gauge*

The maximum circumferentially directed strain is at the centre of the diaphragm and given by:

$$\text{maximum circumferential strain} = \frac{3Pa^2(1-v^2)}{8Et^2}$$

The maximum radially directed strain is at the circumference of the diaphragm and is given by:

$$\text{maximum radial strain} = -\frac{3Pa^2(1-v^2)}{4Et^2}$$

P is the pressure difference between the two sides of the diaphragm, a is its radius, t its thickness, E the Young's modulus for the material and v Poisson's ratio. The sensitivity of the diaphragm sensor is thus determined by the (a/t) ratio.

5.7.3 Capsules and bellows

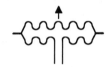

Figure 5.35 *Capsule*

A *capsule* (Figure 5.35) can be considered to be just two diaphragms back to back. When the pressure inside the capsule increases relative to that outside the capsule, both the diaphragms give central displacements outwards and, since the centre of the lower diaphragm is fixed, the centre of the upper diaphragm moves through twice the distance that the single diaphragm does. This movement may be used through linkages and gears to directly move a pointer across a scale or be transformed by some other sensor into an electrical signal.

A *bellows* can be considered as just a stack of capsules (Figure 5.36). Alternatively you can consider it as just a form of spring. When the pressure inside a bellows increases relative to that outside it, the bellows extends. This movement may be used through linkages and gears to directly move a pointer across a scale or be transformed by some other sensor into an electrical signal.

Figure 5.36 *Bellows*

5.7.4 Bourdon tubes

The *Bourdon tube* is a very commonly used form of pressure sensor and is the basis of many mechanical pressure gauges. The Bourdon tube is an elliptical cross-sectional tube made from materials such as stainless steel or phosphor bronze (Figure 5.37(a)). The cross-section must not be circular. The tube depends for its action on the pressure difference between the inside of the tube and the outside of the tube causing the tube to attempt to assume a circular cross-section when the internal pressure is greater than the external pressure. If the tube is in the form of a C (Figure 5.37(b)), this has the result of straightening out the C and hence moving the free end of the C outwards. If the tube is in the form a helix (Figure 5.37(b)), then the free end rotates.

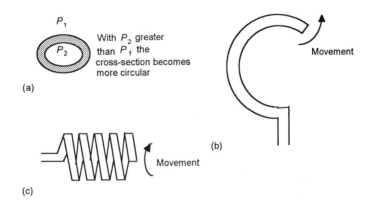

Figure 5.37 *Bourdon tube: (a) cross-section, (b) C type, (c) helical type*

5.8 Piezoelectric sensors

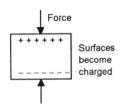

Figure 5.38 *Piezoelectricity*

Figure 5.39 *Capacitor*

With some dielectric materials, when they are stretched or compressed, one face of the material becomes positively charged and the opposite face negatively charged (Figure 5.38). This effect is called *piezoelectricity*. Such materials consist of ions in an orderly array within the crystalline structure. With many crystals, when the crystal is stretched or squashed and the ions moved further apart or closer together, the arrangement of the positive and negative ions is completely symmetrical and the change in the separation of the ions does not change this. With piezoelectric materials, the arrangement of the positive and negative ions is not symmetrical and changing their separation results in a change in the arrangement of the ions to produce a net displacement of charge so that one face of the material becomes positively charged and the other negatively charged. The net charge q on a surface is proportional to the amount x by which the charges have been displaced, and since the displacement is proportional to the applied force F:

$$q = kx = SF$$

where k is a constant and S a constant termed the *charge sensitivity*.

The charge sensitivity depends on the material concerned and the orientation of its crystals. For example, quartz has a charge sensitivity of 2.2 pC/N when cut in one particular direction and the forces applied in a particular direction, −2.0 pC/N when cut in the same way but the forces applied in a right-angled direction. The charge sensitivity of quartz is low compared with other piezoelectric materials. Barium titanate has a much higher charge sensitivity of the order of 130 pC/N and lead zirconate-titanate about 265 pC/N.

In order to measure the charge, metal electrodes are deposited on opposite faces of the piezoelectric material in order to form a capacitor with the piezoelectric material as the dielectric (Figure 5.39). The capacitance C of the arrangement is then:

$$C = \frac{\varepsilon_0 \varepsilon_r A}{t}$$

where ε_r is the relative permittivity of the material, A is area and t its thickness. Since $q = Cv$, where v is the potential difference produced across a capacitor when it has plates with charge q, then:

$$v = \frac{St}{\varepsilon_0 \varepsilon_r A} F$$

Since the force F is applied over an area A, the applied pressure p is F/A and so we can write:

$$v = S_v t p$$

where S_v is a constant $(S/\varepsilon_0 \varepsilon_r)$ and is termed the voltage sensitivity. The resulting voltage is thus proportional to the applied pressure. The voltage sensitivity for quartz is about 0.055 V/m Pa. For barium titanate it is about 0.011 V/m Pa.

The equivalent electrical circuit for such a piezoelectric sensor is a charge generator in parallel with capacitance C_s and in parallel with resistance R_s arising from leakage through the dielectric (Figure 5.40(a)). As a result, when the sensor is connected via a cable, of capacitance C_c, to an amplifier of input capacitance C_A and resistance R_A we have effectively the circuit shown in Figure 5.40(b). We have a total circuit capacitance of $C_s + C_c + C_A$ in parallel with $R_A R_s/(R_A + R_s)$.

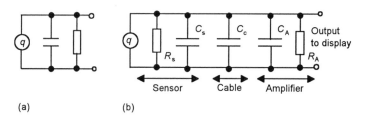

(a)   (b)

Figure 5.40 *(a) Sensor equivalent circuit, (b) sensor connected to charge amplifier*

When the sensor is subject to forces, it becomes charged. Connected across this charge generator is the total circuit capacitance and resistance. The capacitor becomes charged but because of the resistance, the capacitor will discharge with time. The time taken for the discharge will depend on the time constant of the circuit (see Section 3.2), i.e. the value of the product of the total circuit resistance and the total circuit capacitance.

We can think of the charge generator as being a current generator supplying a current i_s where $i_s = dq/dt$. This current then subdivides with

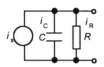

Figure 5.41 *Equivalent circuit*

part being used to charge the circuit capacitance C and part passing through the circuit resistance R (Figure 5.41):

$$i_s = i_C + i_R$$

But the potential difference v across the capacitor, which is the same as that across the resistor and also the output, is given by $q = Cv$ and so, since $q = \int i_C \, dt$, we have:

$$v = \frac{1}{C} \int i_C \, dt$$

This can be written as:

$$v = \frac{1}{C} \int (i_s - i_R) \, dt$$

Hence, since $i_R = v/R$, we can obtain the differential equation:

$$C\frac{dv}{dt} = i_s - \frac{v}{R}$$

$$RC\frac{dv}{dt} + v = Ri_s$$

Since $i_s = dq/dt$ and $q = kx$, where x is the displacement produced by the applied forces, then:

$$RC\frac{dv}{dt} + v = Rk\frac{dx}{dt}$$

Taking the Laplace transform (see Section 3.4):

$$RCsV(s) + V(s) = RksX(s)$$

Hence the transfer function for the system is:

$$G(s) = \frac{V(s)}{X(s)} = \frac{Rks}{RCs+1} = \frac{k}{C}\frac{\tau s}{\tau s+1}$$

where $\tau = RC$ and is the time constant. The term $\tau s/(\tau s + 1)$ means that the system cannot be used for the measurement of steady displacements since the steady-state response to a constant x is zero. What this is saying is that, after sufficient time all the charge produced by a steady displacement will have leaked away and so there is no output voltage for the steady state.

In the above discussion the transfer function was determined between the input of displacement and the output of voltage. However, what we are really concerned with is the input of force, or pressure, and the output of voltage. The transfer function relating the input of force and the output of displacement is effectively that which would be obtained if we considered the piezoelectric material as a mass-spring-damper system (see Sections 3.3 and 3.4.2) and so we have a transfer function of the form:

$$\frac{X(s)}{F(s)} = \frac{b_0\omega_n^2}{s^2 + 2\zeta\omega_n s + \omega_n^2}$$

Hence, the transfer function relating the force input and the output voltage is:

$$\text{transfer function} = \frac{V(s)}{X(s)}\frac{X(s)}{F(s)} = \frac{k}{C}\frac{\tau s}{\tau s + 1}\frac{b_0\omega_n^2}{s^2 + 2\zeta\omega_n s + \omega_n^2}$$

The term $\tau s/(\tau s + 1)$ means that the system cannot be used for the measurement of steady forces since the steady-state response to a constant force is zero. Because of this, piezoelectric sensors are used for the measurement of fluctuating pressures, accelerations and vibration. They cannot be used for the measurement of constant pressures or forces.

Charge amplifiers are frequently used with piezoelectric sensors. These are amplifiers which provide an output voltage proportional to any changes of charge stored on a device connected to the amplifier input terminals. Basically they are operational amplifiers with a capacitor in the feedback path (Figure 5.42). In the figure, the leakage resistance has been assumed to be infinite and so not included.

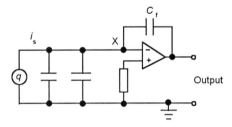

Figure 5.42 *Charge amplifier connected to piezoelectric sensor*

Because point X is a virtual earth, there is no potential difference across the piezoelectric sensor or the cable capacitor and so the current i_s must be the current flowing through the feedback capacitor. Thus we must have, for the feedback capacitor C_f:

$$i_s = \frac{dq_f}{dt} = -C_f\frac{dv}{dt}$$

where q_f is the charge on the feedback capacitor and v the potential difference across it. But the potential difference across the feedback capacitor is the output potential difference and thus v is the output potential difference. But $i_s = dq/dt$, where q is the charge produced on the piezo-electric sensor. Thus:

$$\frac{dq}{dt} = -C_f \frac{dv}{dt}$$

Hence:

$$v = -\frac{q}{C_f}$$

Since $q = kx$, where x is the displacement produced by the forces, or pressure, applied to the sensor, then:

$$v = -\frac{k}{C_f} x$$

The transfer function $V(s)/X(s)$ is thus $-k/C_f$. Hence with a charge amplifier, the transfer function relating the output voltage to the input forces is:

$$\text{transfer function} = -\frac{k}{C_f} \frac{b_0 \omega_n^2}{s^2 + 2\zeta \omega_n s + \omega_n^2}$$

In practice there is also a resistance R_f in parallel with the feedback resistor and this introduces a $\tau_s s/(\tau_s s + 1)$ term, where $\tau_s = R_f C_f$ into the transfer function.

5.9 Photovoltaic sensors

These are p-n junctions which produce a change in current when electromagnetic radiation is incident on the junction. They are widely used where the presence of an object is being detected by it breaking a beam of light.

Figure 5.43 shows the typical form of the current–voltage relationship when in the dark and when illuminated. In the absence of biasing, the effect of the illumination is to give a current which is proportional to the light intensity. In fact, the entire dark characteristic is shifted down by this current.

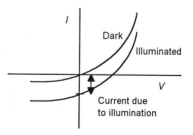

Figure 5.43 *Photo diode characteristic*

5.10 Pyroelectric sensors

Certain materials, e.g. lithium tantalate crystals, exhibit the pyroelectric effect. With such a material, when an electric field is applied across it at a temperature just below the Curie temperature (about 610°C for lithium tantalate) and the material then cooled in the field, the material becomes polarised as a result of electric dipoles within the material lining up (Figure 5.44(a) to (b)). When the field is then removed the material retains its polarisation (the effect is rather like magnetising a piece of iron by exposing it to a magnetic field). When the pyroelectric material is now exposed to infrared radiation, its temperature rises. This reduces the amount of polarisation in the material, the dipoles being shaken up more and losing their alignment (Figure 5.44(c)).

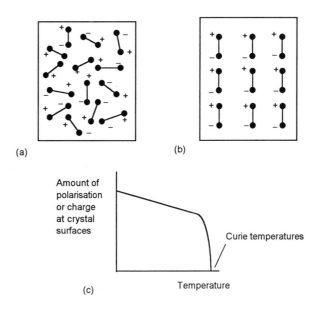

Figure 5.44 *(a) Before polarisation, (b) after polarisation, (c) effect of temperature on the amount of polarisation*

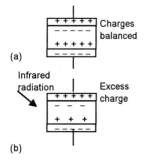

Figure 5.45 *Pyroelectric sensor*

A pyroelectric sensor consists of a pyroelectric crystal with thin metal film electrodes on opposite faces. Because the crystal is permanently polarised, ions are drawn from the surrounding air and electrons from any measurement circuit connected to the sensor to balance the charge at the surfaces of the crystal at the prevailing temperature (Figure 5.45(a)). If radiation is incident on the crystal and changes its temperature, then the polarisation in the crystal is reduced and causes a reduction in the charge at the surfaces of the crystal. As a consequence there is then an excess of charge on the metal electrodes over that needed to balance the charge on the crystal surfaces (Figure 5.45(b)). Thus the effect of the radiation has been to liberate charge. This charge then leaks away through the measurement circuit until the charge on the crystal once again is balanced by the charge on the electrodes. The pyroelectric sensor thus behaves as a charge generator which generates charge when there is a change in the temperature of the sensor as a result of the incidence of infrared radiation, the charge then leaking away through the measurement circuit.

Over the linear part of the graph in Figure 5.44(c) we have the change in charge Δq proportional to the change in temperature Δt:

$$\Delta q = k_p \Delta t$$

where k_p is a sensitivity constant for the crystal. A pyroelectric sensor might be used with a charge amplifier to give an output voltage (Figure 5.46(a)). These are amplifiers which provide an output voltage proportional to any changes of charge stored on a device connected to the amplifier input terminals. Basically they are operational amplifiers with a capacitor in the feedback path. Alternatively the pyroelectric sensor might be considered as a capacitor which is charged by the excess charge and so has a potential difference across it. A field-effect transistor (FET) might then be used (Figure 5.46(b)) to give an output which is a measure of this potential difference.

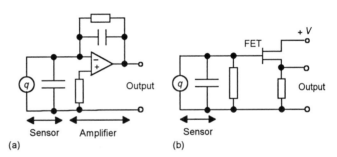

Figure 5.46 *(a) Charge amplifier circuit, (b) voltage measurement circuit*

5.11 Hall effect sensors

With the Hall effect sensor, the action of a magnetic field on a flat plate carrying an electric current generates a potential difference which is a measure of the strength of the field.

A beam of charged particles can be deflected by a magnetic field. A current flowing in a conductor is a beam of moving charges and thus can be deflected by a magnetic field. This effect was discovered by E.R. Hall in 1879 and is now referred to as the *Hall effect*. Consider a plate of a metal, or semiconductor, with a magnetic field applied at right angles to the plane of the plate (Figure 5.47). As a consequence of the magnetic field, the moving charged particles in the plate are acted on by a force which causes them to be deflected to one side of the plate. As a consequence, one side of the plate becomes negatively charged and the other positively charged. This charge separation produces an electric field in the material. The charge separation grows until the forces on the charged particles from the electric field just balance the forces produced by the magnetic field.

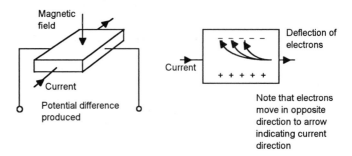

Figure 5.47 *Hall effect*

The resulting electric field across the width of the plate means there is a potential difference across it. This potential difference is given by:

$$V = K_H \frac{BI}{t}$$

where K_H is a constant called the *Hall coefficient*, B is the magnetic flux density, I the current through the plate and t the plate thickness. For a derivation of the above equation see, for example, *Electrical and Magnetic Properties of Materials* by W. Bolton (Longman 1992). Thus if a constant current source is used with a particular thickness sensor, the potential difference is directly proportional to the magnetic flux density.

Hall effect sensors are often integrated with signal processing circuitry on the same chip of material and often used as proximity detectors. Thus a permanent magnet may be attached to a moving object and its proximity to a Hall sensor detected by the changing Hall voltage resulting from the changing magnetic field. A particular application is in brushless d.c. motors where it is necessary to determine when the permanent magnet rotor is correctly aligned with the windings on the stator so that the current through the windings can be switched on at the right instant to maintain the rotor rotation. Hall sensors are used to detect when the alignment is right.

Problems

1 A potentiometer with a uniform resistance per unit length of track has a total resistance of 2 kΩ and a supply voltage of 10 V is connected across it. The output from the potentiometer is connected across a load of resistance 5 kΩ. What will be the error due to this loading when the slider is at the mid-point of the potentiometer track?

2 A potentiometer with a uniform resistance per unit length of track is to have a track length of 100 mm and used with the output being measured with an instrument of resistance 10 kΩ. Determine the resistance

required of the potentiometer if the maximum error is not to exceed 1% of the full-scale reading.

3 A platinum resistance coil has a resistance at 0°C of 100 Ω. Determine the change in resistance that will occur when the temperature rises to 30°C if the temperature coefficient of resistance is 0.0039 K⁻¹.

4 A platinum resistance thermometer has a resistance of 100.00 Ω at 0°C, 138.50 Ω at 100°C and 175.83 Ω at 200°C. What will be the non-linearity error at 100°C if a linear relationship is assumed between 0°C and 200°C?

5 An electrical resistance strain gauge has a resistance of 120 Ω and a gauge factor of 2.1. What will be the change in resistance of the gauge when it experiences a uniaxial strain of 0.0005 along its length?

6 A cantilever has a free length of 250 mm and a strain gauge is attached longitudinally half way along this length. The strain gauge has a resistance of 120 Ω and a gauge factor of 2.1. The cantilever has a rectangular cross-section with a thickness of 3 mm and a width of 60 mm, being made of material with a Young's modulus of 70 GPa. What will be the change in resistance of the strain gauge when a force of 1.0 N is applied at the free end of the cantilever?

7 A capacitive sensor consists of two parallel plates in air, the plates each having an area of 1000 mm² and separated by a distance of 0.3 mm in air. Determine the displacement sensitivity of the arrangement if the dielectric constant for air is 1.0006.

8 A capacitive sensor consists of two parallel plates in air, the plates being 50 mm square and separated by a distance of 1 mm. A sheet of dielectric material of thickness 1 mm and 50 mm square can slide between the plates. The dielectric constant of the material is 4 and that for air may be assumed to be 1. Determine the capacitance of the sensor when the sheet has been displaced so that only half of it is between the capacitor plates.

9 A variable reluctance sensor consists of a ferromagnetic core and armature, as in Figure 5.17, of relative permeability 50 with a variable air gap between the core and armature, the air having a relative permeability of 1. The mean length of the flux path in the ferromagnetic material is 345 mm and the core and armature have cross-sectional areas of 400 mm². Determine the flux in the air gap when a current of 1.2 A passes through a coil of 800 turns which is wrapped round the core and the total length of the air gap is 5 mm.

10 A variable reluctance sensor consists of a ferromagnetic core and armature, as in Figure 5.17, of relative permeability 100 with a variable air gap between the core and armature, the air having a relative

permeability of 1. A coil of 200 turns is wrapped round the core. If the core and the armature have the same cross-sectional area of 25 mm² and the total length of the flux path within the ferromagnetic material is 120 mm, determine an equation relating the inductance of the sensor to the separation of the armature and the core.

11 A variable reluctance tachogenerator consists of a toothed ferro-magnetic wheel which rotates close to a pickup coil (as in Figure 5.22). The wheel has 44 teeth and the pickup coil 100 turns. The flux, in mWb, linked by each turn of the coil is given by:

$$\text{flux} = 0.05 + 0.01 \cos 44\omega t$$

where ω is the angular velocity of the wheel. Determine the amplitude of the signal from the coil when the angular velocity is 1000 revs per minute.

12 A chromel–constantan thermocouple has a cold junction at 20°C. What will be the thermoelectric e.m.f. when the hot junction is at 200°C? Tables give for this thermocouple: 0°C, e.m.f. 0.000 mV; 20°C, e.m.f. 1.192 mV; 200°C, e.m.f. 13.419 mV.

13 An iron–constantan thermocouple has a cold junction at 0°C and is to be used for the measurement of temperatures between 0°C and 400°C. What will be the non-linearity error at 100°C, as a percentage of the full-scale reading, if a linear relationship is assumed over the full range? Tables give for this thermocouple: 0°C, e.m.f. 0.000 mV; 100°C, e.m.f. 5.268 mV; 400°C, e.m.f. 21.846 mV.

14 A force measurement sensor consists of a disc of diameter 40 mm and 1.6 mm thick which is simply supported round its edges. The force is applied centrally and can be considered to be applied round the perimeter of a circle of 4 mm radius. Determine the maximum deflection that will occur if the maximum stress to which the disc is to be subject is 400 MPa. The disc material has a Young's modulus of 200 GPa and Poisson's ratio of 0.3. For a simply supported disc of radius a when the force is applied centrally round the circumference of a circle of radius r:

$$\text{central deflection} = \frac{3F(1-v^2)}{2\pi Et^3}\left[\frac{(a^2-r^2)(3+v)}{2(1+v)} - r^2\ln\left(\frac{a}{r}\right)\right]$$

$$\text{radial stress} = \text{tangential stress} = \frac{3F}{4\pi t^2}\left[2(1+v)\ln\left(\frac{a}{r}\right)\right.$$

$$\left. + (1-v)\left(\frac{a^2-r^2}{a^2}\right)\right]$$

15 A ring-type load cell uses an LVDT sensor to monitor the changes in the ring diameter when it is subject to cyclic forces. The ring element is designed so that it gives a deflection of 1.25 mm when subject to the

maximum forces. The LVDT has a sensitivity of 250 (mV/V input)/mm. What will be the voltage output compared with the voltage input to the LVDT at maximum load?

16 For a circular diaphragm built in round its edges, determine the maximum value of the central deflection to diaphragm thickness ratio (*y/t*) if non-linearity is to be kept below 1% (see the equation given in Section 5.7.2).

17 A cadmium sulphide photo conductive cell has a resistance of 100 kΩ in the dark. This changes to 30 kΩ when in a particular intensity of illumination. If the cell has a time constant of 100 ms, what will the resistance of the cell be 10 ms after it is illuminated?

18 A piezoelectric sensor is connected by a short cable of negligible capacitance and resistance to a voltage measurement device which is purely resistive, having an input resistance of 10 MΩ. Determine the system transfer function given that the charge sensitivity to force of the piezoelectric crystal is 2 pC/N, its capacitance 100 pF, its natural angular frequency 200 krad/s and its damping ratio 0.01.

6 Microsensors

This chapter continues from the previous chapter on sensors with a consideration of the development of microsensors. The term *microsensor* has been used to describe sensors that have physical dimensions less than a millimetre. Such sensors have evolved as a consequence of the development of integrated circuits, often resulting in the integration of sensors and signal conditioning/processing circuitry on the same chip of silicon.

6.1 Silicon-based microsensors

Most microsensors are based on the use of silicon with entire circuits with their components and sensors all being formed within a wafer of silicon. The fabrication of microsensors involves the techniques used for the fabrication of silicon integrated circuits. These are primarily the selective doping and etching of silicon wafers, the techniques used including:

1   Controlled crystal growth to give the basic silicon material.

2   Oxidation to give a surface, insulating, oxide layer.

3   Lithography to imprint a geometric pattern from a mask onto a thin layer of material, termed the resist, which has been used to coat the surface of the oxidised silicon wafer. Ultraviolet light is then directed onto the masked surface and those parts of it exposed to the light undergo a chemical change. This enables the exposed parts of the resist to be dissolved away and leaves an exposed oxide pattern. The exposed oxide can then be etched away to leave the exposed silicon surface.

4   Selective diffusing or implanting dopants in the exposed silicon to change the electrical properties. By selective doping of different layers and different areas, a range of electronic components can be produced.

5   Film deposition to form such elements as electrodes.

6   Depositing of metallic layers and wire bonding to form connections.

7   In addition, there can be micro machining by selective etching to form particular mechanical structures required for sensors, e.g. a cantilever.

Figure 6.1 illustrates how a p-n junction might be made using some of the techniques outlined above. The above list just gives a brief indication of techniques used, for more details the reader is referred to specialist texts, e.g. *Semiconductor Devices: Physics and Technology* by S.M. Sze (Wiley 1985), *Design and Technology of Integrated Circuits* by D. de Cogan (Wiley 1990).

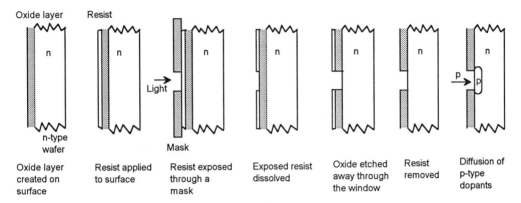

Figure 6.1 *Possible steps in the formation of a diode*

6.2 Thermal microsensors

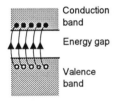

Figure 6.2 *Intrinsic semiconductor*

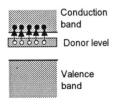

Figure 6.3 *Extrinsic semiconductor*

With intrinsic semiconductors, i.e. pure semiconductors with no doping or impurities, the number of holes in the valence band equals the number of electrons in the conduction band (Figure 6.2). When there is an increase in temperature there is an increase in the number of electrons moving from the valence band to the conduction band and hence an increased number of holes and electrons available for electrical conduction. This increase means a drop in electrical conductivity. The electrical conductivity can be expressed as:

$$\text{electrical conductivity} = A\ e^{-E_g/2kT}$$

where A and k are constants, E_g the energy gap between the valence and conduction bands and T the temperature on the kelvin scale.

While this might suggest that a semiconductor such as silicon could be used to give a temperature-dependent resistor, there are problems in that it requires high purity and usually there is sufficient impurities present for them to affect the conductivity. For this reason, silicon temperature-dependent resistors are based on using doped silicon, i.e. extrinsic silicon. With an n-type semiconductor, the doping has resulted in energy levels containing electrons being inserted into the energy gap (Figure 6.3). At temperatures above about 100 K, the electrons in the donor levels will have all jumped across to the conduction band. The conduction is then due almost entirely to these donated electrons in the conduction band, the number of charge carriers being determined by the amount of doping. As the temperature is increased, though some electrons are able to jump from the valence band to the conduction band, electrical conduction is almost entirely due to the donor electrons. Since the number of donor electrons is constant, increasing the temperature decreases the electrical conductivity because it decreases the mobility of the electrons (think of the electrons moving through a lattice of ions which are oscillating back and forth, the higher the temperature, the greater the amplitude of the oscillations and so the greater the obstacle to the electrons, the shorter distance they can accelerate before suffering a collision and so the lower the average speed of the electrons). It is this extrinsic conduction, when the doped semiconductor is behaving rather like a metal, that is used for temperature sensors. The sensor is then essentially just a piece of doped semiconductor between electrodes.

At yet higher temperatures, significant numbers of electrons are able to jump from the valence band to the conduction band, the higher the temperature the greater the number of electrons than can make the jump, and so the intrinsic form of conduction begins to dominate.

6.2.1 Thermodiodes

A widely used form of thermal microsensor is a semiconductor diode. As outlined in the previous section, when the temperature of doped semiconductors change, the mobility of their charge carriers changes. This affects the rate at which electrons and holes can diffuse across a p-n junction. As a consequence, when a p-n junction has a potential difference V across it, the current I through the junction is given by:

$$I = I_0(e^{eV/kT} - 1)$$

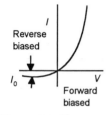

Figure 6.4 *Current – voltage characteristic*

where T is the temperature on the Kelvin scale, e the charge on an electron, k and I_0 are constants (see *Electric and Magnetic Properties of Materials* by W. Bolton (Longman 1992) for a derivation of the above equation). Figure 6.4 shows a graph of the above equation when the temperature is constant.

With reverse bias, V is negative and as V becomes more negative the exponential term becomes smaller and smaller and so the current becomes I_0, the current that occurs when the diode is reverse biased. The effect of changing the temperature is to change the value of the current for a particular potential difference or the potential difference for a particular current. By taking logarithms of the equation we can write the equation in terms of the voltage as:

$$V = \left(\frac{kT}{e}\right) \ln\left(\frac{I}{I_0} + 1\right)$$

For a constant current we have V proportional to the temperature on the kelvin scale. Thus a measurement of the potential difference at constant current can be used as a measure of the temperature.

An electrical circuit used with such a diode involves an operational amplifier. Such a circuit can be incorporated on the same chip as the diode, thus giving a very compact sensor complete with signal conditioning. Such an integrated circuit is the LM3911 (Figure 6.5). The output voltage from such a circuit is proportional to the temperature at the rate of 10 mV/°C. By connecting appropriate external resistors, the scale over which the temperatures are measured can be adjusted.

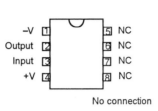
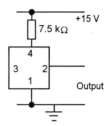

No connection

Figure 6.5 *LM3911*

For a more detailed discussion of the underlying theory of the p-n junction and its dependence on temperature, the reader is referred to texts such as *Electrical and Magnetic Properties of Materials* by W. Bolton (Longman 1992).

6.2.1 Thermotransistors

An alternative to the thermodiode is to use a transistor. In a similar manner to the thermodiode, the voltage across the junction between the base and the emitter depends on the temperature and can be used as a measure of temperature. A common method is to use two transistors with different collector currents and determine the difference in the base-emitter voltages between them, this difference being directly proportional to the temperature on the kelvin scale. Such transistors can be combined with other circuit components on a single chip to give a temperature sensor with its associated signal conditioning. The integrated circuit AD590 is such a device.

6.3 Mechanical microsensors

Single crystal silicon has a value of Young's modulus, about 190 GPa, not too different to that of steel, about 210 GPa, and a yield stress greater than that of steel. It is thus a suitable material for mechanical structures. In addition, when doped silicon is subject to strain its electrical resistance changes (see Section 5.2.3) and so can be used as a strain gauge.

6.3.1 Accelerometer

An accelerometer has been developed in the form of a small silicon mass connected to a silicon cantilever beam (Figure 6.6). The cantilever is about 0.5 mm long with the mass being a piece of silicon with an area of about 1 mm^2 and 200 μm thick. With one form (Figure 6.6(a)), doping has been used in part of the beam to produce a region to act as a semiconductor strain gauge. In another form a capacitive pickup has been used (Figure 6.6(b)).

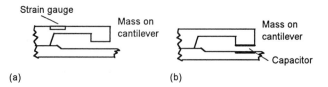

Figure 6.6 *Forms of micro-accelerometers*

6.3.2 Diaphragm pressure gauge

Diaphragm pressure gauges (see Section 5.7.2) can take the form of a circular diaphragm with strain gauges attached to it to give a measure of its displacement when subject to pressure. While strain gauges can be stuck onto a diaphragm, an alternative approach is to create a silicon diaphragm with the strain gauges as specially doped areas of the diaphragm (Figure 6.7).

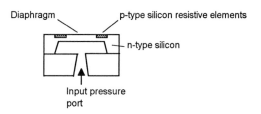

Figure 6.7 *Silicon pressure gauge*

Such sensors typically have a pressure range of 0 to 100 kPa and a total non-linearity error and hysteresis error of about ±0.5%. They have the great merit of being small, sizes less than 1 mm diameter are possible, and cheap. A silicon diaphragm gives a high natural frequency, and so typically can be used for dynamic pressure measurements up to about 80 kHz.

6.4 Magnetic sensors

Silicon exhibits the Hall effect (see Section 5.11), having a high Hall coefficient. With a semiconductor, when we have a current through a slice from either hole movement or electron movement or both the charge carriers are deflected by a magnetic field and give rise to a potential difference across the slice (Figure 6.8). The direction of the potential difference due to hole movement is in the opposite direction to that resulting from electron movement.

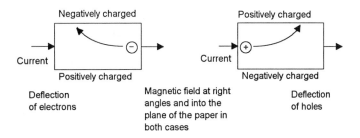

Figure 6.8 *The Hall effect with semiconductors*

Typically a silicon Hall sensor consists of an n-type layer on a p-type substrate with the current passing between two electrodes diffused as highly doped n-type in the n-layer and the resulting potential difference being measured between other highly doped n-type electrodes diffused into the n-layer (Figure 6.9). Highly doped n-material behaves rather like a metal because of the large number of electrons in the conduction band.

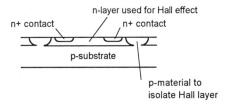

Figure 6.9 *Hall effect sensor*

6.5 Chemical sensors

Chemical sensors are used to detect the presence of chemicals, e.g. the presence of carbon monoxide in air. In this section the topic is only briefly touched on; for more details the reader is referred to specialist texts such as *Principles of Chemical Sensors* by J. Janata (Plenum Press 1989).

6.5.1 Chemoresistors

A simple form of chemical sensor is the resistor which shows a change in resistance as a consequence of the presence of particular chemicals. The electrical conductivity of many metal oxide semiconductors is changed when gas molecules are adsorbed on the surface of the semiconductor. Adsorption results in the formation of bonds between the gas molecules and atoms of the semiconductor by the transfer of electrons and so, by changing the number of charge carriers, changes the electrical conductivity. The change in conductivity is related to the number of gas molecules adsorbed and so to the concentration of the molecules in the surrounding atmosphere. Metal oxides that have been used are SnO_2, ZnO, TiO_2 and In_2O_3. Such a sensor might just consists of a bead, or layer on an inert substrate, of the semiconducting material between two electrodes. Tin oxide sensors have been developed for the measurement of odours due to sulphur compounds, cigarrete smoke in air, organic solvents, methane, hydrogen, ammonia, etc.

6.5.2 Chemocapacitors

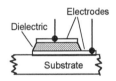

Figure 6.10 *Chemocapacitor*

A parallel plate capacitor with its dielectric exposed to the surrounding atmosphere (Figure 6.10) can be constructed using a dielectric which changes its dielectric constant in response to particular gases. A sensor based on this principle is used for the measurement of the relative humidity, i.e. a measure of the amount of water vapour in the atmosphere. Others have been developed for the detection of carbon monoxide, carbon dioxide, methane and nitrogen.

6.5.3 Chemotransistors

A MOSFET is a field-effect transistor (for a discussion of the principles of such components the reader is referred to texts on electronics, e.g. *Electronics: Circuits, Amplifiers and Gates* by D.V. Bugg (Adam Hilger 1991)). A MOSFET in which the gate is made of a chemically sensitive layer, such as palladium, gives a sensor in which there is a change in the threshold voltage which depends on concentration of the chemicals to which the chemically sensitive layer is exposed. Such a device is sometimes referred to as a CHEMFET (Figure 6.11). With palladium, the sensor is sensitive to hydrogen gas in concentrations less that one part per million.

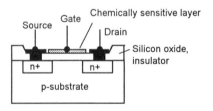

Figure 6.11 *CHEMFET*

Another form, termed an ISFET (ion-sensitive FET), involves a MOSFET with the gate electrode, which is separated from the substrate by an electrolyte, becoming the reference electrode in an electrochemical cell (Figure 6.12). When the ISFET is immersed in an electrolyte, the threshold voltage shifts by an amount which depends on the concentration of hydrogen ions and so gives a measure of the pH. By coating the gate electrode with different membranes, different ions can be selected and their concentration determined.

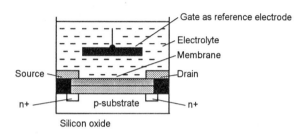

Figure 6.12 *ISFET*

6.6 Radiation microsensors

Photoconductive cells, photodiodes and phototransistors are examples of microsensors that can be used for the detection of ultraviolet, visible and infrared radiation.

Cadmium sulphide photoconductive cells (see Section 5.2.4) are widely used for light in the visible region and can be in the form of flat coils which when encapsulated might have a diameter of about 13 mm and a thickness of 6 mm and response times of the order of 20 ms.

Photodiodes (see Section 5.9) can take a number of forms: the p-n photodiode, the p-i-n photodiode, the Schottky-type photodiode and the avalanche photodiode. Photodiodes have a higher sensitivity and faster response time than photoconductive cells. Response times are typically of the order of microseconds. The type of diode and the materials used determine the range of wavelengths to which it will respond. Silicon photodiodes can be used for wavelengths from the ultraviolet to the near infrared, p-i-n silicon photodiodes from the ultraviolet to just in the infrared, GaAsP Schottky photodiodes from just in the ultraviolet to the red end of the visible spectrum, silicon avalanche photodiodes from the violet end of the visible spectrum to the near infrared. A typical p-i-n photodiode chip might be encapsulated in a 5 mm half round black, infrared transmissive, plastic package and have a peak response at a wavelength of 950 nm. The maximum dark current might be 30 nA and the current 60 μA when exposed to light of wavelength 950 nmm and intensity 1000 lux.

Phototransistors can give greater light sensitivity than photodiodes and are sometimes used as a Darlington pair. An n-p-n silicon phototransistor might be encapsulated in a 3 mm clear resin package, have a peak response at a wavelength of 880 nm, give a maximum dark current of the order of 0.1 μA and a current of between 1 and 20 mA when exposed to light of intensity 1000 lux at a wavelength of 880 nm, the collector–emitter voltage being 3 V.

6.7 Smart sensors

A sensor may be integrated on the same chip with signal conditioning circuits. However, when a sensor is integrated with not only signal conditioning but a microprocessor so that there is also signal processing and control exercised over the sensor, it is generally termed a *smart* or *intelligent sensor*. Because microprocessors require inputs of digital signals, if the output from the sensor is analogue then analogue-to-digital conversion (ADC) is required. Display of the output from the micro-processor, or use in a control circuit, might require a digital-to-analogue conversion (DAC) of the microprocessor. Thus basically, a smart sensor might contain the elements shown in Figure 6.13.

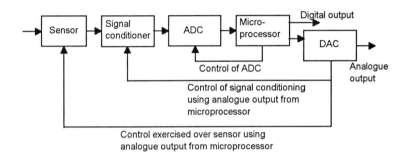

Figure 6.13 *A basic form of smart sensor*

A smart sensor thus not only contains the sensing element but has also a data processing facility. This can allow it to perform automatic calibration, self-testing and compensate for such things as non-linearity, thermocouples not having the cold junction at 0°C, scatter of readings, elimination of abnormal readings, etc.

As an illustration, a smart pressure sensor is available which has the basic sensor element to detect the difference in pressure between the two sides of a diaphragm, but also sensors to detect the static pressure and the temperature. The outputs from the sensors are, after signal conditioning, processed by a microprocessor with the static pressure data and the temperature data being used to correct the value of the differential pressure. The output is then a digital signal representing the value of the pressure difference and this, after digital-to-analogue conversion is available for use in control circuits.

6.9 Microsensor arrays

The above description of a differential pressure system with sensors of the pressure difference, static pressure and temperature is an example of what might be termed a sensor array. Microsensor array devices are concerned with the processing of signals from a number of sources (Figure 6.14). Thus an array of sensors will produce signals, which after some pre-processing will be analysed by the array processor to give an output of some quantity which depends on the inputs to the array of sensors.

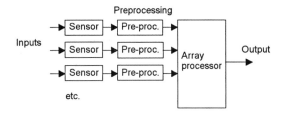

Figure 6.14 *Sensor array system*

There are various levels of integration. Just the array of sensors might be integrated as a single integrated circuit, or perhaps the array of sensors with their associated signal conditioning and pre-processing, or perhaps a completely integrated circuit with sensors, pre-processing and the array processor all on one integrated circuit.

A sensor array might be composed of a number of identical sensors, the outputs from all of them being summed to give a much larger output signal, e.g. an array of thermocouples. Another possibility is the use of an array of identical sensors to form a one-dimensional or two-dimensional map of the signals from some object. An example of this is in the automatic camera where the light input falls on an array of light sensitive sensors. To determine the exposure the camera microprocessor can compare the brightness of different areas of the picture and determine the optimum exposure. Another example is an infrared camera where the array of infrared sensitive sensors gives signals which the microprocessor converts into outputs which can give a visual image on a screen. Another form of array is where the sensors monitor different quantities. Thus an example of this is the pressure gauge where one sensor monitors the pressure difference, another the static pressure and another the temperature, the microprocessors using the data to compute an accurate value for the pressure difference.

Problems

1 A thermodiode is specified as having a time constant of 3 s. What will be the time taken for the thermodiode to respond to a step input and reach 95% of the steady-state value?

2 A silicon thermoresistor has typically a range of −50 to +150°C, a sensitivity of about 7 Ω/°C, and gives a linear output. Compare this with the typical performance of thermistors, commenting on the main differences which would affect selection.

3 Explain the basic principles of the ISFET in its determination of pH.

4 Explain what is meant by a smart sensor.

5 Suggest a form of sensor that might be used to act as a 'nose' for particular odours.

6 Suggest a form of sensor that is small and has a fast response and might be used as a light sensor for the near infrared.

7 Signal conditioning and processing

Signal conditioning is concerned with taking the output from a sensor and converting it into a suitable form for *signal processing* so that it can be displayed or handled by a control system. Signal conditioning might involve transforming a resistance change into a voltage, or perhaps using an amplifier to make a signal bigger. Signal processing might involve taking into account non-linearity before determining the result or combining signals from more than one sensor or averaging the outputs from a single sensor over a period of time.

The following are some examples of common forms of signal conditioning and processing:

1 Transforming the signal from the sensor into the right type of signal. For example, with an electrical resistance strain gauge, the resistance change is converted into a voltage change using a Wheatstone bridge (see Section 7.1). It might mean making the signal digital or analogue (see Section 7.6) or suitable for transmission over a distance (see Section 7.5).

2 Adjusting the level of the signal to the right size. The signal from a thermocouple of a few millivolts might need amplification before further processing can occur. Operational amplifiers are widely used for amplification (See section 7.2).

3 Eliminating or reducing *noise*. Thus filters might be used to eliminate mains noise from a signal (see Section 7.4).

4 Making a non-linear signal from a sensor into a linear one, i.e. one where the output is directly proportional to the variable (see Section 7.2.6).

5 Protection to prevent damage to the next element, e.g. a micro-processor, as a result of high current or voltage (see Section 7.7).

This chapter takes a look at all the above forms of signal conditioning and processing.

7.1 Resistance to voltage conversion

Resistance temperature sensors and electrical strain gauges are examples of sensors where the output is a change in resistance. This generally has to be transformed by signal conditioning into a voltage change. One method by which this can be done is a *potential divider circuit*. Figure 7.1 shows how such a circuit could be used with a thermistor. A constant voltage, of perhaps 6 V, is applied across the thermistor and a series resistor. With a thermistor of nominal resistance 4.7 kΩ at room temperature, the series resistor might be 10 kΩ. The output signal is the voltage across the 10 kΩ resistor. The output voltage is the fraction of the total resistance which is between the output terminals, i.e.

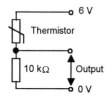

$$\text{output voltage} = \frac{R}{R+R_t}V_s$$

where R is the series resistance of 10 kΩ, R_t the resistance of the thermistor and V_s the supply voltage of 6 V. When the resistance of the thermistor, changes, the fraction of the constant 6 V across the 10 kΩ changes and so the output voltage changes. Note that in the above discussion it has been assumed that any instrument connected across the output has an infinite resistance, otherwise loading effects have to be considered (see Section 5.2.1).

Figure 7.1 *Potential divider circuit*

Another example of a resistance-to-voltage converter, and one which can be used to generate voltages from much smaller resistance changes, is the *Wheatstone bridge*. Figure 7.2 shows the basic form of such a bridge. The sensor resistance element being monitored forms one arm of the bridge, the others being constant resistors.

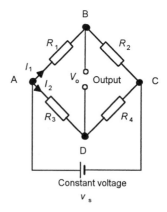

Figure 7.2 *Wheatstone bridge*

When the output voltage V_o is zero, then there is no potential difference between B and D and so the potential at B must equal that at D. Then the potential difference across R_1, i.e. V_{AB}, must equal that across R_3, i.e. V_{AD}. Thus:

$$I_1R_1 = I_2R_2$$

We must also have the potential difference across R_2, i.e. V_{BC}, equal to that across R_4, i.e. V_{DC}. Since there is no potential difference between B and D, there is no current through BD and so the current through R_2 must be the same as that through R_1 and the current through R_4 the same as that through R_3. Thus:

$$I_1R_2 = I_2R_4$$

Dividing these two equations gives:

$$\frac{R_1}{R_2} = \frac{R_3}{R_4}$$

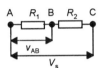

Figure 7.3 *Potential drop across AB*

The bridge is said to be *balanced*.

Now consider what happens when one of the resistances R_1 changes from the value which gives this balanced condition. The supply voltage V_s is connected between points A and C (Figure 7.3) and thus the potential drop across the resistor R_1 is the fraction $R_1/(R_1 + R_2)$ of the supply voltage. Hence

$$V_{AB} = \frac{V_sR_1}{R_1+R_2}$$

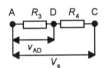

Figure 7.4 *Potential drop across AD*

Similarly, the potential difference across R_3 (Figure 7.4) is:

$$V_{AD} = \frac{V_sR_3}{R_3+R_4}$$

Thus the difference in potential between B and D, i.e. the output potential difference V_o, is:

$$V_o = V_{AB} - V_{AD} = V_s\left(\frac{R_1}{R_1+R_2} - \frac{R_3}{R_3+R_4}\right)$$

This equation gives the balanced condition, obtained earlier, when $V_o = 0$.

Consider resistance R_1 to be a sensor, e.g. a strain gauge, which has a resistance change. When the resistance changes from R_1 to $R_1 + \delta R_1$ then the output will change from V_o to $V_o + \delta V_o$, where

$$V_o + \delta V_o = V_s\left(\frac{R_1+\delta R_1}{R_1+\delta R_1 + R_2} - \frac{R_3}{R_3+R_4}\right)$$

Hence:

$$(V_o + \delta V_o) - V_o = V_s\left(\frac{R_1+\delta R_1}{R_1+\delta R_1 + R_2} - \frac{R_1}{R_1+R_2}\right)$$

If δR_1 is much smaller than R_1 then the above equation approximates to:

$$\delta V_o \approx V_s\left(\frac{\delta R_1}{R_1 + R_2}\right)$$

With this approximation, the change in output voltage is proportional to the change in the resistance of the sensor. Note that this gives the output voltage when the load resistance across the output is infinite. If there is a finite resistance then the loading effect has to be considered (see Section 7.1.1).

To illustrate the above, consider a platinum resistance temperature sensor which has a resistance at 0°C of 100 Ω. It is connected as one arm of a Wheatstone bridge and the bridge is balanced at this temperature with each of the other arms also being 100 Ω. The variation of the resistance of the platinum with temperature can be represented by

$R_t = R_0(1 + \alpha t)$

where R_t is the resistance at t°C, R_0 the resistance at 0°C and α the temperature coefficient of resistance. If the temperature coefficient of resistance of platinum is 0.0039 K^{-1}, then:

Change in resistance $= R_t - R_0 = R_0 \alpha t$

$$= 100 \times 0.0039 \times 1 = 0.39 \ \Omega/K$$

Suppose the supply voltage, with negligible internal resistance, to be 6.0 V. Since the resistance change is small compared to the 100 Ω, the approximate equation for the out-of-balance voltage can be used. Hence, for a change in temperature of 1°C, the change in voltage is given by:

$$\delta V_o \approx V_s\left(\frac{\delta R_1}{R_1 + R_2}\right) = \frac{6.0 \times 0.39}{100 + 100} = 0.012 \text{ V}$$

The load across the output is assumed to be infinite.

7.1.1 Loading of a Wheatstone bridge

Thévenin's theorem states that a network involving a source of voltage can be replaced by an equivalent network consisting of an ideal independent voltage source in series with an internal resistance, the voltage of the source being the open-circuit voltage, i.e. the voltage with infinite resistance loading, and the internal resistance being that measured at the output terminals when the independent voltage source is replaced by a short circuit (see textbooks on electrical circuits, e.g. *Electrical Circuit Principles* by W. Bolton (Longman 1992)).

With the Wheatstone bridge circuit shown in Figure 7.2, the potential difference V_o between B and D when the bridge is unbalanced is (see the previous section for the derivation of this equation):

$$V_{Th} = V_o = V_s\left(\frac{R_1}{R_1 + R_2} - \frac{R_3}{R_3 + R_4}\right)$$

This is the open-circuit voltage.

The Thévenin series resistance is that of a pair of parallel resistors R_1 and R_2 in series with the parallel resistor pair R_3 and R_4, as illustrated in Figure 7.5.

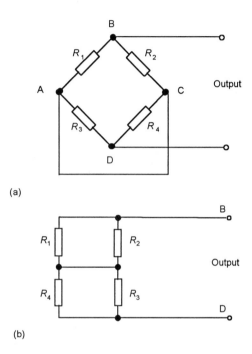

(a)

(b)

Figure 7.5 *Thévenin resistance*

The Thévenin resistance is thus:

$$R_{Th} = \frac{R_1 R_2}{R_1 + R_2} + \frac{R_3 R_4}{R_3 + R_4}$$

The Thévenin equivalent circuit for the Wheatstone bridge is thus as shown in Figure 7.6. The potential difference V_L across the load resistance R_L is the output. Thus the output voltage V_L is:

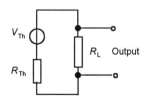

Figure 7.6 *Equivalent circuit*

$$V_L = IR_L = \frac{V_{Th} R_L}{R_L + R_{Th}} = \frac{R_L V_s\left[\dfrac{R_1}{R_1 + R_2} - \dfrac{R_4}{R_3 + R_4}\right]}{R_L + \left[\dfrac{R_1 R_2}{R_1 + R_2} + \dfrac{R_3 R_4}{R_3 + R_4}\right]}$$

This can be simplified to give:

$$V_L = \frac{R_L V_s (R_1 R_3 - R_2 R_4)}{R_L (R_1 + R_2)(R_3 + R_4) + R_1 R_2 (R_3 + R_4) + R_3 R_4 (R_1 + R_2)}$$

We can compare this value of the output with that which would occur with infinite resistance load, namely:

$$V_o = V_s \left(\frac{R_1}{R_1 + R_2} - \frac{R_3}{R_3 + R_4} \right) = \frac{V_s (R_1 R_3 - R_2 R_4)}{(R_1 + R_2)(R_3 + R_4)}$$

Hence:

$$\frac{V_L}{V_o} = \frac{1}{1 + (1/R_L)\left[R_1 R_2 / (R_1 + R_2) + R_3 R_4 / (R_3 + R_4) \right]}$$

If $R_L = \infty$, then the equation gives $V_L = V_o$. If R_L is not infinity then its effect is to reduce the output signal, V_L becoming less than V_o. The size of this reduction depends on the size of the $[R_1 R_2/(R_1 + R_2) + R_3 R_4/(R_3 + R_4)]$ term. But this term is the Thévenin equivalent resistance of the bridge, thus if we have a load which has a resistance which is, say, ten times the equivalent resistance then:

$$\frac{V_L}{V_o} = \frac{1}{1 + (1/10)} = 0.91$$

There is thus a 9% drop in the output voltage because of the load resistance.

7.1.2 Temperature compensation with a resistance temperature sensor

In measurements of temperature involving a resistive sensor, e.g. a platinum resistance temperature sensor, the actual sensing element may have to be at the end of long leads. Not only will the sensor resistance be affected by changes in temperature but also the resistance of these leads. What is required is just the changes in the resistance of the sensor as a result of temperature and so some means has to be employed to compensate for the changes in the resistance of the leads to the sensor.

One method of doing this is to use three leads to the sensor, each being the same length, as shown in Figure 7.7. The sensor is connected into the Wheatstone bridge in such a way that lead 1 is in series with the R_3 resistor while lead 3 is in series with the platinum resistance coil R_1. Lead 2 is the connection to the power supply. Any change in the resistances of the leads is likely to affect all three leads equally, since they are of the same material, diameter and length and wired so that they are close together. The result is that changes in lead resistance occur equally in two arms of the bridge and cancels out if R_1 and R_3 are the same resistance.

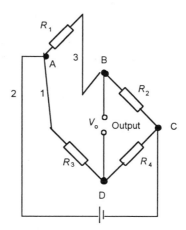

Figure 7.7 *Three-lead form of temperature compensation*

7.1.3 Temperature compensation with strain gauges

The resistance strain gauge changes resistance when the strain applied to it changes. Unfortunately, it also changes resistance if the temperature changes. Thus, when used for the measurement of strain, the effects of temperature have to be eliminated. One way of achieving this is to use a *dummy strain gauge*. This is a strain gauge which is identical to the one used for the measurement of the strain, the *active gauge*, and is mounted on the same material but is not subject to the strain. It is positioned close to the active gauge so that it suffers the same temperature changes. As a consequence, a temperature change will cause both gauges to change resistance by the same amount. The active gauge is mounted in one arm of a Wheatstone bridge and the dummy gauge in another arm (Figure 7.8) so that the effects of temperature-induced resistance changes cancel out.

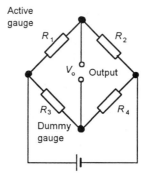

Figure 7.8 *Use of a dummy gauge for temperature compensation*

Where strain gauges are used with such items as load cells or diaphragm pressure gauges it is customary to employ four identical resistance strain gauges so that one gauge can be in each arm of a Wheatstone bridge. Figure 7.9 shows gauges attached to a cantilever and the associated bridge circuit. Two of the gauges are attached so that when the force is applied they are in tension and the other two gauges are in compression. The gauges that are in tension will increase in resistance while those in compression will decrease in resistance. As the gauges are connected as the four arms of a Wheatstone bridge, then since all will be equally affected by any temperature changes the arrangement is temperature compensated. The arrangement also gives a much greater output voltage than would occur with just a single active gauge.

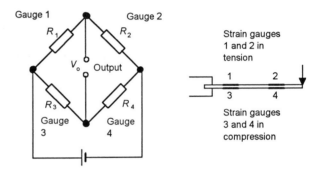

Figure 7.9 *Four active strain gauges*

With a four active arm bridge, the out-of-balance potential difference is given by:

$$V_o = V_s\left(\frac{R_1}{R_1+R_2} - \frac{R_3}{R_3+R_4}\right) = V_s\left(\frac{R_1R_4 - R_2R_3}{(R_1+R_2)(R_3+R_4)}\right)$$

Each of the resistors will change its resistance when the force is applied to the cantilever. The changes in resistance in relation to the denominator terms where we have the sum of resistances is insignificant and can be neglected. Thus

$$V_o = V_s\left(\frac{(R_1+\delta R_1)(R_4+\delta R_4) - (R_2+\delta R_2)(R_3+\delta R_3)}{(R_1+R_2)(R_3+R_4)}\right)$$

We have an initially balanced bridge with $R_1R_4 = R_2R_3$, then, neglecting products of δ terms:

$$V_o = \frac{V_sR_1R_4}{(R_1+R_2)(R_3+R_4)}\left(\frac{\delta R_1}{R_1} - \frac{\delta R_2}{R_2} - \frac{\delta R_3}{R_3} + \frac{\delta R_4}{R_4}\right)$$

For a system where we have four identical strain gauges and $R_1 = R_2 = R_3 = R_4 = R$, then:

$$V_o = \frac{V_s}{4R}(\delta R_1 - \delta R_2 - \delta R_3 + \delta R_4)$$

With a cantilever we have gauges 1 and 2 in tension and gauges 3 and 4 in compression. Thus if G is the gauge factor we have $\delta R_1/R = G\varepsilon_1$, $\delta R_2/R = G\varepsilon_2$, $\delta R_3/R = G\varepsilon_3$ and $\delta R_4/R = G\varepsilon_1$, where ε_1, ε_2, ε_3 and ε_4 are the strains.

$$V_o = \tfrac{1}{4}GV_s(\varepsilon_1 - \varepsilon_2 - \varepsilon_3 + \varepsilon_4)$$

The tensile strains will be positive and so give an increase in resistance, the compressive strains will be negative and give a decrease in resistance.

Consider a load cell with four identical strain gauges arranged as shown in Figure 5.32 so that gauges 1 and 2 are longitudinal to the applied force and subject to the same tension but gauges 3 and 4 are transverse to the applied force and so subject to the compression. The change in resistance of a gauge subject to the tensile strain is $\delta R/R = G\varepsilon$ and a gauge subject to the compressive strain is $\delta R/R = -G\nu\varepsilon$, where ν is Poisson's ratio. Thus:

$$V_o = \tfrac{1}{2}GV_s\varepsilon(1 + \upsilon)$$

7.1.4 Thermocouple compensation

A thermocouple gives an e.m.f. which depends on the temperature of its two junctions (see Section 5.6). With one junction kept at 0°C, tables can be used to obtain the value of the temperature relating to a particular e.m.f. However, the cold junction is often allowed to be at the ambient temperature. Compensation it thus required because not only is the cold junction not at 0°C but it is at a temperature which may vary. To compensate for this we can use the law of intermediate temperatures (see Section 5.6) and so a potential difference must be added to the thermocouple, the size of this e.m.f. being that e.m.f. which would be generated by the thermocouple with one junction at 0°C and the other at the ambient temperature. Such a potential difference can be produced by using a resistance temperature sensor in a Wheatstone bridge with the bridge balanced at 0°C and the out-of-balance output voltage from the bridge providing the correction potential difference at other temperatures.

The resistance of a metal resistance temperature sensor R_t at a temperature t can be described, over a small range of temperature, by (see Section 5.2.2):

$$R_t = R_0(1 + \alpha t)$$

where R_0 is the resistance at 0°C and α the temperature coefficient of resistance. Thus, when there is a change in temperature:

change in resistance $= R_t - R_0 = R_0\alpha t$

Using the simplified equation for the output voltage for the bridge, taking R_1 to be the resistance temperature sensor:

$$\delta V_o \approx V_s\left(\frac{\delta R_1}{R_1 + R_2}\right) = \frac{V_s R_0 \alpha t}{R_0 + R_2}$$

The thermocouple e.m.f. e is likely to vary with temperature t in a reasonably linear manner over the small temperature range being considered, i.e. from 0°C to the ambient temperature. Thus we can write $e = at$, where a is a constant, i.e. the e.m.f. produced per degree change in temperature. Hence for compensation we must have:

$$at = \frac{V_s R_0 \alpha t}{R_0 + R_2}$$

and so:

$$aR_2 = R_0(V_s\alpha - a)$$

For an iron–constantan thermocouple giving 51 μV/°C, compensation can be provided by a nickel resistance element with a resistance of 10 Ω at 0°C and a temperature coefficient of resistance of 0.0067 K^{-1}, a supply voltage for the bridge of 1.0 V, and R_2 as 1304 Ω.

7.2 Operational amplifiers

The *operational amplifier* is a very high gain d.c. amplifier, the gain typically being of the order of 100 000 or more, which is supplied as an integrated circuit on a silicon chip. It has two inputs, known as the inverting input (–) and the non-inverting input (+). In addition there are inputs for a negative voltage supply, a positive voltage supply and two inputs termed offset null, these being to enable corrections to be made for the non-ideal behaviour of the amplifier. Figure 7.10 shows the pin connections for a 741 type operational amplifier with the symbol for the operational amplifier shown superimposed. On the symbol the + sign indicates the non-inverting input and the – sign the inverting input.

The operational amplifier is a very widely used element in signal conditioning and processing circuits and the following indicates common examples of such circuits. For more details the reader is referred to more specialist texts such as *Feedback Circuits and Op Amps* by D.H. Horrocks (Chapman and Hall 1990), *Analysis and Design of Analog Integrated Circuits* by P.R. Gray and R.G. Meyer (Wiley 1993) and *Analog Electronics with Op Amps* by A.J. Peyton and W. Walsh (CUP 1993).

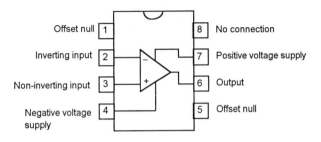

Figure 7.10 *Pin connections for a 741 operational amplifier*

7.2.1 Inverting amplifier

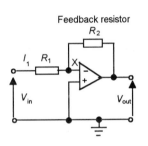

Figure 7.11 *Inverting amplifier*

Figure 7.11 shows the connections made to the amplifier when it is used as an *inverting amplifier*, such a form of amplifier giving an output which is an inverted form of the input, i.e. it is out of phase by 180°. The input is taken to the inverting input through a resistor R_1 with the non-inverting input being connected to ground. A feedback path is provided from the output, via the resistor R_2 to the inverting input. The operational amplifier has a very high voltage gain of about 100 000 and the change in output voltage is limited to about ±10 V. Thus the input voltage at point X must be between +0.0001 V and −0.0001 V. This is virtually zero and so point X is at virtually earth potential and hence is termed a *virtual earth*. With an ideal operational amplifier with an infinite gain, point X is at zero potential. The potential difference across R_1 is $(V_{in} - V_X)$. Hence, for an ideal operational amplifier with an infinite gain, and hence $V_X = 0$, the input potential V_{in} can be considered to be across R_1. Thus:

$$V_{in} = I_1 R_1$$

The operational amplifier has a very high impedance between its input terminals, for a 741 this is about 2 MΩ. Thus virtually no current flows through X into it. For an ideal operational amplifier the input impedance is taken to be infinite and so there is no current flow through X into the amplifier input. Hence, since the current entering the junction at X must equal the current leaving it, the current I_1 through R_1 must be the current through R_2. The potential difference across R_2 is $(V_X - V_{out})$ and thus, since V_X is zero for the ideal amplifier, the potential difference across R_2 is $-V_{out}$. Thus:

$$-V_{out} = I_1 R_2$$

Dividing these two equations gives the ratio of the output voltage to the input voltage, i.e. the voltage gain of the circuit. Thus:

$$\text{voltage gain of circuit} = \frac{V_{out}}{V_{in}} = -\frac{R_2}{R_1}$$

The voltage gain of the circuit is determined solely by the relative values of R_2 and R_1. The negative sign indicates that the output is inverted, i.e. 180° out of phase, with respect to the input.

To illustrate the above, consider an inverting operational amplifier circuit which has a resistance of 10 kΩ in the inverting input line and a feedback resistance of 100 kΩ. The voltage gain of the circuit is:

$$\text{voltage gain of circuit} = \frac{V_{out}}{V_{in}} = -\frac{R_2}{R_1} = -\frac{100}{10} = -10$$

7.2.2 Non-inverting amplifier

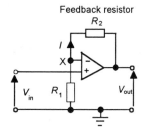

Feedback resistor
R_2

Figure 7.12 *Non-inverting amplifier*

Figure 7.12 shows the operational amplifier connected as a non-inverting amplifier. Since the operational amplifier has a very high input impedance, there is virtually no current flowing into the inverting input. The inverting voltage input voltage is V_{in}. Since there is virtually no current through the operational amplifier between the two inputs there can be virtually no potential difference between them. Thus, with the ideal operational amplifier, we must have $V_X = V_{in}$. The output voltage is generated by the current I which flows from earth through R_1 and R_2. Thus:

$$I = -\frac{V_{out}}{R_1 + R_2}$$

But X is at the potential V_{in}, thus the potential difference across the feedback resistor R_2 is $(V_{in} - V_{out})$ and we must have:

$$I = \frac{V_{in} - V_{out}}{R_2}$$

Thus:

$$-\frac{V_{out}}{R_1 + R_2} = \frac{V_{in} - V_{out}}{R_2}$$

Hence:

$$\text{voltage gain of circuit} = \frac{V_{out}}{V_{in}} = \frac{R_1 + R_2}{R_1} = 1 + \frac{R_2}{R_1}$$

A particular form of this amplifier which is often used has the feedback loop as a short circuit, i.e. $R_2 = 0$. The voltage gain is then just 1. However, the input voltage to the circuit is across a large resistance, the input resistance of an operational amplifier such as the 741 typically being 2 MΩ. The resistance between the output terminal and the ground line is, however, much smaller, typically 75 Ω. Thus the resistance in the circuit that follows is a relatively small one compared with the resistance in the input circuit and so affects that circuit less. Such a form of amplifier circuit is referred to as a *voltage follower* and is typically used for sensors which require high impedance inputs such as piezoelectric sensors.

7.2.3 Current-to-voltage converter

Figure 7.13 shows how the standard inverting amplifier can be used as a current-to-voltage converter. Point X is the virtual earth. Thus any input current has to flow through the feedback resistor R_2. The voltage drop across R_2 must therefore produce the output voltage. Thus:

$$V_{out} = -IR_2$$

So the output voltage is just the input current multiplied by the scaling factor R_2. The advantage of this method of converting a current to a voltage, compared with just passing a current through a resistor and taking the potential difference across it, is that there is a high impedance across the input and so there is less likelihood of loading problems.

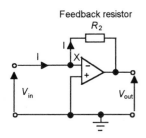

Figure 7.13 *Current-to-voltage converter*

7.2.4 Voltage-to-current converter

Situations often arise where the output needs to be a current in order to drive perhaps an electromechanical device such as a relay or possibly give a display on a moving coil meter. A voltage-to-current converter is provided by the basic inverting amplifier circuit with the device through which the current is required being the feedback resistor. Since X is a virtual earth, the potential difference across R_1 is V_{in} and the current through it I_1. Hence $I_1 = V_{in}/R_1$. The current through R_2 is I_1. Thus the input voltage has been converted to the current I_1 through the feedback resistor, with the current being V_{in}/R_1.

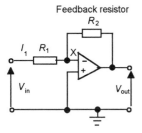

Figure 7.14 *Voltage-to-current converter*

7.2.5 Summing amplifier

In some data processing elements there is a need to add signals from more than one sensor in order to give an output which is proportional to the sum of the input signals. Figure 7.15 shows how an operational amplifier can be used as a summing amplifier for this purpose. As with the standard inverting amplifier, X is a virtual earth. Thus the sum of the currents entering X must equal that leaving it. Hence:

$$I = I_A + I_B + I_C$$

But $I_A = V_A/R_A$, $I_B = V_B/R_B$ and $I_C = V_C/R_C$. Because of the high input impedance of the operational amplifier, the same current I must pass through the feedback resistor R_2. Since the potential difference across R_2 is

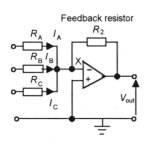

Figure 7.15 *Summing amplifier*

$(V_X - V_{out})$ and since V_X can be assumed to be zero, we have $I = -V_{out}/R_2$. Thus:

$$-\frac{V_{out}}{R_2} = \frac{V_A}{R_A} + \frac{V_B}{R_B} + \frac{V_C}{R_C}$$

The output is thus the scaled sum of the inputs, i.e.

$$V_{out} = -\left(\frac{R_2}{R_A}V_A + \frac{R_2}{R_B}V_B + \frac{R_2}{R_C}V_C\right)$$

If $R_A = R_B = R_C = R_1$ then:

$$V_{out} = -\frac{R_1}{R_2}(V_A + V_B + V_C)$$

To illustrate the above, consider a circuit that is required to produce an output voltage which is the average of the input voltages from three sensors. If an inverted output is acceptable, a circuit of the form shown in Figure 7.15 can be used. Each of the three inputs must be scaled to 1/3 to give an output which is the average and so a voltage gain of the circuit of 1/3 for each of the input signals is required. Hence, if the feedback resistance is 15 kΩ the resistors in each input arm will be 5 kΩ.

7.2.6 Differential amplifier

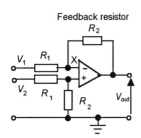

Figure 7.16 *Differential amplifier*

Figure 7.16 shows how an operational amplifier can be used as a differential amplifier, amplifying the difference between two input signals. Since the operational amplifier has high impedance between its two inputs, there will be virtually no current through the operational amplifier between the two input terminals. There is thus no potential difference between the two inputs and therefore both will be at the same potential, that at X.

The voltage V_2 is across resistors R_1 and R_2 in series. We thus have a potential divider circuit with the potential at the non-inverting input, which must be the same as that at X of V_X, as:

$$\frac{V_X}{V_2} = \frac{R_2}{R_1 + R_2}$$

The current through the feedback resistance must be equal to that from V_1 through R_1. Hence:

$$\frac{V_1 - V_X}{R_1} = \frac{V_X - V_{out}}{R_2}$$

This can be rearranged to give:

$$\frac{V_{out}}{R_2} = V_X\left(\frac{1}{R_2} + \frac{1}{R_1}\right) - \frac{V_1}{R_1}$$

Hence substituting for V_X using the earlier equation:

$$V_{out} = \frac{R_2}{R_1}(V_2 - V_1)$$

The output is thus proportional to the difference between the two input voltages.

Such a circuit might be used with a thermocouple to amplify the difference in e.m.f.s between the hot and cold junctions. Figure 7.17 shows how it might be used in such a situation. Suppose we require there to be an output of 1 mV/°C. With an iron–constantan thermocouple with the cold junction at 0°C, the e.m.f. produced between the hot and cold junctions is about 53 μV/°C. For the circuit we thus have, for a 1°C temperature difference between the junctions:

$$V_{out} = \frac{R_2}{R_1}(V_2 - V_1)$$

$$1 \times 10^{-3} = \frac{R_2}{R_1} \times 53 \times 10^{-6}$$

Hence we must have $R_2/R_1 = 18.9$. Thus if we take for R_1 a resistance of 10 kΩ then R_2 must be 189 kΩ.

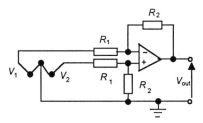

Figure 7.17 *Thermocouple input to a differential amplifier*

A differential amplifier might be used with a Wheatstone bridge, perhaps one used with strain gauges, to amplify the out-of-balance potential difference. With the bridge, when it is balanced both the outputs are at the same potential, there being thus no output potential difference. Both the inputs might thus be at, say, 5.00 V. When the bridge is no longer balanced we might have one input at 5.01 V and the other at 4.99 V. The amplifier then amplifies this difference in voltages of 0.02 V between the two inputs. The original 5.00 V signal which is common to both inputs is termed the *common mode voltage* V_{CM}. For the amplifier only to amplify the difference between the two signals assumes that the two input channels are perfectly matched and the operational amplifier has the same, high, gain for both of them. In practice this is not perfectly achieved and thus the output is not perfectly proportional to just the difference between the two input voltages. We then have:

$$V_{out} = G_d \Delta V + G_{CM} V_{CM}$$

where G_d is the gain for the voltage difference ΔV, G_{CM} the gain for the common mode voltage V_{CM}. The smaller the value of G_{CM} the smaller the effect of the common mode voltage on the output. The extent to which an operational amplifier deviates from the ideal situation is specified by the *common mode rejection ratio* (CMRR):

$$CMRR = \frac{G_d}{G_{CM}}$$

To minimise the effect of the common mode voltage on the output, a high CMRR is required. Common mode rejection ratios are generally specified in decibels (dB). Thus, on the decibel scale a CMRR of, say, 10 000 would be 20 lg 10 000 = 80 dB. A typical operational amplifier might have a CMRR between about 80 and 100 dB.

The differential amplifier is the simplest form of what is often termed an *instrumentation amplifier*. A more usual form involves three operational amplifiers (Figure 7.18). Such a circuit is available as a single integrated circuit. The first stage involves the amplifiers A_1 and A_2. These amplify the two input signals without any increase in the common mode voltage before amplifier A_3 is used to amplify the differential signal. The differential amplification produced by A_1 and A_2 is $(R_1 + R_2 + R_3)/R_1$ and that produced by A_3 is R_5/R_4 and so the overall amplification is the product of these two amplifications. The overall gain is usually set by varying the value of R_1. Normally the circuit has $R_2 = R_3$, $R_4 = R_6$ and $R_5 = R_7$.

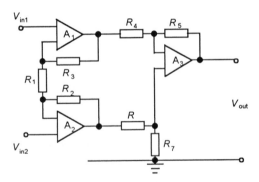

Figure 7.18 *Three operational amplifier form of an instrumentation amplifier*

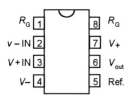

Figure 7.19 *INA114*

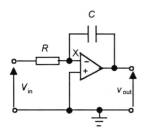

Figure 7.20 *Integrating amplifier*

Figure 7.19 shows the pin connections for an integrated circuit instrumentation amplifier involving three operational amplifiers. The gain is set by connecting a single external resistor R_G between pins 1 and 8, the differential gain then being, when R_G is in kΩ:

$$\text{gain} = 1 + \frac{50}{R_G}$$

7.2.7 Integrating amplifier

Consider an inverting operational amplifier circuit with the feedback being via a capacitor (Figure 7.20). Current is the rate of movement of charge q and since, for a capacitor the charge $q = Cv$, where v is the voltage across it, then the current through the capacitor $i = \mathrm{d}q/\mathrm{d}t = C\,\mathrm{d}v/\mathrm{d}t$. The potential difference across C is $(v_X - v_{out})$ and since v_X is effectively zero, being the virtual earth, it is $-v_{out}$. Thus the current through the capacitor is $-C\,\mathrm{d}v_{out}/\mathrm{d}t$. But, because of the high resistance between the input terminals of the operational amplifier, this is also the current through the input resistance R. Hence:

$$\frac{v_{in}}{R} = -C\frac{\mathrm{d}v_{out}}{\mathrm{d}t}$$

Rearranging this gives:

$$\mathrm{d}v_{out} = -\left(\frac{1}{RC}\right)v_{in}\,\mathrm{d}t$$

Integrating both sides:

$$v_{out}(t_2) - v_{out}(t_1) = -\frac{1}{RC}\int_{t_1}^{t_2} v_{in}\,\mathrm{d}t$$

$v_{out}(t_2)$ is the output voltage at time t_2 and $v_{out}(t_1)$ is the output voltage at time t_1. The output is proportional to the integral of the input voltage, i.e. the area under a graph of input voltage with time. Thus a constant input voltage V_{in} would produce an output $V_{in}t$, i.e. a voltage which is proportional to time (such a voltage is termed a ramp voltage).

7.2.8 Differentiating amplifier

The differentiating amplifier is similar to the integrating amplifier but with the capacitor and resistor reversed (Figure 7.21). Then, in a derivation similar to that given above for the integrating amplifier:

$$v_{out} = -RC\frac{\mathrm{d}v_{in}}{\mathrm{d}t}$$

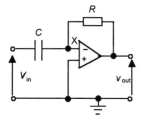

Figure 7.21 *Differentiating amplifier*

7.2.9 Logarithmic amplifier

Some sensors have outputs which are non-linear and thus there might be a need to linearise the output from such a sensor. This can be done using an operational amplifier circuit which is designed to have a non-linear relationship between its input and output such that when its input is non-linear the output is linear. This can be achieved by a suitable choice of component for the feedback loop.

The logarithmic amplifier shown in Figure 7.22 is an example of a non-linear amplifier. The feedback loop contains a diode (or a transistor with a grounded base). The diode has a non-linear characteristic with the voltage V across it related to the current I through it by $V = C \ln I$, where C is a constant. Then, since the current through the feedback loop is the same as the current through the input resistance R_1 and the potential difference across the diode is $-V_{out}$:

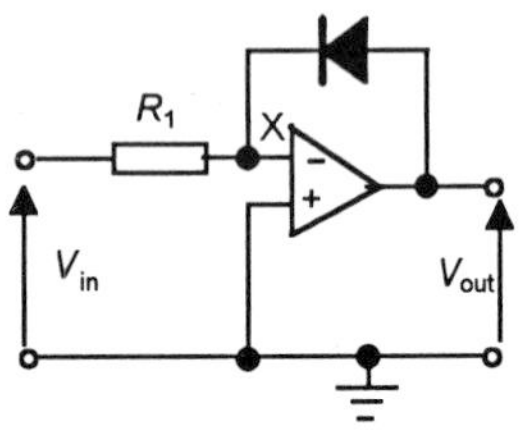

$$V_{out} = -C \ln(V_{in}/R) = K \ln V_{in}$$

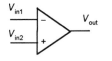

Figure 7.22 *Logarithmic amplifier*

where K is some constant. However, if we have a non-linear sensor giving an input which varies with some variable x in the form $V_{in} = A\ e^{ax}$, with A and a being constants, then:

$$V_{out} = K \ln V_{in} = K \ln(A\ e^{ax}) = K \ln A + Kax$$

The result is a linear relationship between V_{out} and x.

7.2.10 Offset voltage

The operational amplifier is basically a differential amplifier, amplifying the difference between its two inputs (Figure 7.23). It is not affected by the levels of the two voltages, only the difference between them. Thus:

$$V_{out} = G_v(V_{in1} - V_{in2})$$

Figure 7.23 *Operational amplifier*

where G_v is the voltage gain. Thus if the two inputs are shorted we might expect to obtain no output. However, in practice this does not occur and an output voltage might be detected. This effect is produced by imbalances in the internal circuitry in the operational amplifier. The output voltage can be made zero by applying a suitable voltage between the input terminals. This is known as the *offset voltage*. Many operational amplifiers are provided with arrangements for applying such an offset voltage via a potentiometer. With the 741 this is done by applying a voltage between pins 1 and 5 (see Figure 7.10).

7.3 Noise   The term *noise* is used, in this context, for the unwanted signals that may be picked up by a measurement system and interfere with the signals being measured. There are two basic types of electrical noise:

1   *Interference*
    This is due to the interaction between external electrical and magnetic fields and the measurement system circuits, e.g. the circuit picking up interference from nearby mains power circuits.

2   *Random noise*
    This is due to the random motion of electrons and other charge carriers in components and is determined by the basic physical properties of components in the system.

7.3.1 Interference

The three main types of interference are:

1   *Inductive coupling*
    A changing current in a nearby circuit produces a changing magnetic field which can induce e.m.f.s, as a result of electromagnetic induction, in conductors in the measurement system.

2   *Capacitive coupling*
    Nearby power cables, the earth, and conductors in the measurement system are separated from each other by a dielectric, air. There can thus be capacitance between the power cable and conductors, and between the conductors and earth. These capacitors couple the measurement system conductors to the other systems and thus signals in the other systems affecting the charges on these capacitors can result in interference in the measurement system.

3   *Multiple earths*
    If the measurement system has more than one connection to earth, there may be problems since there may be some difference in potential between the earth points. If this occurs, an interference current may arise in the measurement system.

Methods of reducing interference are:

1   *Twisted pairs of wires*
    This involves the elements of the measurement system being connected by twisted wire pairs (Figure 7.24). A changing magnetic field will induce e.m.f.s in each loop, but because of the twisting the directions of the e.m.f.s in a wire will be in one direction for one loop and in the opposite direction for the next loop and so cancel out.

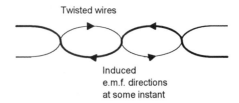

Twisted wires

Induced
e.m.f. directions
at some instant

Figure 7.24 *Twisted pairs*

2 *Electrostatic screening*
Capacitive coupling can be avoided by completely enclosing the system in an earthed metal screen. Problems may occur if there are multiple earths. Coaxial cable gives screening of connections between elements, however, the cable should only be earthed at one end if multiple earths are to be avoided.

3 *Single earth*
Multiple earthing problems can be avoided if there is only a single earthing point.

4 *Differential amplifiers*
A differential amplifier can be used to amplify the difference between two signals. Thus if both signals contain the same interference, then the output from the amplifier will not have amplified any interference signals.

5 *Filters*
A filter can be selected which transmits the measurement signal but rejects interference signals.

7.3.2 Random noise

Random noise can arise in a number of ways:

1 *Thermal noise*
This is sometimes referred to as *Johnson noise*. This noise is generated by the random motion of electrons and other charge carriers in resistors and semiconductors. It is spread over an infinite range of frequencies and is thus often referred to as *white noise*. The r.m.s. noise voltage for a bandwidth of frequency f_1 to f_2 is:

$$\text{r.m.s. noise voltage} = \sqrt{4kRT\,(f_2 - f_1)}$$

where k is Boltzmann's constant, R the resistance and T the temperature on the kelvin scale. Thus a wide band amplifier will generate more noise than a narrow band one. High resistances and high temperature will also result in more noise.

2 *Shot noise*
This is noise due to the random fluctuations in the rate at which charge carriers diffuse across potential barriers, such as in a p-n junction. The r.m.s. noise voltage for a bandwidth from frequency f_1 to f_2 is:

$$\text{r.m.s. noise voltage} = \sqrt{2kTr_d(f_2 - f_1)}$$

where k is Boltzmann's constant, T the temperature on the kelvin scale, r_d the differential diode resistance and equal to kT/qI, with q being the charge on an electron and I the d.c. current through the junction. Because this noise is spread across the frequency spectrum it is also referred to as white noise.

3 *Flicker noise*
This is noise due to the flow of charge carriers in a discontinuous medium, e.g. a carbon composite resistor. The r.m.s. noise voltage is approximately inversely proportional to the frequency. It is sometimes referred to as *pink noise* because it tends to be restricted to the lower frequencies below about 10 Hz (red is the lower frequency end of the visible spectrum).

4 *Poor connections*
Noise can result from fluctuation contact resistances due to bad soldering or dirt on switch contacts.

7.3.3 Noise rejection

The term *normal mode noise* is used to describe all the noise occurring within the signal source. To a measurement system, this noise is indistinguishable from the actual quantity it is being used to measure. The ability of a system to reject normal mode noise is called the *normal mode rejection ratio* (N.M.R.R.) or *series mode rejection ratio*. This is defined in decibels as:

$$\text{N.M.R.R.} = 20\lg\left(\frac{V_n}{V_e}\right)$$

where V_n is the peak value of the normal mode noise and V_e the peak value of the error it produces in the measurement at a particular frequency.

The term *common mode noise* is used to describe the noise occurring between the earth terminal of a system and its lower potential terminal. The ability of a system to prevent common mode noise introducing an error in the measurement is called the *common mode rejection ratio* (C.M.R.R.). This is defined in decibels as:

$$C.M.R.R. = 20 \lg\left(\frac{V_{cm}}{V_e}\right)$$

where V_{cm} is the peak value of the common mode noise and V_e the peak value of the error it produces in the measurement at a particular frequency.

7.3.4 Signal to noise ratio

The signal to noise ratio (S/N ratio) is the ratio of the signal power to the noise power:

$$\text{S/N ratio} = \frac{\text{signal power}}{\text{noise power}}$$

This is usually expressed in decibels as:

$$\text{S/N ratio} = 10 \lg \frac{\text{signal power}}{\text{noise power}}$$

Since the power is V^2/R, then if V_s is the signal voltage and V_n the noise voltage:

$$\text{S/N ratio} = 10 \lg\left(\frac{V_s}{V_n}\right)^2 = 20 \lg\left(\frac{V_s}{V_n}\right)$$

Averaging can be used to improve the signal to noise ratio for a repetitive signal. For the same point in the signal waveform cycle, samples are taken for a number of cycles and the average value obtained. Because of the random nature of the noise signal superimposed on the measurement signal, the noise component in each sample will differ, sometimes being negative and sometimes positive. The result of the averaging is thus to give a reduced average noise. This averaging can be used for a number of points in the waveform cycle and the signal then reconstructed with this lower noise level.

As an illustration of the above, consider an amplifier which has a bandwidth of 20 kHz and an input resistance of 1 kΩ. The input signal to the amplifier comes from a sensor which gives a signal of 1 mV and has an internal resistance of 1 kΩ. The thermal noise generated in the source resistance at a temperature of 20°C (293 K) will be ($k = 1.38 \times 10^{-23}$ J/K):

$$\text{r.m.s. noise voltage} = \sqrt{4kRT(f_2 - f_1)}$$

$$= \sqrt{4 \times 1.38 \times 10^{-23} \times 10^3 \times 293 \times 20 \times 10^3}$$

$$= 0.57 \ \mu V$$

This thermal noise voltage generated in the sensor resistance can be considered as a noise voltage source in series with the sensor resistance.

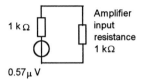

1 kΩ

Amplifier
input
resistance
1 kΩ

0.57 μ V

Figure 7.25 *Input circuit*

Thus we can represent the input circuit in the way shown in Figure 7.25. The r.m.s. noise current in the input circuit due to the noise originating in the sensor is then:

$$\text{r.m.s. noise current } I_n = \frac{0.57 \times 10^{-6}}{10^3 + 10^3} = 0.29 \times 10^{-12} \text{ A}$$

The noise power developed in the sensor is thus:

$$\text{noise power} = I_n^2 R_s = (0.29 \times 10^{-12})^2 \times 10^3 = 8.41 \times 10^{-23} \text{ W}$$

The input power from the sensor is:

$$\text{input power} = I^2 R_s = \left(\frac{1 \times 10^{-3}}{10^3 + 10^3}\right)^2 \times 10^3 = 2.5 \times 10^{-10} \text{ W}$$

Thus the input signal to noise ratio is:

$$\text{S/N ratio} = \frac{2.5 \times 10^{-10}}{8.41 \times 10^{-23}} = 2.97 \times 10^{-12}$$

Expressed in decibels, this is $10 \lg(2.97 \times 10^{-12}) = 125$ dB.

7.3.5 Noise figure

The *noise figure* or *noise factor*, F, is, when expressed in decibels, defined as:

$$F = 10 \lg \frac{\text{total noise power outp}}{\text{total input noise powe}}$$

The noise figure is thus a measure of the amount of noise introduced into the output by the measurement system. If the input noise power is P_{ni} and the system has an overall transfer function G, then the output noise power due to the input noise is GP_i. But G is the output signal power P_{so} divided by the input signal power P_{si}. Hence we can write:

$$F = 10 \lg \left[\frac{(P_{si}/P_{ni})}{(P_{so}/P_{no})} \right]$$

Hence:

$$F = \frac{\text{input S/N ratio}}{\text{output S/N rati}}$$

where the S/N ratios are expressed in decibels.

7.4 Filters

The term *filtering* is used to describe the process of removing a certain band of frequencies from a signal and permitting others to be transmitted. The range of frequencies passed by a filter is known as the *pass band*, the range not passed as the *stop band* and the boundary between stopping and passing as the *cut-off frequency*. Filters are classified according to the frequency ranges they transmit or reject:

1 A *low-pass filter* (Figure 7.26(a)) has a pass band which allows all frequencies from 0 up to some frequency to be transmitted.

2 A *high-pass filter* (Figure 7.26(b)) has a pass band which allows all frequencies from some value up to infinity to be transmitted.

3 A *band-pass filter* (Figure 7.26(c)) allows all the frequencies within a specified band to be transmitted.

4 A *band-stop filter* (Figure 7.26(d)) stops all frequencies with a particular band from being transmitted.

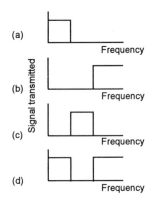

(a) Frequency
(b) Frequency
(c) Frequency
(d) Frequency

igure 7.26 *Characteristics of ideal filters*

Filters are described as *passive* if they are composed only of resistors, capacitors and inductors, and *active* when the filter also involves an operational amplifier.

Figure 7.27 shows the form a simple low pass passive filter can take and its frequency response curve. The circuit is a potential divider with the output being taken across the reactance of the capacitor. Thus:

$$V_{out} = \frac{(1/j\omega C)V_{in}}{R + (1/j\omega C)} = \frac{1}{1 + j\omega RC}V_{in}$$

Thus the ratio of the size of the amplitude of the output signal to the amplitude of the input signal, i.e. its magnitude, varies with the angular frequency ω according to:

$$\text{ratio} = \frac{1}{\sqrt{1 + (\omega RC)^2}}$$

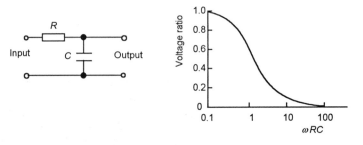

Figure 7.27 *Simple passive low-pass filter*

The phase ϕ between the output and input signals is given by:

$$\phi = \tan^{-1}(\omega RC)$$

Since noise tends to occur at higher frequencies, a low-pass filter can be used to block it off and thus low-pass filters are widely used as part of signal conditioning. A low-pass filter might be selected with a cut-off frequency of 40 Hz, thus blocking off any inference signals from the a.c. mains supply and noise in general.

Figure 7.28 shows the circuit for a simple high-pass filter and how the amplitude ratio varies with frequency. The circuit is a potential divider with the output being taken across the resistance. Thus:

$$V_{out} = \frac{RV_{in}}{R + (1/j\omega C)} = \frac{j\omega RC}{1 + j\omega RC}V_{in}$$

Thus the ratio of the size of the amplitude of the output voltage compared with that of the input voltage is given by:

$$ratio = \frac{\omega RC}{\sqrt{1 + (\omega RC)^2}}$$

and the phase difference ϕ between the output and input by:

$$\phi = \frac{\pi}{2} - \tan^{-1}\omega RC$$

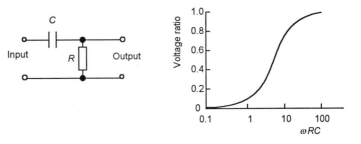

Figure 7.28 *Simple passive high-pass filter*

The above filters are examples of first order filters in that the differential equation describing their behaviour is first order. This is because there is only one element, the single capacitor, whose behaviour is frequency dependent. By increasing the number of orders, by for example including inductance and increasing the number of stages of the circuits used, the sharpness of the cut-off can be improved. One such form is termed *Butterworth filter*. The amplitude ratio for such a filter is given by:

$$\text{ratio} = 1 + \left(\frac{\omega}{\omega_c}\right)^{2n}$$

where ω is the angular frequency, ω_c is the cut-off angular frequency and n is the order of the filter. The higher the order, i.e. the greater the number of frequency-dependent elements, the sharper the cut-off.

Passive filters suffer from the problems that their frequency response can be affected by the load across the output, they are not easily tuned to respond over a wide frequency range, and, because all the output energy has to be taken from that supplied by the input, they usually attenuate, i.e. reduce the amplitude, signals even in their pass band. These problems can be overcome by the use of operational amplifiers to give active filters. Figure 7.29 shows a simple form of an active low-pass filter. The ratio of the amplitudes of the output to the input is given by:

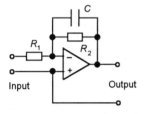

Figure 7.29 *Active low-pass filter*

$$\text{ratio} = -\frac{R_2}{R_1}\frac{1}{\sqrt{1 + (\omega/R_2 C)}}$$

Thus the gain of the filter can be adjusted independently of the frequency at which the cut off occurs by changing R_2/R_1. The cut-off frequency is varied by changing $R_2 C$.

For a more detailed discussion of filters, both passive and active, see *Filter Handbook* by S. Niewiadomski (Heinemann Newnes 1989).

7.5 Modulation

When transmitting analogue alternating signals over some distance, modulation is often used. This enables data transmission at much higher frequencies and so allows the use of high-pass filters to eliminate the noise signals that usually occur at much lower frequencies. Modulation techniques used are:

1 *Amplitude modulation*

With amplitude modulation the amplitude of a carrier wave, of much higher frequency than the input, is varied according to the size of the voltage input, i.e. the wave carrying the signal from the sensor. Thus for a carrier wave that can be represented by:

$$v = V \sin(\omega t + \phi)$$

the amplitude term V is varied according to the way the voltage input varies.

2 *Phase modulation*

For a carrier wave that can be represented by:

$$v = V \sin(\omega t + \phi)$$

phase modulation involves varying the phase ϕ of the carrier wave according to the size of the voltage input.

3   *Frequency modulation*
For a carrier wave that can be represented by:

$$v = V \sin(\omega t + \phi)$$

frequency modulation involves varying the angular frequency ω of the carrier wave according to the size of the voltage input.

After transmission, the modulated signal can be demodulated so that an output can be obtained which is related to the original signal before it was modulated.

When transmitting d.c. signals, they are often converted into alternating signals and modulation then used. This is because in the transmission of low-level d.c. signals from sensors, the gain of an operational amplifier used to amplify them may drift and so the output drifts. This problem can be overcome if the signal is alternating rather than direct and, in addition, it becomes easier to eliminate interference.

One way this conversion of d.c. to alternating signal can be achieved is by electronically chopping the d.c. signal to convert it into a number of pulses with amplitude related to the size of the d.c. signal (Figure 7.30). The output from the chopper is a chain of pulses with heights which vary according to the way the level of the d.c. signal varied. This process is called *pulse amplitude modulation*. After amplification and any other signal conditioning, the modulated signal can be demodulated to give a d.c. output. An alternative to this is *pulse width modulation* where the width, i.e. duration, of a pulse depends on the size of the voltage (Figure 7.31).

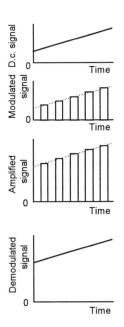

Figure 7.30 *Pulse amplitude modulation*

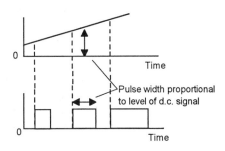

Figure 7.31 *Pulse width modulation*

7.6 Analogue and digital conversions

The output from most sensors tends to be in analogue form. However, where a microprocessor or computer is used as part of the measurement or control system, the analogue signal has to be converted into a digital form before it can be used as an input to the microprocessor or computer. Many of the devices actuated by the output require analogue inputs and so the digital output from a microprocessor or computer has often to be converted into an analogue form before it can be used.

Digital systems operate on the principle of elements which, like switches, can be turned on or off. Since such elements have only two states, on or off, digital signals consists of binary elements, called *bits*, which are either 0, for off, or 1 for on. By combining bits it is possible to define numbers greater than 1, such a combination being termed a *word*. When a number is represented by this system, the digit position in the number indicates the weight attached to it, the weight increasing by a factor of 2 as we proceed from right to left. Thus we have:

2^4	2^3	2^2	2^1	2^0
bit 4	bit 3	bit 2	bit 1	bit 0

The bit 0 is termed the *least significant bit* (LSB) and the highest bit the *most significant bit* (MSB), in the above example this is bit 4. For example, the binary number 1010 has 0 as the least significant bit and 1 has the most significant bit. When converted to a decimal number we have:

bit 4	bit 3	bit 2	bit 1
1	0	1	0
$2^3 = 8$	0	$2^1 = 2$	0

Thus the decimal equivalent is 10. The conversion of a binary number to a decimal number thus involves the addition of the powers of 2 indicated by the digits in the word.

The conversion of a decimal number to a binary number involves looking for the appropriate powers of two. We can do this by successive divisions by 2, noting the remainder at each division. Thus if we have the decimal number 15, then:

$15 \div 2 = 7$, remainder 1. This is the LSB
$7 \div 2 = 3$, remainder 1
$3 \div 2 = 1$, remainder 1. This is the MSB.

The binary number is 111.

A 4-bit binary word permits bits 2^0, 2^1, 2^2, 2^3, i.e. a number from 0 to 15. The maximum count is thus given by:

$$\text{maximum count} = 2^n - 1$$

where n is the number of bits. For an 8-bit word we have a maximum count of $2^8 - 1 = 255$.

The least significant bit is the smallest number that is indicated by a digital word. It thus indicates the *resolution*. Hence with an n-bit word, the resolution is:

$$\text{resolution} = \frac{2^0}{2^n - 1} = \frac{1}{2^n - 1} = \frac{1}{\text{maximum count}}$$

Hence with an 8-bit word, the resolution that can be achieved is 1/255 or 0.39% of the full scale.

7.6.1 Analogue-to-digital conversions

Analogue-to-digital conversion involves transforming an analogue signal into a digital signal. It is done by taking samples of the analogue signal (Figure 7.32(a)) at regular intervals, a clock supplying regular time signal pulses (Figure 7.32(b)) to the converter so that every time it receives a pulse it samples the analogue signal. The result of the sampling is a series of narrow pulses with heights which, at the instant the sample is taken, represents the analogue signal (Figure 7.32(c)). A sampled signal is held until the next pulse occurs (Figure 7.32(d)) so that conversion can take place to a digital signal at the following analogue-to-digital converter. Analogue-to-digital conversion thus involves a *sample and hold* unit followed by an analogue-to-digital converter (Figure 7.33).

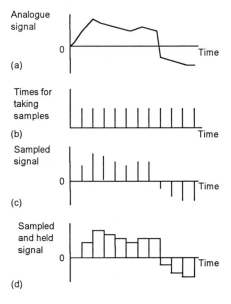

Figure 7.32 *Sampling and holding*

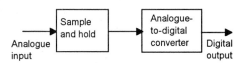

Figure 7.33 *Analogue-to-digital conversion*

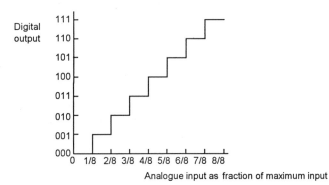

Figure 7.34 *Digital output from an analogue-to-digital converter*

The relationship between the sampled and held input V_A and the output from an analogue-to-digital converter is illustrated by the graph shown in Figure 7.34 for a 3-bit digital output. Each rise in the analogue voltage of (1/8) of the full-scale value results in a further bit being generated. The analogue voltages are rounded either up or down to one of a number of levels, these being termed *quantisation levels*. The resolution of the converter is the analogue voltage required to change the digital output by one bit, any change smaller than this has no effect on the output. Because there is a range of analogue voltage that will give the same digital output, there is an error associated with the conversion which is termed the *quantisation error*. The maximum quantisation error is ± half the resolution, i.e. half the size of one quantum step.

Consider a thermocouple giving an output of 50 μV/°C and being used over a range of 0 to 200°C with a change of one digit on a digital display being for a temperature change of 0.5°C. The resultion required is thus 0.5/200. The resolution given by a digital display with a word length n is $1/(2^n - 1)$. Thus we must have $0.5/200 = 1/(2^n - 1)$. Hence $n = 8.6$ and so a 9-bit word length is required to give the necessary resolution.

7.6.2 Analogue-to-digital converters

The input to an analogue-to-digital converter is an analogue signal and the output is a binary word that represents the level of the input signal. There are a number of forms of analogue-to-digital converter, the most common being successive approximations, ramp, dual ramp and flash. The term *conversion time* is used to specify the time it takes a converter to generate a complete digital word when supplied with the analogue input.

Successive approximations is probably the most commonly used analogue-to-digital conversion method. This method is based on comparing the input analogue voltage with another analogue voltage until the two are equal or as close as it is possible to set them. Figure 7.35(a) illustrates the subsystems involved. A clock is used to emit a regular sequence of pulses, i.e. 0 and 1 bits, which are stored in a register (a location where a word can be stored) and becomes the input to a digital-to-analogue converter (DAC) . At the start of the conversion process the input to the DAC is set with the most significant bit at 1 and the other bits at 0. For a 4- bit converter this would mean 1000. Since this represents the mid-point of the digital range, it gives an analogue output voltage from the DAC of 8/16 of the full-scale analogue voltage (Figure 7.35(b)). This first approximation is then compared with the input analogue voltage. If it is smaller than the input voltage, a second approximation is made and the next bit is turned on to give an input to the DAC of 1100, i.e. an analogue voltage from the DAC of 12/16 of the full-scale analogue voltage. If this results in a voltage which is larger than the input voltage, the bit is turned off and the next bit tried, i.e. an input of 1010 to the DAC and so 10/16 of the full-scale analogue voltage. If this is too low then the next bit is added to give 1011, i.e. 11/16 of the input voltage. If this is too high then the bit is turned off and the nearest digital equivalent to the input analogue voltage given as 1010. Because each of the bits in the word is tried in sequence, with an n-bit word it only takes n steps to make the comparison. Thus if the clock has a frequency f, the time between pulses is $1/f$. Hence the time taken to generate the word, i.e. the conversion time, is n/f.

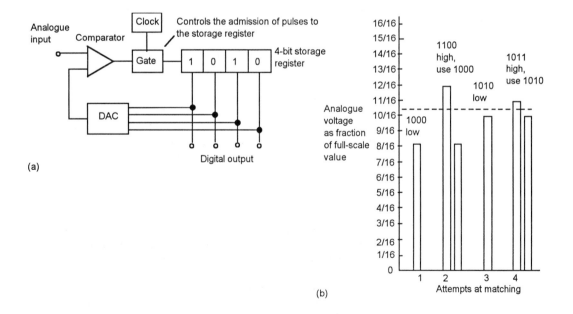

Figure 7.35 *4-bit successive approximations ADC*

The *ramp* form of analogue-to-digital converter involves an analogue voltage which is increased at a constant rate, a so-called ramp voltage. This ramp voltage is generated by a capacitor being charged from a constant current source. The charge q on the capacitor is It, where I is the constant current and t the time for which the current charges the capacitor. Since the potential difference across the capacitor is q/C, then the potential difference is proportional to the time for which the charging occurs and so a ramp voltage is produced. The ramp voltage is applied to a comparator where it is compared with the input analogue voltage. The time taken for the ramp voltage to increase to the value of the input analogue voltage will depend on the size of the sampled analogue voltage. When the ramp voltage starts, a gate is opened which starts a binary counter counting the regular pulses from a clock. When the two voltages are equal, the gate closes and the word indicated by the counter is the digital representation of the sampled analogue voltage. Figure 7.36 indicates the subsystems involved in the ramp form of analogue-to-digital converter. Ramp converters are cheap but relatively slow.

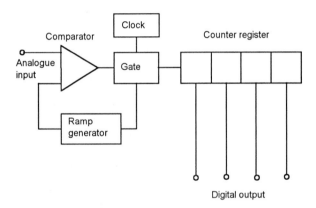

Figure 7.36 *Ramp analogue-to-digital converter*

The accuracy of the single ramp method can be improved by using the *dual ramp* method. In this method the measurement is accomplished in two steps. In the first step the input analogue voltage is applied to an integrating circuit which integrates it with respect to time and produces a voltage which increases at a constant rate with time, a ramp voltage (Figure 7.37), across the capacitor of the integrator. After a fixed time this integration ceases, leaving the capacitor of the integrator charged. Since $i = dq/dt = q/t_1$ for a constant rate of charging, the charge Q_{in} on the capacitor after a time t_1 is $it_1 = v_{in}t_1/R$, where R is the resistance in the input line of the integrator. The second step then involves the integration of a reference voltage to produce a second ramp voltage. This voltage is in the opposite direction to the voltage produced by the integration to the input voltage and so reduces the output of the integrator. The time taken for the output of the integrator to reach zero is used to control a gate admitting pulses from a clock to a binary register and so give a digital output which is a measure of the analogue input. If t_2 is the time taken for the charge to reach zero, then we must have $v_{in}t_1/R = v_{ref}t_2/R$ and so t_2 is a measure of v_{in}. Dual ramp converters are relatively slow but have the advantage of being able to integrate over time and so reject noise.

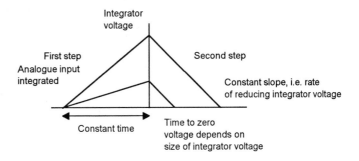

Figure 7.37 *Dual ramp ADC*

The *flash analogue-to-digital converter* is a very fast, but more costly, form of ADC. For an *n*-bit converter, $2^n - 1$ separate voltage comparators are used in parallel (Figure 7.38). Each comparator has the input analogue voltage as one of its inputs. A reference voltage is applied to a ladder of resistors so that the reference voltage applied to each comparator is one bit higher in size than the reference voltage applied to the lower previous comparator. When the analogue input voltage is applied, all those comparators for which the analogue voltage is greater than the reference voltage element will go high and those for which the analogue voltage is less than the reference voltage element will go low. The resulting outputs of the comparators are fed in parallel to a gate system which translates them into the digital output. Because all the bits of the output digital word are simultaneously produced, this converter is very fast.

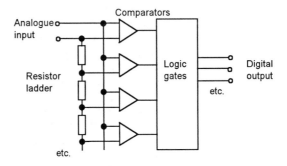

Figure 7.38 *Flash ADC*

Analogue-to-digital converters, with internal clock, sample and hold circuit and interfaces for direct connection to microprocessors, are available as integrated circuits. For example, the ADS7806 is a 12-bit analogue-to-digital converter of the successive approximations form, complete with sample and hold, internal clock and interfaces for microprocessors. It has a conversion time of 20 μs. The 8-bit ZN439 operating in the same way has a conversion time of 5 μs. Figure 7.39 shows how the ZN439 might be used to convert the output from a sensor supplying an analogue voltage into a digital signal for a microprocessor, with the microprocessor controlling the ADC. The ADC is first selected by taking the chip select pin low. Upon receipt of a negative-going pulse on the start conversion pin the status output goes high and the successive approximations process starts. At the end of the conversion the status pin goes low. The digital output is sent to internal buffers where the data is held until read by taking the output enable pin low.

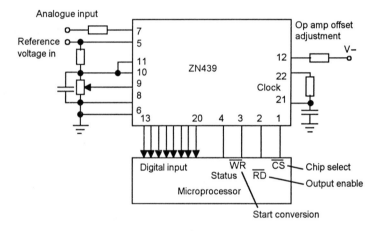

Figure 7.39 *ZN439 connected to a microprocessor*

7.6.3 Digital-to-analogue conversion

The input to a digital-to-analogue converter is a binary word and the output is an analogue signal that represents the weighted sum of the non-zero bits in that word. Thus, for example, an input of 0010 must give an analogue output which is twice that given by an input of 0001. An input of 0100 must give an analogue output which is twice that given by 0010 and four times that given by 0001.

7.6.4 Digital-to-analogue converters

One form of digital-to-analogue converter uses a summing amplifier to form the weighted sum of all the non-zero bits in the input word, the weighted sum being obtained by the use of a *weighted resistor network* (Figure 7.40). The reference voltage is connected to the resistors by means of switches which respond to binary 1, each such switch having an input from one of the bits in the digital word. Thus an input of 1001 would close the switches attached to the least significant bit and the most significant bit. The input resistances depend on which bit in the word a switch is responding to. Hence the sum of the voltages switched on is a weighted sum of the digits in the word.

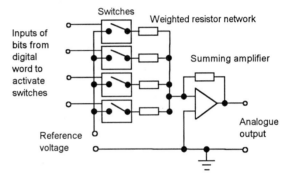

Figure 7.40 *Weighted resistor network DAC*

Another, more commonly used version uses an *R-2R ladder network* (Figure 7.41). Each successive 2R resistor in the ladder is switched on or off, and thus the reference voltage switched across the 2R and R resistors, according to whether there is a 1 or 0 in the digital input to that resistor.

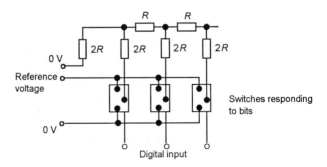

Figure 7.41 *R-2R ladder network DAC*

Digital-to-analogue converters are available as integrated circuits. For example ZN429E is an 8-bit converter using the *R-2R* ladder. The *settling time*, i.e. the time taken for the analogue output voltage to settle within a specified band, usually ±½ the least significant bit, about its final value when the input digital word is suddenly changed, is 1 μs.

7.6.5 Sampling theorem

Analogue-to-digital converters sample analogue signals on a regular basis and convert the signal at each sample time into a digital value. The greater the number of samples taken in each cycle of the analogue signal, the more representative the sample is of the analogue signal. In order to obtain meaningful results, the analogue input signal must be sampled at a rate which is twice that of the highest frequency in the analogue signal. This condition is known as the *Nyquist criterion*. Failure to meet this criterion results in a sample which represents some other analogue signal, this effect being termed *aliasing* (see Appendix E). Figure 7.42 shows waveforms that might be constructed when a sinusoidal analogue signal is sampled at a rate which fails to meet the Nyquist criterion and one that does. It is obvious from the sample points with the one failing to meet the criterion that a completely different frequency analogue signal could give the same sample.

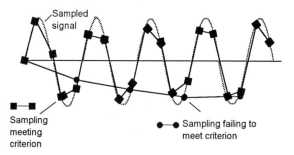

Figure 7.42 *Sampling*

7.7 Digital signal processing

Before digital electronics became so widespread, all the processing of signals was carried out by analogue components. The emergence of digital signal processing (DSP) has changed that and now digital processing of signals is widely used. With digital signal processing the sequence of operations is typically of the form shown in Figure 7.43. The signals from the sensor are converted from analogue to digital by an analogue-to-digital converter and then digital signal processing is used to process the signal for storage, display or transmission for use in some control process or passing through a digital-to-analogue converter to reconstitute an analogue waveform.

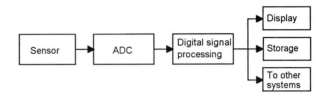

Figure 7.43 *Digital system*

Once a signal has been digitised, digital electronics can perform a wide range of tasks. For example, such operations as filtering, modulation and spectral analysis can be carried out digitally by means of software. It is necessary to recognise that digital signal processing is different from data processing. Data processing involves the manipulation of data which is already digitised and which is not a function of real time. Digital signal processing involves real-time analogue signals which are processed digitally. See Appendix E for further discussion.

Digital signal processing has advantages over the conventional analogue processing of signals. Because it is just concerned with sequences of binary arithmetic the outcomes can be entirely predictable and so give precision and stability. It is easy to modify since changes are often just software changes; the hardware is simpler and some functions can be implemented which are not practical with analogue systems.

7.7.1 Hardware and software

Digital signal processing has required the development of fast computing algorithms which permit real-time processing. An *algorithm* is a sequence of logical steps of operations that has to be completed to produce the required result. High computational intensive tasks are involved in carrying out digital processing. Even though most microprocessors can perform such computations, they are generally too slow. Thus, in order to give the required speed, special microprocessor integrated circuits tend to be used, or add-on microprocessors or plug-in boards with personal computers.

With traditional microprocessors the basic operations are fetch and then excute. Thus an instruction is fetched during one cycle. During the next cycle that data is fetched and the instruction executed. Thus to carry out the operation of multiplying two numbers A and B the sequence is:

Fetch instruction to read A.
Read A.
Fetch instruction to read B.
Read B.
Fetch instruction to multiply A and B.
Multiply A and B.
Fetch instruction to write the product.

Write the product.

This involves four instruction fetches and four executions and so a minimum of eight cycles. To speed things up, chips are used which permit operations to be carried out in parallel. This is termed *pipelining*. While one instruction is being fetched and executed, the remaining instructions can be fetched, thus considerably reducing the number of cycles required. Thus we might have:

Fetch instruction to read *A*.
Read *A*. Fetch instruction to read *B*.
Read *B*. Fetch instruction to multiply *A* and *B*.
Multiply *A* and *B*. Fetch instruction to write the product.
Write the product.

In the case of a digital filter being used in place of an analogue filter, we have the analogue signal being sampled and converted to a number of digital signals by an analogue-to-digital converter. The algorithm representing the required filter action, e.g. low pass, is then used to carry out arithmetic on the numbers representing the digital signals. This output of new numbers may be then used in further digital processing or converted from digital to analogue, producing a filtered version of the original analogue signal.

Software is required to implement algorithms. The trend is now to use the more user-friendly graphical interfaces. In Chapter 8 the LabView software is discussed, this providing a graphical interface to enable the user to easily and quickly develop the required processing by the selection and organisation of icons on the computer screen rather than the writing out of program instructions. Thus to use a digital filter in a program all that needs to be done is for the filter icon to be selected.

The above is just a very sketchy outline of digital signal processing. For a more detailed consideration the reader is referred to texts such as *Digital Signal Processing Design* by A. Bateman and W. Yates (Pitman 1988), *Introductory Digital Signal Processing with Computer Applications* by P.A. Lynn and W. Fuerst (Wiley 1994), *Introduction to Digital Signal Processing* by M.J.T. Smith and R.M. Merserau (Wiley 1992), *Digital Signals, Processors and Noise* by P.A. Lynn (Macmillan 1992).

7.8 Interfacing   The term *interface* is given to the item used to make connections between input and output devices and a microprocessor. There could be inputs from sensors, switches, and keyboards and outputs to displays and actuators. The interface generally has to contain signal conditioning and protection. The protection is to prevent damage to the microprocessor system from excessive voltages, excessive currents or signals of the wrong polarity. The signal conditioning is because signals from sensors are generally not suitable for direct processing by a microprocessor. Signal conditioning may include amplification, analogue-to-digital conversion and often shaping of signals to make them the right shape.

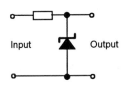

Figure 7.44 *Zener diode protection circuit*

A high current can be protected against by the incorporation in the input line of a series resistor to limit the current to an acceptable level and a fuse to break if the current does exceed a safe level. High voltages, and wrong polarity, may be protected against by the use of a Zener diode circuit (Figure 7.44). Zener diodes behave like ordinary diodes up to some breakdown voltage when they become conducting. Their resistance drops to a very low value. The result is that the voltage across the diode, and hence that outputted to the next circuit, drops. Because the Zener diode is a diode with a low resistance for current in one direction through it and a high resistance for the opposite direction, it also provides protection against wrong polarity.

Often it is necessary to completely isolate circuits and remove all electrical connections between them. This can be done using an *optoisolator* (Figure 7.45). This converts an electrical signal into an optical signal, transmits it to a detector which then converts it back into an electrical signal. The input signal is fed through an infrared light-emitting diode (LED) and the resulting infrared signal is detected by a photo transistor. To prevent the LED having the wrong polarity or too high an applied voltage, it is likely to be protected by a Zener diode circuit.

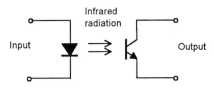

Figure 7.45 *Optoisolator*

Problems

1   Show that the output voltage for a Wheatstone bridge with a single strain gauge in one arm of the bridge and the other arms all having the same resistance as that of the unstrained strain gauge is given by:

$$\text{output voltage} = \frac{V_s}{4} G\varepsilon$$

where V_s is the supply voltage to the bridge, G the gauge factor of the strain gauge and ε the strain acting on the gauge.

2   Show that the output voltage for a Wheatstone bridge, which has two identical platinum resistance thermometers connected into arms 1 and 3 and resistors of the same resistance R but much greater resistance than the zero temperature resistance R_0 of the resistance thermometers, is related to the difference in temperature between the two thermometer elements by:

$$\text{output voltage} = \frac{R_0}{R} V_s a \times \text{temperature difference}$$

where V_s is the supply voltage to the bridge and α the temperature coefficient of resistance for platinum.

3 A Wheatstone bridge has a platinum resistance temperature sensor with a resistance of 120 Ω at 0°C in one arm of the bridge. At this temperature the bridge is balanced with each of the other arms being 120 Ω. What will be the output voltage from the bridge for a change in temperature of 20°C? The supply voltage to the bridge is 6.0 V and the temperature coefficient of resistance of the platinum is 0.0039 K^{-1}.

4 A diaphragm pressure gauge employs four strain gauges to monitor the displacement of the diaphragm. A differential pressure applied to the diaphragm results in two of the gauges on one side of the diaphragm being subject to a tensile strain of 1.0×10^{-5} and the two on the other side a compressive strain of 1.0×10^{-5}. The gauges have a gauge factor of 2.1 and resistance 120 Ω and are connected in the bridge with the gauges giving subject to the tensile strains in arms 1 and 3 and those subject to compressive strain in arms 2 and 4. If the supply voltage for the bridge is 10 V, what will be the voltage output from the bridge?

5 A Wheatstone bridge is being used with a strain gauge load cell, the out-of-balance voltage being measured by a voltmeter. What will be the drop in the output voltage, compared with that given by a voltmeter of infinite resistance, when using a voltmeter which has a resistance which is 20 times that of the equivalent Thévenin resistance of the bridge?

6 A thermocouple gives an e.m.f. of 820 μV when the hot junction is at 20°C and the cold junction at 0°C. Explain how a Wheatstone bridge incorporating a metal resistance element can be used to compensate for when the cold junction is at the ambient temperature rather than 0°C and determine the parameters for the bridge if a nickel resistance element is used with a resistance of 10 Ω at 0°C and a temperature coefficient of resistance of 0.0067 K^{-1} and the bridge voltage supply is 2 V.

7 An operational amplifier circuit is required to produce an output that ranges from 0 to −5 V when the input goes from 0 to 100 mV. By what factor is the resistance in the feedback arm greater than that in the input?

8 A summing amplifier circuit is required to produce an output that ranges from 0 to −5 V when the input from each of two inputs goes from 0 to 100 mV. If the feedback resistance is 1 kΩ, what will be the resistances in the input lines?

9 What will be the feedback resistance required for an inverting amplifier which is to have a voltage gain of 50 and an input resistance of 10 kΩ?

10 What will be the feedback resistance required for a non-inverting amplifier which is to have a voltage gain of 50 and an input resistance of 10 kΩ?

11 A differential amplifier is to have a voltage gain of 100 and input resistances of 1 kΩ. What will be the feedback resistance required?

12 A differential amplifier is to be used to amplify the voltage produced between the two junctions of a thermocouple. The input resistances are to be 1 kΩ. What value of feedback resistance is required if there is to be an output of 10 mV for a temperature difference between the thermocouple junctions of 100°C with a copper–constantan thermocouple. The thermocouple can be asssumed to give an output of 43 μV/°C?

13 An *RC* circuit is to be used as a passive low-pass filter. What will be the time constant, i.e. the value of *RC*, required if the output voltage is to be attenuated by 3 dB at a frequency of 100 Hz?

14 An *RC* circuit is to be used as a passive high-pass filter. What will be the time constant, i.e. the value of *RC*, required if the output voltage is to be attenuated by 3 dB at a frequency of 200 Hz?

15 What is the resolution of an analogue-to-digital converter with a word length of 12 bits?

16 A sensor gives a maximum analogue output of 5 V. What word length is required for an analogue-to-digital converter if there is to be a resolution of 10 mV?

17 A successive approximations form of a 12-bit ADC has a clock of frequency 1 MHz. What will be the conversion time?

18 For a weighted resistor DAC of the form shown in Figure 7.40, how should the values of the input resistors be weighted for a 4-bit DAC?

19 What is the voltage resolution of an 8-bit DAC when it has a full-scale input of 5 V?

20 A 12-bit ADC has an input range of 0 to 5 V. What will be the maximum percentage quantisation error?

8 Data acquisition and processing systems

The term *data acquisition system* is used for that part of a measurement system that quantifies and stores the data. Processing then involves manipulating the data to obtain the required result. While the system could be a scientist or engineer reading data from instruments, recording it in a notebook, and then manipulating the data to calculate the required quantity, the term is generally used for automated data acquisition involving micro-processors or computers for the data processing. This chapter is about such automated systems.

8.1 Automated systems

Automated data acquisition systems take the signals from sensors, use signal conditioning to convert the signal from perhaps a resistance change to a voltage, convert an analogue signal to a digital signal and store it for processing by a microprocessor. Thus the basic form of such a system may be as shown in Figure 8.1. Such a system may be provided by an instrument termed a *data logger* or by a personal computer (PC) using *input/output plug-in boards*, both perhaps supplemented by separate signal conditioning.

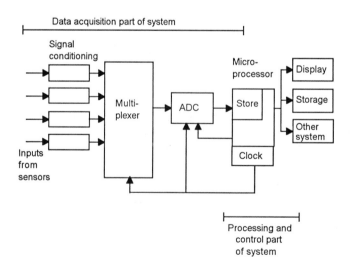

Figure 8.1 *Automated data acquisition and processing system*

8.1.1 Data loggers

A *data logger* is a microprocessor-based device which collects and stores data, being programmed from its front panel. Figure 8.2 shows a basic system. A large number of sensors can be connected to the data logger and its microprocessor permits simple programming to determine which sensors to use, when and how often to use them to obtain data, and has the ability to carry out simple arithmetic operations in order to determine the result from a number of measurements. Generally the results are then presented in the form of a digital display, for recording by a printer or for storage on a floppy disk which can then be transferred for reading and analysis by a computer or directly downloaded to a computer or other controller.

Because thermocouples are often used as inputs to data loggers, compensation circuits for the cold junction temperature being the ambient temperature and linearisation are frequently included. A multiplexer is used to select the inputs from the sensors and process it through a single amplifier and analogue-to-digital converter. Typically the multiplexer will scan through five input channels per second and thus if there are, say, fifty sensors connected to the data logger it will take ten seconds to complete a scan. Thus applications involving fast changing signals can present problems. The gain of the amplifier is controlled by the microprocessor so that it can be automatically adjusted to bring all the input signals to the common ±10 V range required by the ADC. The analogue-to-digital converters are generally the dual ramp form since the conversion time required is not particularly fast and the interference rejection properties of this form of ADC can be of benefit.

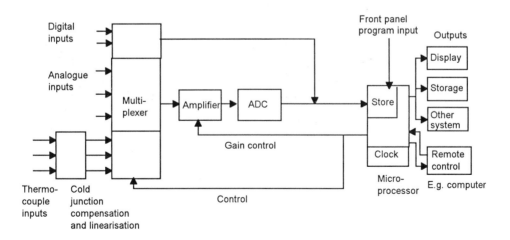

Figure 8.2 *Basic data logger system*

8.1.2 Computer-based systems

Personal computer-based systems involving the interface between the computer and the sensors being by means of plug-in boards/cards (see Section 8.3 for a discussion of boards) are very widely used. These boards carry the components for the data acquisition so that the computer can store data, carry out data processing, control the data acquisition process and issue signals for use by actuators. The plug-in board is controlled by the computer and the digitised data transferred from the board to the memory of the computer. The processing of the data is then carried out within the computer according to how it is programmed. Commercially available software is available for this purpose.

Personal computer-based systems may also involve intelligent instruments being connected through a plug-in board/card to the computer so that their operation can be controlled by it (see Section 8.3).

8.2 Microprocessor systems

Computers have three basic sections:

1 *Central processing unit* (CPU)
  This recognises and carries out program instructions.

2 *Memory*
  This is memory to store program instructions and also memory to store data.

3 *Input/output interfaces*
  These enable the computer to handle communications to and from the outside world.

Digital signals move from one section to the other along paths called *buses*. A bus is just a number of conductors along which digital signals can travel. Data is carried by the *data bus*, information concerning the address of the memory location for the accessing of stored data is carried by the *address bus* and the signals related to control actions by the *control bus*. Figure 8.3 shows the basic system.

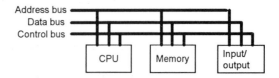

Figure 8.3 *Basic form of a computer system*

The buses have been drawn as wide bands because they each contain a number of parallel wires, a word being transmitted by each bit of the word being simultaneously transmitted, one per wire, along the wires. Thus, for example, the 4-bit word 1010 would be transmitted with the least significant bit (0) along one wire, the next significant bit (1) along the next wire, the next significant bit (0) along the next wire and the most significant bit (1) along the fourth wire.

8.2.1 The buses

All the information in a microprocessor system is represented by a sequence of bits. An 8-bit sequence is called a *byte*, a 4-bit sequence a *nibble*. The term *word* is used for the number of bits used to contain a piece of information. The word length thus might be 8, 16, 32 or even 64 bits.

The data bus is used to transmit data words between the memory and the CPU and between the input/output interfaces and the CPU. Word lengths may be 8, 16 or 32 bits. For an 8-bit word it needs eight parallel wires, for a 16-bit word 16 wires and for a 32-bit word 32 wires. The word length determines the number of values that can be represented by the data. With a word length of 4 bits, the number of values possible is $2^4 = 16$, with an 8-bit word $2^8 = 256$, with a 16-bit word $2^{16} = 65\ 536$ and with a 32-bit word $2^{32} = 4\ 294\ 960\ 000$.

The address bus carries signals to indicate where data is to be found. The memory is configured as a set of storage cells, each cell having its own address. A particular address is selected by the address being placed on the address bus. Then only that location is open to communication with the CPU. A computer with an 8-bit address bus would be able to address $2^8 = 256$ locations, this being memory locations from 0 to $2^8 - 1 = 255$. A 16-bit wide address bus enables $2^{16} = 65\ 536$ locations to be addressed. Memory sizes, i.e. the number of stage cells and so separate addresses they contain, are usually written in terms of so many K, where K is equal to $2^{10} = 1024$. Thus a 16-bit address bus addresses a memory of size $65\ 536/1024$ or 64K.

The control bus is the means by which signals are sent to synchronise the separate elements. Thus the system clock signals are carried by the control bus. The CPU sends some control signals to other elements to indicate the type of operation being performed, e.g. whether it needs to receive (the term used is READ) a signal or send (the term used is WRITE) a signal.

To store data in a memory cell, the CPU places the address of the cell on the address bus and the data to be stored on the data bus. The CPU then uses the control bus to indicate that the data is being written, i.e. sent.

8.2.2 The central processing unit

The central processing unit (CPU) is the section of the system which processes the data, fetching instructions from memory, decoding them and executing them. It consists of a control unit, an arithmetic and logic unit (ALU) and registers. The control unit determines the timing and sequencing of operations. The arithmetic and logic unit is responsible for performing the data manipulation. The internal data that the CPU is currently using is temporarily stored in a group of registers. There are a number of types of registers; the number, size and types varying from one microprocessor to another.

The processing speed of the CPU depends on both its clock speed and the bus size. The clock provides the timing for each operation, the number of clock ticks required for an operation depending on the microprocessor concerned. Bus size affects the processing speed because it determines the amount of data that can be transferred per operation, the larger the bus size the greater the amount.

8.2.3 RAM and ROM

Computer memory devices are in two basic types, RAM and ROM. ROM stands for *read-only memory* and is for memory where data is permanently stored and can only to be read by the computer system. No new information can be stored, i.e. written, into that memory. ROM memory is programmed by the manufacturer and contains the computer operating system and standard routines that might be used. This memory is not lost when the power to the computer is switched off. RAM stands for *random access memory* and is for memory which the computer system can use for temporary storage, being able to write information into it and read information from it. RAM memory is used for storing the user's programs. A user would normally save the contents of RAM on floppy or hard disc before turning the power off, since all the information in RAM is lost when the power is switched off.

8.2.4 Input/output interfaces

Interface circuits are used between the microprocessor and items connected to it. The actions required of interface circuits include:

1 *Electrical buffering/isolation*
   This is required when the items connected to the microprocessor operate at a different voltage or current to that used by the microprocessor and isolation is required to prevent damage to the microprocessor.

2 *Putting the signal in the right form*
   The input from a sensor might be analogue and thus need to be converted to digital. Likewise the digital output from the micro-processor might need to be converted to analogue for use by a connected peripheral item. There might also be a need to change the incoming or outgoing voltage or current levels to more suitable values.

3 *Timing control*
   This is needed when the data transfer rates of the microprocessor differ from those used by the connected item. For this reason, special lines are used to send signals to control the rate of data transfer. These are

referred to as *handshake lines* and the process as *handshaking*. Thus the connected item might send a signal DATA READY along a handshake line when it has put data on the data bus to the input interface. The CPU then determines that the DATA READY signal is active, reads the data from the input and then sends the signal INPUT ACKNOWLEDGED to the connected item. This signal indicates that the transfer has taken place and thus the sequence can be repeated with more data to be sent. For an external item to receive an output from the microprocessor, it can send a PERIPHERAL READY signal along a handshake line. The CPU determines that the PERIPHERAL READY signal is present on the line and then sends an OUTPUT READY signal and the data. The next PERIPHERAL READY signal might be used to indicate that the transfer has been completed.

4   *Serial to parallel and vice versa*
When a word is transmitted with each bit simultaneously along parallel data paths it is said to be transmitted by *parallel data transfer*. Such a form is used within the computer system. External devices may, however, only use one data path and thus a word has to be transmitted sequentially one bit at a time. Such a form of transfer is termed *serial data transfer*. Serial data transfer is a slower method of data transfer than parallel data transfer. Thus if an external item is transmitting data by serial data transfer, then it needs to be converted to parallel data transfer for use by a microprocessor.

8.3 Data acquisition boards

Data acquisition boards are available for many computers and offer, on a plug-in board, various combinations of analogue, digital and timing/counting inputs and outputs for interfacing sensors with computers. The board is a printed circuit board that is inserted into an expansion slot in the computer, plugging into the computer bus system. Figure 8.4 shows the basic elements of a simple data acquisition board. Analogue inputs from the sensors are accessed through a multiplexer. An analogue-to-digital converter then converts the amplified sampled signal to a digital signal. The control element can be set up so that each of the inputs is sequentially sampled or perhaps samples are taken at regular time intervals or perhaps just a single sensor signal is used. The other main element is the bus interface element which contains two registers, the control and status register and the data register.

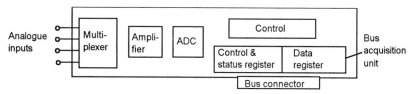

Figure 8.4 *Basic elements of a data acquisition card*

Consider now what happens when the computer is programmed to take a sample of the voltage input from a particular sensor. The computer first activates the board by writing a control word into the control and status register. This word indicates the type of operation that is to be carried out. The result of this is that the multiplexer is switched to the appropriate channel for the sensor concerned and the signal from that sensor passed, after amplification at an instrumentation amplifier, to the analogue-to-digital converter. The outcome from the ADC is then passed to the data register and the word in the control and status register changed to indicate that the conversion is complete. Following that signal, the computer sets the address bus to select the address of where the data is stored and then issues the signal on the control bus to read the data and so take it into storage in the computer for processing. In order to avoid the computer having to wait and do nothing while the board carries out the acquisition, an interrupt system can be used whereby the board signals the computer when the acquisition is complete and then the computer microprocessor can interrupt any program it was carrying out and jump from that program to a subroutine which stores its current position in the program, reads the data from the board and then jumps back to its original program at the point where it left it (Figure 8.5).

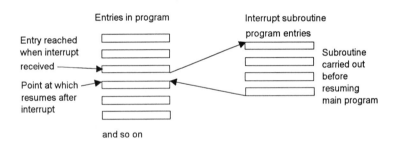

Figure 8.5 *The interrupt subroutine*

With the above system, the acquired data moves from the board to the microprocessor which has to interrupt what it is doing, set the memory address and then direct the data to that address in the memory. The microprocessor thus controls the movement. A faster system is to transfer the acquired data directly from the board to memory without involving the microprocessor. This is termed *direct memory access* (DMA). To carry this out, a DMA controller is connected to the bus. This controller supplies the memory address locations where the data is to be put and so enables the data to be routed direct to memory.

Data acquisition board specifications for the analogue inputs include items such as:

1   *Sampling rate for analogue inputs*
    A board may be specified as having a sampling rate of, say, 100 kS/s (100 thousand samples per second). This determines the number of samples the analogue-to-digital converter takes per second. According to the Nyquist sampling criterion, the sample rate must be at least twice the maximum frequency component in the analogue signal. Thus the 100 kS/s board can sample a maximum frequency of 50 kHz.

2   *Number of channels*
    Multiplexing is used to sample across more than one input. The analogue-to-digital converter samples one channel, switches to the next channel, samples it, switches to the next channel, samples it, and so on. Because the analogue-to-digital converter is sampling many channels instead of just one, the sampling rate must be considered as applied to the sum of the channels. Thus if the sampling rate is 100 kS/s and there are 10 channels then the sampling rate for one of the channels is 100/10 = 10 kS/s.

3   *Resolution*
    This is often specified as the number of bits that the analogue-to-digital converter uses to represent the analogue signal or the fraction of the analogue signal that represents a change of 1 bit. For example, the resolution of a board might be specified as 12 bits, 1 in 4096 (i.e. 1 in 2^{12}).

4   *Range*
    This is the minimum and maximum analogue voltage levels that the analogue-to-digital converter can handle. For example, a board might be specified as having a range of 0 to 10 V. Note that this is after amplification by the built-in amplifier on the board. Multifunction data acquisition boards offer a number of ranges from which the range to be used can be selected.

5   *Gain*
    The gain available from the amplifier on the board is also selectable and this, coupled with the range and resolution, determines the smallest detectable change in input analogue voltage. Thus if the gain is 20 and the range 0 to 10 V for a 16-bit board, the smallest detectable input voltage is $10/(20 \times 2^{16})$ or 7.6 μV.

Data acquisition board specification for analogue outputs to provide stimuli for the data acquisition system include items such as:

1   *Settling time*
    This is the time taken for the digital-to-analogue converter to give an analogue output which settles to within the specified accuracy. Thus a board might have a settling time to 0.012% of full-scale reading for a 10 V step of 7 μs.

2   *Slew rate*
    This is the maximum rate of change that the digital-to-analogue converter can produce on the output signal. A board might have a DAC with a slew rate of 10 V/μs. A digital-to-analogue converter with a small settling time and a high slew rate can thus produce high-frequency output signals.

3   *Range*
    This is the minimum and maximum analogue voltage levels that the digital-to-analogue converter can produce. For example, a board might be specified as having an analogue output range of 0 to 10 V.

Data acquisition board specification for digital inputs/outputs include items such as:

1   *Number of channels*
    This specifies the number of digital lines in or out.

2   *Digital logic levels*
    This specifies the maximum and minimum voltages.

3   *Handshaking*
    In transferring digital data between a computer and equipment such as data loggers and printers, the handshaking circuitry for communication synchronisation has to be specified.

Data acquisition boards include timers and counters which are available for use in such events as timed control of laboratory events and counting digital events. For such applications the specification include items such as:

1   *Clock frequency*
    This is the frequency of the internal clock and determines how fast the counter can increment. A board might have a clock frequency of, say, 20 MHz.

2   *Resolution*
    This is specified as the number of bits the counter uses to count with. A higher resolution thus indicates that the counter can count up to a larger number.

As an example of a data acquisition board, consider the National Instruments NB-MIO-16. This is a multifunction board which plugs directly into a Macintosh NuBus connector (other versions are available for IBM computers). Figure 8.6 shows the general appearance of such a board and Figure 8.7 a block diagram of the board's contents. It has 16 analogue input channels, a 12-bit analogue-to-digital, two 12-bit digital-to-analogue converters, two analogue output channels, an 8-bit digital interface and three timers/counters. The analogue-to-digital converter has a resolution of 12 bits, 1 in 4096 and a maximum sampling rate of 100 kS/s. The amplifier gain is software selectable. For one particular mode, the analogue input signal ranges which can be selected with a gain of 1 are ±10 V, ±5 V or 0 to 10 V; with a gain of 2 they are ±5 V, ±2.5 V and 0 to 5 V; with a gain of 4 they are ±2.5 V, ±1.25 V or 0 to 2.5 V; with a gain of 8 they are ±1.25 V, ±0.63 V or 0 to 1.25 V. The two analogue output channels have a resolution of 12 bits, 1 in 4096, and a range which can be selected as ±10 V or 0 to 10 V.

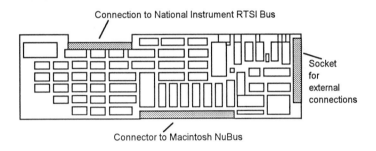

Figure 8.6 *General form of the NB-MIO-16 card*

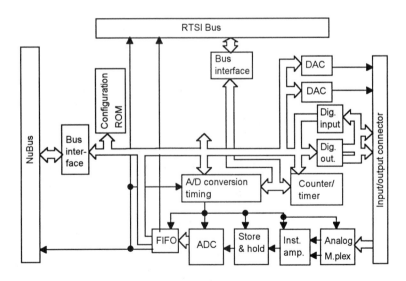

Figure 8.7 *NB-MIO-16*

The RTSI Bus (real-time system integration bus) is the National Instruments timing bus used to connect boards directly, by means of a connector on the board, so that precise synchronisation can take place. The NuBus is the bus system used in the Macintosh computers. The FIFO element on the board is the *first-in, first-out* element. This temporarily stores samples from the analogue-to-digital converter pending a transfer to the computer memory. Data acquisition can be triggered by software commands, external pulses or the onboard clock.

The above is a description of a multifunction input/output board which can be used to acquire data from sensors and communicate with a computer or take data from a computer and communicate with external devices. For specialised analogue applications dedicated analogue boards can be used and for purely digital and timing applications dedicated digital and timing boards can be used.

8.4 Digital communication

There are a number of possibilities for the connections of instruments to computers. These include:

1 *Serial connection*
   Serial connection offers the possibility of longer distances for the transfer of data but typically at a slower transfer speed. Commonly used serial instruments, e.g. data loggers, output data in a form that can easily be manipulated by a computer program and often can be connected directly to the standard serial port on the computer.

2 *General purpose interface bus*
   This is a standard method widely used to connect programmable instruments to a computer so that they can be controlled by the software run by the computer.

3 *VXIbus*
   This method of connecting instruments to computers was introduced in 1987 and its use is increasing. It uses a mainframe chassis to hold modular instruments on plug-in boards. These may include data acquisition boards and general purpose bus instruments.

Digital data can be communicated between devices by serial or parallel communication. With serial communication the data is sent sequentially bit by bit; with parallel communication the groups of bits are sent simultaneously. Certain standards are used for the manner in which digital information is communicated between devices. With serial communications the main standard is *RS-232C*. With parallel communications the standard used is the *general purpose interface bus* (GPIB) (IEEE-488). Parallel communication is a faster method of communication than serial but is restricted to a maximum transmission rate of 1 M bytes/s up to a maximum transmission distance of 15 m and is used for high speed, short distance, communicating in situations where there is little interference. Serial data communication can be used over much larger distances.

8.4.1 Serial communication

Serial data communication can be divided into three categories, depending on whether the communication is one way or two way.

1 *Simplex*
   This involves communication from A to B where B is not capable of transmitting back to A. This form of communication involves two wires.

2 *Half duplex*
   This involves transmissions from A to B and B to A but not simultaneously. This form of communication involves a two- or four-wire cable.

3 *Full duplex*
   This involves simultaneous transmissions from A to B and B to A. This involves a two- or four-wire cable.

With each of the above forms of communication, it is necessary for a receiver to be ready to receive and identify each set of data as it is received from the transmitter. This can be done in two possible ways. *Asynchronous transmission* involves each frame of data being preceded by a start bit and terminated by a stop bit. The receiving device then knows exactly where data starts and ends. Because of the need to check for start and stop bits, such a form of transmission is restricted to less than 1200 bits/s. Asynchronous transmission tends to be used for intermittent transmission. With *synchronous transmission* there is no need for start and stop bits since the transmitter and receiver are synchronised to a common clock. The transmitting data is preceded by a synchronising clock signal and then each character frame is recognised as a block of 7 or 8 bits. Periodic resynchronisation occurs. This mode of transmission can be used with more than 1200 bits/s and is commonly used where there is an almost constant stream of data, e.g. computer files. The term *protocol* is used for a set of rules by which two stations may transfer data.

Most computer systems have a serial communication interface built into the system so that longer range communication can occur than is feasible with the parallel interface. A commonly used standard serial interface is the RS-232, the letter used after the number indicating which edition of the standard is being employed. This interface can be used for transmission rates up 20 000 bits/s over distances up to about 15 m, with lower bit rates at longer distances. RS232 signals can be grouped into three categories:

1 *Data*
   RS-232 provides two independent serial data channels, termed primary and secondary. Both these channels are used for full duplex operation.

2   *Handshake control*
Handshaking signals are used to control the flow of serial data over the communication path.

3   *Timing*
For synchronous operation it is necessary to pass clock signals between transmitters and receivers.

Table 8.1 gives the RS-232 connector pin numbers and signals for which each is used. Not all the pins and signals are necessarily used in a particular set-up. The signal ground wire allows for a return path. Data termination equipment (DTE) is normally fitted with a male connector and data communications equipment (DCE) with a female connector. Data terminating equipment, such as a computer, is capable of sending and/or receiving data; data communications equipment is generally a device such as a modem which can facilitate communication.

Table 8.1 *RS-232 pins*

Pin	Abbreviation	Direction	Signal/function
1	FG		Frame/protective ground
2	TXD	To DCE	Transmitted data
3	RXD	To DTE	Received data
4	RTS	To DCE	Request to send
5	CTS	To DTE	Clear to send
6	DSR	To DTE	DCE ready
7	SG		Signal ground/common return
8	DCD	To DTE	Received line detector
9			
10			
11			
12	SDCD	To DTE	Secondary received line signal detector
13	SCTS	To DTE	Secondary clear to send
14	STD	To DCE	Secondary transmitted data
15	TC	To DTE	Transmit signal timing
16	SRD	To DTE	Secondary received data
17	RC	To DTE	Received signal timing
18		To DCE	Local loop back
19	SRTS	To DCE	Secondary request to send
20	DTR	To DCE	Data terminal ready
21	SQ	To DEC/DTE	Remote loop back/signal quality detector
22	RI	To DTE	Ring indicator
23		To DEC/DTE	Data signal rate selector
24	TC	To DCE	Transmit signal timing
25		To DTE	Test mode

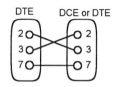

Figure 8.8 *Minimum serial connections*

For the simplest bi-directional link only the two lines 2 and 3 for transmitted data and received data, with signal ground (7) for the return path of these signals, are needed (Figure 8.8). Thus the minimum connection is via a three-wire cable. A configuration that is widely used with interfaces involving computers is pins 1, 2, 3, 4, 5, 6, 7 and 20. The signals sent through pins 4, 5, 6 and 20 are used to check that the receiving end is ready to receive a signal; the transmitting end is ready to send and the data is ready to be sent.

RS232 is limited concerning the distance over which it can be used, noise limiting the transmission of high numbers of bits per second when the length of cable is more than about 15 m. Other standards such as RS422 and RS423 are similar to RS232; both being capable of being used for higher transmission rates and over longer distances.

8.4.2 General purpose instrument bus

The standard interface most commonly used for parallel communications is the *General purpose instrument bus*, the IEEE-488 standard, originally developed by Hewlett Packard to link their computers and instruments and known as the *Hewlett Packard Instrumentation Bus*. This bus provides a means of making interconnections so that parallel data communications can take place between listeners, talkers and controllers. Listeners are devices that accept data from the bus, talkers place data, on request, on the bus and controllers manage the flow of data on the bus and provide processing facilities. There is a total of 24 lines, of which eight bi-directional lines are used to carry data and commands between the various devices connected to the bus: five lines are used for control and status signals, three are used for handshaking between devices and eight are ground return lines (Figure 8.9). Table 8.2 lists the functions of all the lines and their pin numbers in a 25-way D-type connector. Up to 15 devices can be attached to the bus at any one time, each device having its own address.

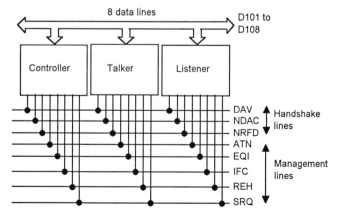

Figure 8.9 *GPIB bus structure*

Table 8.2 *IEEE-488 bus system*

Pin	Signal group	Abbreviation	Signal/function
1	Data	D101	Data line 1
2	Data	D102	Data line 2
3	Data	D103	Data line 3
4	Data	D104	Data line 4
5	Management	EOI	End Or Identify. This is used either to signify the end of a message sequence from a talker device or is used by the controller to ask a device to identify itself.
6	Handshake	DAV	Data Valid. When the level is low on this line then the information on the data bus is valid and acceptable.
7	Handshake	NRFD	Not Ready For Data. This line is used by listener devices taking it high to indicate that they are ready to accept data.
8	Handshake	NDAC	Not Data Accepted. This line is used by listeners taking it high to indicate that data is being accepted.
9	Management	IFC	Interface Clear. This is used by the controller to reset all the devices of the system to the start state.
10	Management	SRQ	Service Request. This is used by devices to signal to the controller that they need attention.
11	Management	ATN	Attention. This is used by the controller to signal that it is placing a command on the data lines.
12		SHIELD	Shield.
13	Data	D105	Data line 5.
14	Data	D106	Data line 6.
15	Data	D107	Data line 7.
16	Data	D108	Data line 8.
17	Management	REN	Remote Enable. This enables a device to indicate that it is to be selected for remote control rather than by its own control panel.
18		GND	Ground/common. (Twisted pair with DAV)
19		GND	Ground/common. (Twisted pair with NRFD)
20		GND	Ground/common. (Twisted pair with NDAC)
21		GND	Ground/common. (Twisted pair with IFC)
22		GND	Ground/common. (Twisted pair with SRG)
23		GND	Ground/common. (Twisted pair with ATN)
24		GND	Signal ground.

The 8-bit parallel data bus can transmit data as one 8-bit byte at a time. The management lines each have an individual task in the control of information. For example, commands from the controller are signalled by taking the Attention Line (ATN) low, otherwise it is high, and thus indicating that the data lines contain data. The commands can be directed to individual devices by placing addresses on the data lines. Each device on the bus has its own address. Device addresses are sent via the data lines as a parallel 7-bit word, the lowest 5-bits providing the device address and the other 2 bits control information. If both these bits are 0 then the commands are sent to all addresses; if bit 6 is 1 and bit 7 a 0 the addressed device is switched to be a listener; if bit 6 is 0 and bit 7 is 1 then the device is switched to be a talker.

The handshake lines are used for controlling the transfer of data. The three lines ensure that the talker will only talk when it is being listened to by listeners. Initially DAV is high, indicating that there is no valid data on the data bus, NRFD and NDAC being low. When a data word is put on the data lines, NRFD is made high to indicate that all listeners are ready to accept data and DAV is made low to indicate that new data is on the data lines. When a device accepts a data word it sets NDAC high to indicate that it has accepted the data and NRFD low to indicate that it is now not ready to accept data. When all the listeners have set NDAC high, then the talker cancels the data valid signal, DAV going high. This then results in NDAC being set low. The entire process can then be repeated for another word being put on the data bus.

This interplay between the controller and the devices on the bus is controlled by software and thus the user has rarely to worry about the commands since the software package has routines which automatically provide the appropriate commands.

The GPIB is a bus which can be used to interface a wide range of instruments, via a plug-in board, to a computer. It might thus be used to connect to a computer such instruments as digital multimeters and digital oscilloscopes and computer peripherals such as printers and plotters. The GPIB hardware (Figure 8.10) consists of a bus controller board that plugs into a computer with standard cables used to link the board to the instruments via interfaces.

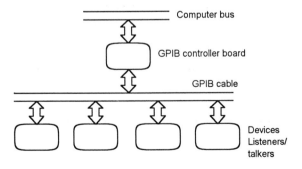

Figure 8.10 *GPIB hardware*

An example of such a board is the National Instruments AT-GPIB/TNT board. This is a board for use with the IBM PC AT and compatible computers. There are two versions of the board, one for which DIP switches and jumpers have to be set to configure the board, and another, termed plug and play, for which configuration is achieved through software. Configuration involves setting input/output addresses, memory buffer address, DMA channel and the interrupt level. Other boards are available for other types of computers.

8.4.3 Personal computer buses

Computer buses and instrumentation buses are the two main forms of buses. A computer bus is the bus used to connect the CPU to the input/output ports or other devices. The form of the bus used depends on the microprocessor, there being different forms of bus for different microprocessors. The next paragraph lists some of the more commonly used computer buses.

The *XT computer bus* was introduced in 1983 for 8-bit data transfers with IBM PC/XT and compatible computers. The *AT bus*, also referred to as the *industry standard architecture (ISA) bus* was later introduced for use with 16-bit transfers with IBM PC and other compatible computers using 80286 and 80386 microprocessors. The AT bus is compatible with the XT bus so that plug-in XT boards can be used in AT bus slots. The *extended industry standard architecture (EISA) bus* was developed to cope with 32-bit data transfers with IBM PC and other compatible computers using 80386 and 80486 microprocessors. The *Micro Channel architecture (MCA) bus* is a 16-bit or 32-bit data transfer bus designed for use with IBM personal System/1 (PS/2) computers. Boards for use with this bus are not, however, compatible with PC/XT/AT boards. The *NuBus* is the bus used in Apple's Macintosh II computers. It is a 32-bit bus. The *S-bus* is the 32-bit bus used in Sun Microsystem's SPARC stations, the 32-bit *TURBOchannel* being the bus used in DECstation 5000 workstations. The *VME bus* was designed by Motorola for use with its 32-bit 68000-microprocessor-based system. Such a bus is now, however, widely used with other computer systems as the bus for use with instrumentation systems.

The above are all termed *backplane buses*, the term backplane being for the printed circuit board on which connectors are mounted with each connector being capable of accepting a printed circuit card containing a particular function, e.g. memory. Figure 8.11 illustrates this. The backplane provides the data, address and control bus signals to each card, so enabling systems to be easily expanded by the use of off-the-shelf cards. The other basic form of bus is the *cable bus* such as RS232 and GPIB.

It is into such computer buses that data acquisition boards and boards used for interfacing instruments and other peripherals have to interface. Because of the range of computer buses in use, data acquisition and instrument boards are usually available in various configurations, depending on the computer with which they are to be used.

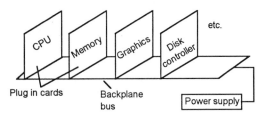

Figure 8.11 *Backplane bus*

8.4.4 VXIbus

The *VMEbus* was designed as a parallel means of connection supporting 8-, 16- and 32-bit microprocessors. The *VXIbus* (VME Extensions for Instrumentation) is an extension of the specification of the VMEbus which has been designed for instrumentation applications such as automatic test equipment where higher speed communications are required than possible with the GPIB bus. It also provides better synchronisation and triggering and has been developed by a consortium of instrument manufacturers so that interoperability is possible between the products of different companies.

The system involves VXI boards plugging into a mainframe, the main frame having up to 13 slots (Figure 8.12). Four basic sizes have been specified for VXI boards: size A 100 mm high by 160 mm deep with a P1 connector, size B 233.35 mm high by 160 mm deep with a P1 and possibly also a P2 connector, size C 233.35 mm high and 340 mm deep with a P1 and possibly a P2 connector, size D 366.7 mm high and 340 mm deep with a P1 and possibly a P2 and a P3 connector. The size A and B boards are thus specified for the VME bus, the C and D boards being additions for the VXI bus. The slot 0 module is responsible for system resources such as clocks, trigger signals and configuration signals. The other slots can contain VXI instruments, plug-in data acquisition boards and boards for GPIB instruments.

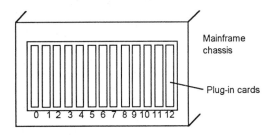

Figure 8.12 *A VXI mainframe*

There are a number of possible system configurations that can be used with the VXI system. One type involves a VXI mainframe linked to an external controller, a computer, via a GPIB link (Figure 8.13(a)). The controller talks across this link using GPIB protocol to an interface board in the chassis which translates the GPIB protocol into the VXI protocol. This makes VXI instruments appear to the controller to be GPIB instruments and enables them to be programmed using GPIB methods. A second option is to place the complete computer in the VXI chassis (Figure 8.13(b)). This option offers the smallest possible physical size for the system and allows the computer to directly use the VXI backplane bus. The third option is to use a special high-speed system-bus-on-cable, the MXIbus, to link a computer and the VXI chassis (Figure 8.13(c)). The MXI is 20 times faster than the GPIB.

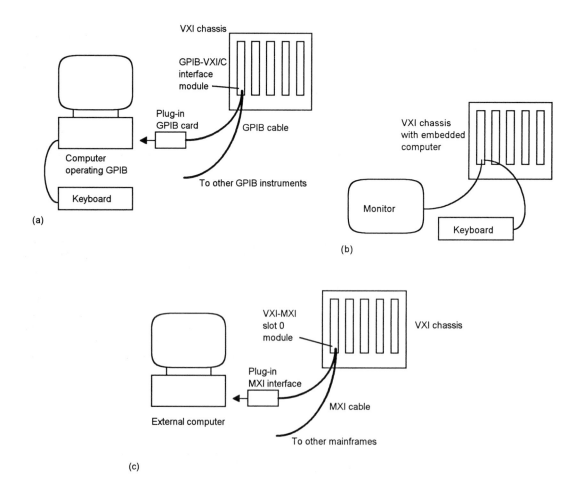

Figure 8.13 *VXI configuration options*

8.5 Software   The software used in automated data acquisition and instrument control systems includes:

1 *Instrument driver software*
This is the software used to control specific hardware interfaces. It provides a tool for the development of a measurement system involving instruments without requiring the operator to write programs for the instruments. For example, there are GPIB, VXI and serial drivers for instrument control. Such drivers contain the hardware set-up and communication protocols so making it easier to access an instrument. For example, there are GPIB drivers for the Hewlett Packard 54100A digitising oscilloscope, the Solartron 7061 digital multimeter and the Philips PM6680 counter/timer.

2 *Data analysis software*
Such software can be used to filter a signal and remove noise, correct for non-linearity, compensate for environmental changes and, in general, correct data for faults and corruption. Statistical analysis can be carried out. For example, the mean, standard deviation and fitting data to straight lines of curves carried out.

3 *Application software*
Application software is used to take care of all the software requirements of an instrumentation system, e.g. data acquisition, instrument control, data analysis and data presentation.

8.5.1 LabVIEW

As an illustration of the type of application software available, consider LabView. LabVIEW is a programming language and set of subroutines developed by National Instruments for data acquisition and scientific programming. It does not require the program to be set out in a sequence of steps that have to be executed consecutively but is a graphical programming language which has each subprogram and program structure represented by icons. All the programming is then done graphically by drawing lines between connection points on the various icons. The resulting picture shows the data flow. Figure 8.14 illustrates this for the simple program of getting input A, getting input B, adding A and B and then displaying the result.

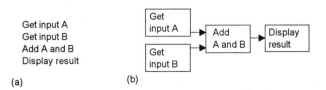

Figure 8.14 *(a) Sequential step programming, (b) graphical programming*

Programs that use one or more LabVIEW functions are called *virtual instruments* or VIs. VIs may be combined to make a new virtual instrument and an icon made for it. Each virtual instrument has a front panel and a diagram. The front panel is the place for user interaction, including boxes for each input variable and is where the results are displayed. The front panel may be thought of as representing the front panel of an instrument with the instrument controls and display. The diagram contains the actual program and is thus basically of the form shown in Figure 8.14(b).

Figure 8.15 shows a simple example of a front panel display, being that for a data acquisition board used for analogue-to-digital conversion for a single input channel, a description of the program which can be selected using the Help window and the program.

The device is the slot number in the computer where the board is installed. The channel is the selected input channel being used. The high and low limits are the maximum and minimum voltages that are required. These voltages automatically set the amplifier gain on the board to achieve these values. The sample is the output.

This simple application might be extended to give an output which displays how the output varies with time, so essentially the board together with the computer providing a form of oscilloscope. Alternatively, the output may be sampled a specified number of times and the average value obtained and given as the output (Figure 8.16).

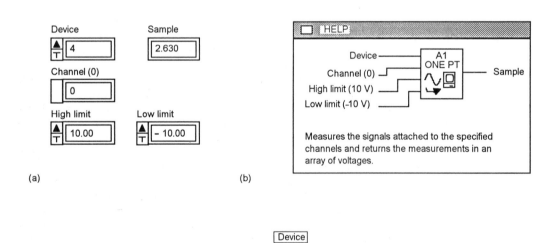

Figure 8.15 *Example of (a) front display, (b) Help window, (c) program*

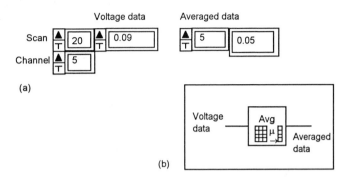

Figure 8.16 *(a) Front panel, (b) description*

To create an application, the program can be developed in this way and a virtual instrument developed. Driver programs are available for intelligent instruments and can be incorporated directly into the application program. For details concerning the development of programs for LabView, the reader is referred to the manual accompanying the software and *LabView Graphical Programming* by G.W. Johnson (McGraw-Hill 1994).

8.6 Programmable logic controller

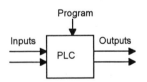

Figure 8.17 *Programmable logic controller system*

A *programmable logic controller* (PLC) is a microprocessor-based controller that uses a programmable memory to store instructions and to implement functions such as logic, sequencing, timing, counting and arithmetic in order to control machines and processes (Figure 8.17). The term *logic* is used because programming is primarily concerned with implementing logic and switching operations. Input devices, e.g. sensors such as switches, and output devices in the system being controlled, e.g. motors, valves, etc., are connected to the PLC. The operator then enters a sequence of instructions, i.e. a program, into the memory of the PLC. The controller then monitors the inputs and outputs according to this program and carries out the control rules for which it has been programmed.

PLCs are designed to be operated by engineers with perhaps a limited knowledge of computers and computing languages. They are not designed so that only computer programmers can set up or change the programs. Thus, the designers of the PLC have pre-programmed it so that the control program can be entered using a simple, rather intuitive, form of language termed *relay ladder logic*. They have the great advantage that the same basic controller can be used with a wide range of control systems. To modify a control system and the rules that are to be used, all that is necessary is for an operator to key in a different set of instructions. There is no need to rewire. The result is a flexible, cost-effective, system which can be used with control systems which vary quite widely in their nature and complexity.

The first PLC was developed in 1969. They are now widely used and extend from small self-contained units for use with perhaps 20 digital inputs/outputs to modular systems which can be used for large numbers of inputs/outputs, handle digital or analogue inputs/outputs, and also carry out proportional-integral-derivative control modes.

8.6.1 Functional elements of a PLC

Typically, a PLC system has five basic components. These are the processor unit, memory, the power supply unit, input/output interface section and the programming device (Figure 8.18).

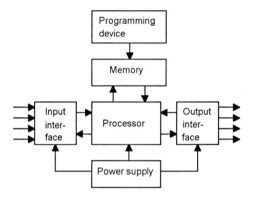

Figure 8.18 *The PLC system*

The *processor unit* or *central processing unit* (CPU) is the unit containing the microprocessor. It interprets the input signals and carries out the control actions, according to the program stored in its memory, communicating the decisions as action signals to the outputs. The *power supply unit* is needed to convert the mains a.c. voltage to the low d.c. voltage (5 V) necessary for the processor and the circuits in the input and output interface modules. The *programming device* is used to enter the required program into the memory of the processor. The *memory unit* is where the rules are stored that are to be used for the control actions by the microprocessor. The *input and output sections* are where the processor receives information from external devices and communicates information to external devices.

8.6.2 Mechanical design of PLC systems

There are two common types of mechanical design; a *single box*, and the *modular* and *rack types*. The single box type is commonly used for small programmable controllers and is supplied as an integral compact package complete with power supply, processor, memory, and input/output units

Socket for cable
from program console

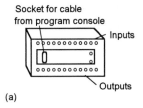

Inputs

Outputs

(a)

Input modules

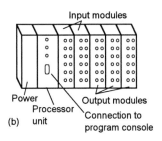

Power
Processor
(b)   unit

Output modules
Connection to
program console

Figure 8.19 *(a) Single box,*
(b) modular/rack types

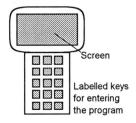

Screen

Labelled keys
for entering
the program

Figure 8.20 *Hand-held*
programmer

(Figure 8.19(a)). Typically such a PLC might have 40 input/output points and a memory which can store some 300 to 1000 instructions. The modular type consists of separate modules for power supply, processor, etc., which are often mounted on rails within a metal cabinet. The rack type can be used for all sizes of programmable controllers and has the various functional units packaged in individual modules which can be plugged into sockets in a base rack (Figure 8.19(b)). The mix of modules required for a particular purpose is decided by the user and the appropriate ones then plugged into the rack. Thus it is comparatively easy to expand the number of input/output connections by just adding more input/output modules or to expand the memory by adding more memory units.

Programs are entered into a PLC's memory using a program device which is usually detachable and can be moved from one controller to the next without disturbing operations. It is not necessary for the programming device to be connected to the PLC. Programming devices can be a hand-held device, a desktop console or a computer. Hand-held systems incorporate a small keyboard and liquid crystal display, Figure 8.20 showing a typical form. Desktop devices are likely to have a visual display unit with a full keyboard and screen display. Personal computers are widely configured as program development workstations. Some PLCs only require the computer to have appropriate software, others special communication cards. A major advantage of using a computer is that the program can be stored on the hard disk or a floppy disk and copies easily made. The disadvantage is that the programming often tends to be not so user-friendly. Hand-held programming consoles will normally contain enough memory to allow the unit to retain programs while being carried from one place to another. Only when the program is ready is it transferred to the PLC.

8.6.3 Internal architecture of a PLC

Figure 8.21 shows the basic internal architecture of a PLC. It consists of a central processing unit (CPU) containing the system microprocessor, memory, and input/output circuitry. The CPU controls and processes all the operations within the PLC. It is supplied with a clock with a frequency of typically between 1 and 8 MHz. This frequency determines the operating speed of the PLC and provides the timing and synchronisation for all elements in the system. The information within the PLC is carried by means of digital signals along buses. The CPU uses the *data bus* for sending data between the constituent elements, the *address bus* to send the addresses of locations for accessing stored data and the *control bus* for signals relating to internal control actions. The *system bus* is used for communications between the input/output ports and the input/output unit.

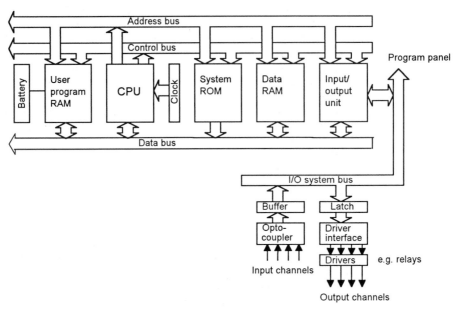

Figure 8.21 *Architecture of a PLC*

There are several memory elements:

1 System *read-only-memory* (ROM) to give permanent storage for the operating system and fixed data used by the CPU.

2 *Random-access memory* (RAM) for the user's program.

3 *Random-access memory* (RAM) for data. This is where information is stored on the status of input and output devices and the values of timers and counters and other internal devices. The data RAM is sometimes referred to as a *data table* or *register table*. Part of this memory, i.e. a block of addresses, will be set aside for input and output addresses and the states of those inputs and outputs. Part will be set aside for preset data and part for storing counter values, timer values, etc.

4 Possibly, as a bolt-on extra module, *erasable and programmable read-only memory* (EPROM) for ROMS that can be programmed and then the program made permanent.

The programs and data in RAM can be changed by the user. All PLCs will have some amount of RAM to store programs that have been developed by the user and program data. However, to prevent the loss of programs when the power supply is switched off, a battery is used in the PLC to maintain the RAM contents for a period of time. After a program has been developed in RAM it may be loaded into an EPROM memory chip, often a bolt-on module to the PLC, and so made permanent. In addition, there are temporary *buffer* stores for the input/output channels.

The processor memory consists of a large number of locations which can store words. So that each word can be located, every memory location is given a unique *address*. Just like houses in a town are each given a distinct address so that they can be located, so each word location is given an address so that data stored at a particular location can be accessed by the CPU either to read data located there or put, i.e. write, data there. It is the address bus which carries the information indicating which address is to be accessed. If the address bus consists of only 8 lines, then the number of 8-bit words, and hence number of distinct addresses, is $2^8 = 256$. With 16 address lines, 65 536 addresses are possible.

The input/output unit provides the interface between the system and the outside world, allowing for connections to be made through input/output channels to input devices such as sensors and output devices such as motors and solenoids. It is also through the input/output unit that programs are entered from a program panel. Every input/output point has a unique address which can be used by the CPU.

The input/output channels provide signal conditioning and isolation functions so that sensors and actuators can often be directly connected to them without the need for other circuitry. Electrical isolation from the external world is usually by means of *optoisolators* (the term *optocoupler* is also often used). The digital signal that is generally compatible with the microprocessor in the PLC is 5 V d.c. However, signal conditioning in the input channel, with isolation, enables a wide range of input signals to be supplied to it. A range of inputs might be available with a larger PLC, e.g. 5 V, 24 V, 110 V and 240 V digital/discrete, i.e. on-off, signals. A small PLC is likely to have just one form of input, e.g. 24 V. Figure 8.22 shows the basic form a d.c. input channel might take.

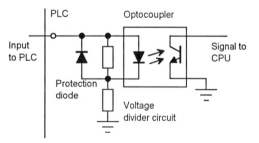

Figure 8.22 *Basic d.c. input circuit*

Outputs are often specified as being of relay type, transistor type or triac type. With the *relay type*, the signal from the PLC output is used to operate a relay and so is able to switch currents of the order of a few amperes in an external circuit. The relay not only allows small currents to switch much

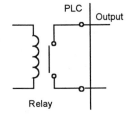

Figure 8.23 *Relay output*

larger currents but also isolates the PLC from the external circuit. Relays are, however, relatively slow to operate. Relay outputs are suitable for a.c. and d.c. switching. They can withstand high surge currents and voltage transients. Figure 8.23 shows the basic feature of a relay output. The *transistor type* of output uses a transistor to switch current through the external circuit. This gives a considerably faster switching action. It is, however, strictly for d.c. switching and is destroyed by overcurrent and high reverse voltage. As a protection, either a fuse or built-in electronic protection is used. Optoisolators are used to provide isolation. Figure 8.24 shows the basic form of such a transistor output channel. Triac outputs, with optoisolators for isolation, can be used to control external loads which are connected to the a.c. power supply. It is strictly for a.c. operation and is very easily destroyed by overcurrent. Fuses are virtually always included to protect such outputs.

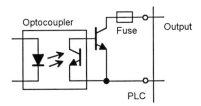

Figure 8.24 *Basic form of transistor output*

8.6.4 Programming

As an introduction to ladder diagrams, consider the simple wiring diagram for an electrical circuit in Figure 8.25(a). The diagram shows the circuit for switching on or off an electric motor. We can redraw this diagram in a different way, using two vertical lines to represent the input power rails and stringing the rest of the circuit between them. Figure 8.25(b) shows the result. Both circuits have the switch in series with the motor and supplied with electrical power when the switch is closed. The circuit shown in Figure 8.25(b) is termed a *ladder diagram*.

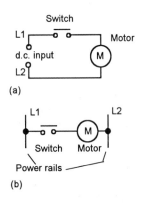

Figure 8.25 *Ways of drawing an electrical circuit*

With such a diagram the power supply for the circuits is always shown as two vertical lines with the rest of the circuit as horizontal lines. The power lines, or rails as they are often termed, are like the vertical sides of a ladder with the horizontal circuit lines like the rungs of the ladder. The horizontal rungs show only the control portion of the circuit, in the case of Figure 8.25 it is just the switch in series with the motor. Circuit diagrams often show the relative physical location of the circuit components and how they are actually wired. With ladder diagrams no attempt is made to show the actual physical locations and the emphasis is on clearly showing how the control is exercised.

A very commonly used method of programming PLCs is based on the use of *ladder diagrams*. Writing a program is then equivalent to drawing a switching circuit. The ladder diagram consists of two vertical lines representing the power rails. Circuits are connected as horizontal lines, i.e. the rungs of the ladder, between these two verticals. In drawing a ladder diagram, certain conventions are adopted:

1   The vertical lines of the diagram represent the power rails between which circuits are connected.

2   Each rung on the ladder defines one operation in the control process.

3   A ladder diagram is read from left to right and from top to bottom, Figure 8.26 showing the scanning motion employed by the PLC. The top rung is read from left to right. Then the second rung down is read from left to right and so on. When the PLC is in its run mode, it goes through the entire ladder program to the end, the end rung of the program being clearly denoted, and then promptly resumes at the start. This procedure of going through all the rungs of the program is termed a *cycle*.

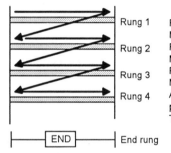

Figure 8.26 *Scanning the ladder program*

4   Each rung must start with an input or inputs and must end with at least one output. The term input is used for a control action, such as closing the contacts of a switch, used as an input to the PLC. The term output is used for a device connected to the output of a PLC, e.g. a motor.

5   Electrical devices are shown in their normal condition. Thus a switch which is normally open until some object closes it is shown as open on the ladder diagram. A switch that is normally closed is shown closed.

6   A particular device can appear in more than one rung of a ladder. For example, we might have a relay which switches on one or more devices. The same letters and/or numbers are used to label the device in each situation.

7   The inputs and outputs are all identified by their addresses, the notation used depending on the PLC manufacturer. This is the address of the input or output in the memory of the PLC. The Mitsubishi F series of PLCs precedes input elements by an X and output elements by a Y and uses the following numbers:

Inputs   X400–407, 410–413, 500–507, 510–513
              (24 possible inputs)

Outputs   Y430–437, 530–537
              (16 possible outputs)

Toshiba also uses an X and Y with inputs such as X000 and X001, outputs Y000 and Y001. Siemens precedes input numbers by I and outputs by Q, e.g. I0.1 and Q2.0. Sprecher+Schuh precedes input numbers by X and output numbers by Y, e.g. X001 and Y001. Allen Bradley uses I and O, e.g. I:21/01 and O:22/01. See Section 3.5 for a discussion of such addresses.

Figure 8.27 shows some of the standard symbols that are used for input and output devices.

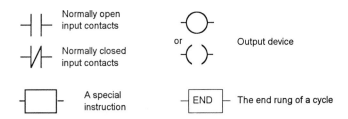

Figure 8.27 *Basic Standard symbols*

To illustrate the drawing of the rung of a ladder diagram, consider a situation where the energising of an output device, e.g. a motor, depends on a normally open start switch being activated by being closed. The input is thus the switch and the output the motor. Figure 8.28 shows the ladder diagram, a number of different address notations being shown.

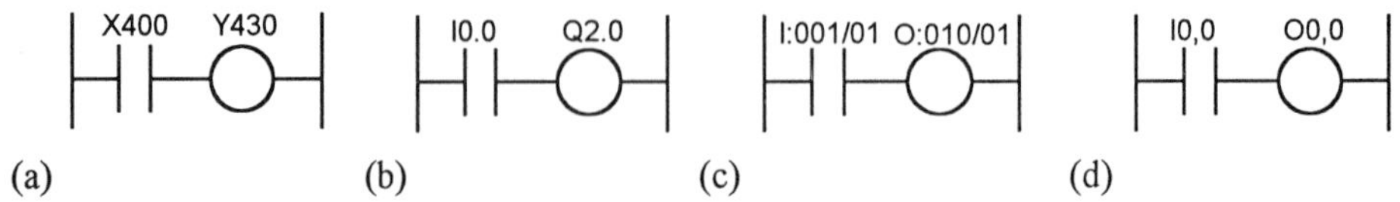

    X400   Y430   I0.0   Q2.0   I:001/01 O:010/01   I0,0   O0,0

      (a)   (b)   (c)   (d)

Figure 8.28 *Notation: (a) Mitsubishi, (b) Siemens, (c) Allen Bradley, (d) Telemecanique*

Starting with the input, we have the normally open symbol ‖ for the input contacts. There are no other input devices and the line terminates with the output, denoted by the symbol O. When the switch is closed, i.e. there is an input, the output of the motor is activated.

Figure 8.29(a) shows an AND gate system on a ladder diagram. The ladder diagram starts with ‖, a normally open set of contacts labelled input X400, and in series with it ‖, another normally open set of contacts labelled input X401. The line then terminates with O to represent the output, labelled Y430. For there to be an output from Y430, both input X400 and input X401 have to occur. Figures 8.29(b) and (c) show the same program with the address notations of other manufacturers.

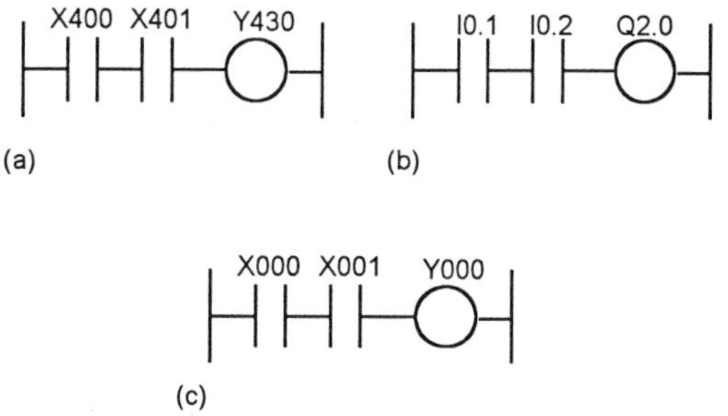

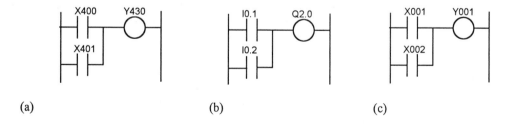

(a)   (b)

(c)

Figure 8.29 *AND gate: (a) Mitsubishi, (b) Siemens, (c) Toshiba notations*

Figure 8.30(a) shows an OR logic gate system on a ladder diagram. The ladder diagram starts with ‖, normally open contacts labelled input X400 and in parallel with it ‖, normally open contacts labelled input X401. Either input X400 <u>or</u> input X401 have to be activated for the output to be energised. The line then terminates with O to represent the output Y430. Figures 8.30(b) and (c) show how such a gate could appear with Siemens and Sprecher+Schuh notations.

(a)   (b)   (c)

Figure 8.30 *OR gate: (a) Mitsubishi, (b) Siemens, (c) Sprecher+Schuh notations*

Figure 8.31(a) shows a system where the input X400 contacts are shown as being normally closed. This is in series with the output O. With no input to input X400, the contacts are closed and so there is an output from Y430. When there is an input to input X400, it opens and there is then no output. This system can be termed a NOT gate.

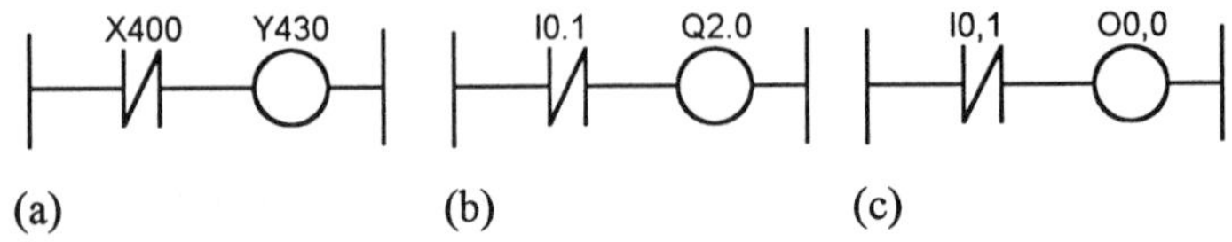

(a)　　　　　　　　　(b)　　　　　　　　　(c)

Figure 8.31 *Program with normally closed input contacts: (a) Mitsubishi, (b) Siemens, (c) Telemecanique notations*

Figure 8.32(a) shows a ladder diagram which gives a NAND gate. Both sets of input contacts are normally closed. When there are no inputs to input X400 and input X401 then there is an output from Y430. When there are inputs to input X400 and input X401, or to just one of them, then there is an output from Y430.

(a)　　　　　　　　　　　　　(b)

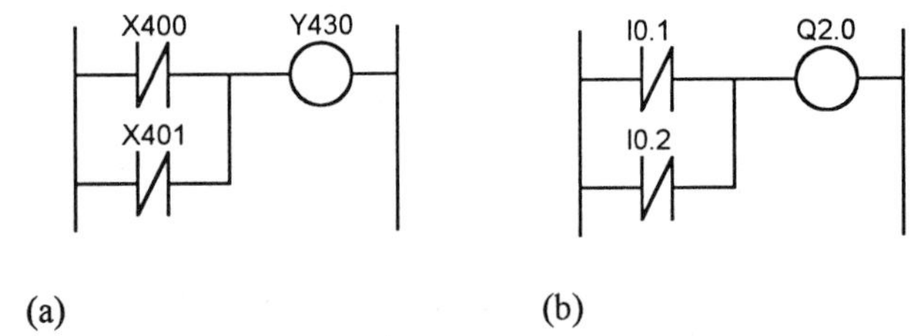

Figure 8.32 *NAND gate: (a) Mitsubishi, (b) Siemens notations*

Figure 8.33 shows a ladder diagram of a NOR system. When input X400 and input X401 are both not activated, there is an output from Y430. When either X400 or X401 are activated there is no output.

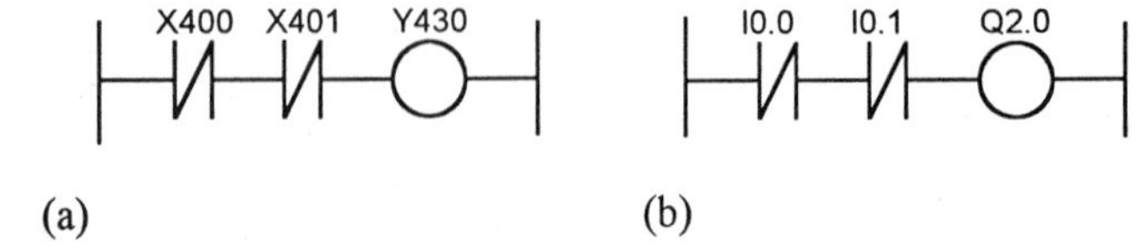

(a)　　　　　　　　　　　　　(b)

Figure 8.33 *NOR gate: (a) Mitsubishi, (b) Siemens notations*

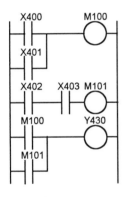

Figure 8.34 *XOR gate: (a) Mitsubishi, (b) Siemens, (c) Toshiba*

Figure 8.34 shows a ladder diagram for an XOR gate system. When input X400 and input X401 are not activated then there is no output from Y430. When just input X400 is activated, then the upper branch results in an output. When just input X401 is activated, then the lower branch results in an output. When both inputs X400 and X401 are activated, there is no output. In this example of a logic gate, input X400 and input X401 have two sets of contacts in the circuits, one set being normally open and the other normally closed. With PLC programming, each input may have as many sets of contacts as necessary.

The above are just some examples of ladder programs involving logic gates with inputs that are either on or off. However, there are many control tasks which can involve time delays and event counting. These requirements are met by internal relays, times and counters which are supplied as a feature of PLCs.

Consider the use of an internal relay, sometimes termed auxiliary relay or marker. This behaves just like a relay with a set of associated contacts, but in reality is not an actual relay but a simulation by the software of the PLC. Figure 8.35 shows an example, M100 and M101 being the addresses of two internal relays with their output coils O and associated contacts ||. The first rung of the program shows an input arrangement involving X400 and X401 being used to control the output of the internal relay. The second rung shows the other input arrangement involving X402 and X403 controlling the output of internal relay M101. The third rung shows the output Y430 controlled by the two internal relays. The result is that when one set of input conditions or the other is met, there is an output.

Consider the use of a timer. There may be a situation where an output has to occur some time after a sensor detects an event occurring or an output, such as a motor running, being activated for a specified amount of time after an input occurs from some sensor. Figure 8.36 shows a simple program involving a timer. In the timer format shown in the program, being that used by Mitsubishi, the timer is considered as an output which gives a delayed time reaction to contacts. Thus when input X400 is energised the timer T450 starts. Suppose it has been preset for a time of 5 s. After 5 s its contacts T450 are activated and close, so starting output Y430 after a delay of 5 s from the input to X400. By using a number of such timers a sequence of events can be made to occur after some initial input.

Figure 8.35 *Program with internal relays*

igure 8.36 *Simple timer program*

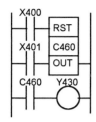

Figure 8.37 *Counter program*

A counter is set to some preset value and when this value of input pulses has been received, it operates its contacts. Figure 8.37 shows a basic counting program. When input X400 occurs the counter is reset. Impulses to input X401 are counted and when the count reaches the set value the contacts for the counter close. Thus for the counter C460 with a set value of 10, following the input to X400 to reset the counter, after 10 pulses to input X401 the contacts C460 close and there is an output from Y430. This arrangement might be for the counting of pulses of reflected light from items moving along a conveyor belt, with each block of ten items being diverted into a carton.

With a program, each horizontal rung on the ladder represents an instruction in the program to be used by the PLC. The entire ladder gives the complete program. There are several methods that can be used for keying in the program into a programming terminal. Whatever method is used to enter the program into a programming terminal or computer, the output to the memory of the PLC has to be in a form that can be handled by the microprocessor. This is termed *machine language* and is just binary code, e.g. 0010100001110001. One method of entering the program into the programming terminal involves using keys with symbols depicting the various elements of the ladder diagram and keying them in so that the ladder diagram appears on the screen of the programming terminal. Where a computer is involved it might mean pressing particular keys such as F1, F2, F3, etc. The terminal or computer then translates this into machine language.

8.6.5 Application example

As an illustration of the way a PLC might be used, consider the task of using a PLC as an on-off controller for a heater in the control of temperature in some enclosure. The heater is to be switched on when the temperature falls below the required temperature and switched off when the temperature is at or above the required temperature. The sensor used for the temperature might be a thermocouple or a thermistor. When connected in an appropriate circuit, the sensor will give a suitable voltage signal related to the temperature. This voltage can be compared, using an operational amplifier, with the voltage set for the required temperature with the result that a high output signal is given when the temperature is above the required temperature and a low output signal when it is below. Thus when the temperature falls from above the required temperature to below it, the signal switches from a high to a low value. This transition can be used as the input to a PLC. The PLC can then be programmed to give an output when there is a low input and this output used to switch on the heater. Figure 8.38 shows the arrangement that might be used and a Mitsubishi ladder program. The input from the operational amplifier has been connected to the input port with the address X400. This input has contacts which are normally closed. When the input goes high, the contacts open. The output is taken from the output port with the address Y430. Thus there is an output when the input is low and no output when the input is high.

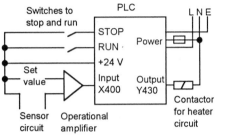

Figure 8.38 *Temperature control*

Now consider a more complex temperature control task involving a domestic central heating system (Figure 8.39). The central heating boiler is to be thermostatically controlled and supply hot water to the radiator system in the house and also to the hot water tank to provide hot water from the taps in the house. Pump motors have to be switched on to direct the hot water from the boiler to either, or both, of the radiator and hot water systems according to whether the temperature sensors for the room temperature and the hot water tank indicate that the radiators or tank need heating. The entire system is to be controlled by a clock so that it only operates for certain hours of the day. Figure 8.40 shows how a Mitsubishi PLC might be used.

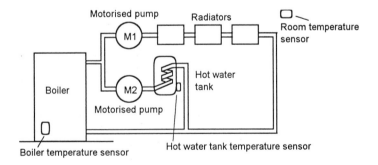

Figure 8.39 *Central heating system*

For further discussion of PLCs the reader is referred to more specialist texts such as *Programmable Logic Controllers* by W. Bolton (Butterworth-Heinemann 1996), *Programmable Controllers* by I.G. Warnock (Prentice-Hall 1988), *Programmable Controllers* by E.A. Parr (Newnes 1993).

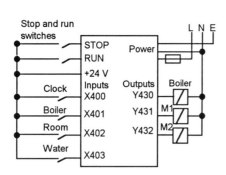
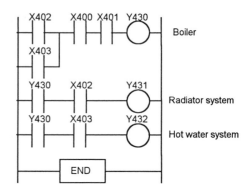

Figure 8.40 *Central heating system*

Problems

1 A data acquisition board has a 12-bit analogue-to-digital converter and is set for input signals in the range 0 to 10 V with the amplifier gain at 10. What is the resolution in volts?

2 A load cell has a sensitivity of 25 mV/kN and is connected to a digital acquisition board which has a 0 to 10 V, 12-bit analogue-to-digital converter. What amplifier gain should be used if the cell is to give an output for forces in the range 0.1 kN to 10 kN?

3 An analogue sensor gives a signal of nominally 1 V and is fed into a 12-bit analogue-to-digital converter system having an input impedance of 1 MΩ. If the output impedance of the analogue sensor is 600 Ω, will loading present any problems?

4 A data acquisition board has an analogue-to-digital converter with a sampling rate of 80 kS/s. What will be the maximum frequency of a analogue input with (a) one channel, (b) 5 channels?

5 Explain the basic principles of a data logger.

6 Draw a block diagram of a PLC showing the main functional items and how buses link them, explaining the function of each block.

7 State the characteristics of the relay, transistor and triac types of PLC output channels.

8 For the PLC ladder programs shown in Figure 8.41, state whether there is an output when (i) there is no input to X400 and (ii) when there is.

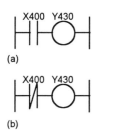

Figure 8.41 *Problem 8*

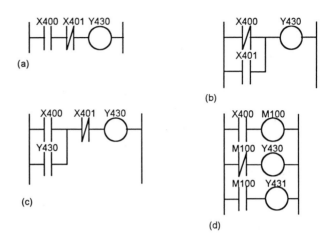

Figure 8.42 *Problem 9*

9 For the PLC ladder programs shown in Figure 8.42, state what input conditions have to be met in each case for there to be an output from Y430 or outputs from Y430 and Y431.

10 Design a PLC-controlled system which could be used to maintain the level of a liquid at a constant height in a container.

9 Data presentation

Elements which can be used to present data can be broadly classified as indicators/displays or recorders. *Indicators* and *displays* give just an instantaneous, but non-permanent, visual display of data while *recorders* record the data over a period of time and automatically give a permanent record. The data by either method can be presented in analogue or digital form. This chapter is a discussion of commonly used indicators/displays and recorders.

9.1 Indicators

The basic indicating devices are:

1 *Pointer moving against a scale*
   Figure 9.1(a) shows a number of basic forms where a pointer moves across a scale. The moving coil meter is an example of such an indicator.

2 *Scale moving against a fixed pointer*
   Figure 9.1(b) shows a number of basic forms of scales moving past fixed pointers.

3 *Bar graph*
   In the bar graph indicator a semitransparent ribbon moves over a scale, the top of the ribbon indicating the value (Figure 9.1(c)).

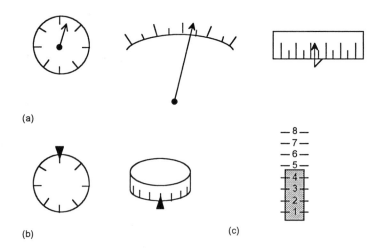

(a)

(b)   (c)

Figure 9.1 *Examples of indicators*

9.1.1 Moving coil meter

The *moving coil meter* is an analogue indicator in which a pointer moves across a scale with the amount of movement being related to the value of the input to the meter. The basic instrument movement is a micro ammeter with shunts, multipliers and rectifiers being used to convert it to other ranges and measurements. The instrument movement consists of a coil situated in a constant magnetic field which is always at right angles to the sides of the coil, no matter what angle the coil has rotated through (Figure 9.2). As a consequence of this, the force experienced by the sides of the coil is, for the same current, the same at all angular deflections of the coil from its initial position. For a coil carrying a current I with vertical sides of length L and horizontal sides of length b in a magnetic field of uniform flux density B which is always at right angles to the coil sides, the force acting on each turn of the coil's vertical sides is BIL. The forces acting on the vertical sides result in a turning moment, the force acting on the horizontal sides giving no such moment.

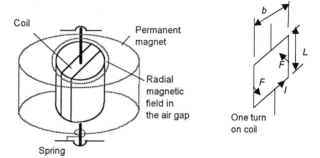

Figure 9.2 *Basic moving coil meter movement*

The turning moment, i.e. torque, about the central vertical axis of the coil for a single turn is:

$$\text{torque} = BIL \times \tfrac{1}{2}b + BIL \times \tfrac{1}{2}b = BIL$$

Since Lb is the area A of the face of the coil:

$$\text{torque} = BIA$$

For a coil with N turns:

$$\text{torque} = NBIA$$

For a particular meter N, B and A will be constant. This torque causes the coil to rotate against springs. The opposing torque developed by the springs is proportional to the angular deflection θ and so:

torque due to springs = $K_s\theta$

where K_s is a constant for the springs. The coil thus rotates until the torque developed by the current through the coil is balanced by that produced by the springs. Then:

$K_s\theta = NBIA$

The angular deflection of the pointer, which moves with the coil, is thus proportional to the current through the coil.

Moving coil meters generally have resistances of the order of a hundred ohms. The accuracy is generally about ±0.1% to ±5%, this being determined by such factors as the way the meter is mounted, bearing friction, the presence of nearby magnetic fields or ferrous materials, inaccuracies in the marking of the scales, and reading errors due to parallax. The meter takes time to respond to changes in the current, typically taking of the order of a few seconds to reach the steady-state deflection.

9.1.2 Dynamic response of the moving coil movement

When the current through a moving coil movement changes then the coil experiences a changing torque due to this current. If the current at some instant of time has risen to i, then:

torque due to current = $NBAi = K_c i$

where $K_c = NBA$ and is a constant for a particular movement. This torque starts the coil rotating and an opposing torque is generated by the springs, or if the coil is suspended from a wire or a ribbon by the twisting of the suspension. The opposing torque is $K_s\theta$, where θ is the angle through which the coil has rotated. The net torque acting on the coil is thus:

net torque = $K_c i - K_s\theta$

This net torque gives rise to an angular acceleration. Angular velocity is the rate at which the angle changes and so is $d\theta/dt$ and angular acceleration is the rate at which angular velocity changes and so is $d^2\theta/dt^2$. Since the net torque is equal to the product of the moment of inertia J of the coil and its angular acceleration, then:

$$K_c i - K_s\theta = J\frac{d^2\theta}{dt^2}$$

We can consider the meter coil as being supplied by the current from a voltage source in series with resistance, i.e. the Thévenin equivalent circuit shown in Figure 9.3(a). This, however, neglects the fact that the coil is rotating in a magnetic field and so there is an induced e.m.f. produced in the coil. The circuit is thus of the form shown in Figure 9.3(b).

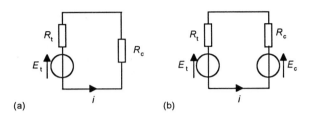

Figure 9.3 *Equivalent circuit*

The induced e.m.f. produced by a conductor of length L carrying a current I and moving at right angles through a magnetic field with a velocity v is given by $e = Blv$. Thus for the coil, where we have a vertical turn of the coil moving in a circular path, we have $v = r\omega$, where ω is the angular velocity and r the radius of the path, and so the induced e.m.f. in one vertical wire is $Blr\omega$. The radius of the path is $b/2$ and since there are two vertical sides and N turns:

$$E_c = Nblb\omega = NBA\frac{\mathrm{d}\theta}{\mathrm{d}t} = K_c\frac{\mathrm{d}\theta}{\mathrm{d}t}$$

Thus the net voltage in the equivalent circuit is:

$$\text{net voltage} = E_t - K_c\frac{\mathrm{d}\theta}{\mathrm{d}t}$$

Hence the current i in the circuit is:

$$i = \frac{E_t - K_c\dfrac{\mathrm{d}\theta}{\mathrm{d}t}}{R_t + R_c}$$

and we can write:

$$K_c i - K_s\theta = J\frac{\mathrm{d}^2\theta}{\mathrm{d}t^2}$$

$$K_c\frac{E_t - K_c\dfrac{\mathrm{d}\theta}{\mathrm{d}t}}{R_t + R_c} - K_s\theta = J\frac{\mathrm{d}^2\theta}{\mathrm{d}t^2}$$

This can be rearranged to give:

$$\frac{\mathrm{d}^2\theta}{\mathrm{d}t^2} + \frac{K_c^2}{J(R_t + R_c)}\frac{\mathrm{d}\theta}{\mathrm{d}t} + \frac{K_s}{J}\theta = \frac{K_c}{J(R_t + R_c)}E_t$$

This second order differential equation describes the motion of the meter coil when subject to a change in current (see Section 3.3 for an analysis of a similar differential equation and Appendix B for a general discussion of differential equations). The system has a natural angular frequency ω_n of:

$$\omega_n = \sqrt{\frac{K_s}{J}}$$

and a damping factor of:

$$\zeta = \frac{K_c^2}{2\sqrt{K_sJ}\,(R_t + R_c)}$$

At damping factors less than 1, i.e. the critical damping condition, the deflection overshoots the steady-state value before finally settling down. A damping factor of 0.91 gives a percentage overshoot of 0.1; 0.82 a percentage overshoot of 1.2; 0.72 a percentage overshoot of 4.0; 0.62 a percentage overshoot of 8.4; and 0.5 a percentage overshoot of 16.5. With damping factors greater than 1, the coil just takes a long time to reach the steady-state value. Moving coil meters typically have a damping factor of about 0.7, this giving a reasonably fast response without too great an overshoot. The natural frequency of a meter coil is typically about 0.5 to 1.5 Hz, this being low because of the relatively high moment of inertia of the coil and pointer.

The steady-state value occurs when $d\theta/dt$ and $d^2\theta/dt^2$ equal 0 and so we then have:

$$\frac{K_s}{J}\theta = \frac{K_c}{J(R_t + R_c)}E_t$$

Hence:

$$\text{steady-state value of } \theta = \frac{K_c}{K_s(R_t + R_c)}E_t$$

The steady-state voltage sensitivity θ/E_t is thus:

$$\text{steady-state voltage sensitivity} = \frac{K_c}{K_s(R_t + R_c)}$$

The steady-state voltage sensitivity can thus be changed by adding resistors in series or parallel with the input circuit.

See Section 9.3 for a discussion of the dynamic response of the moving coil meter movement in relation to the galvanometric recorder.

9.2 Illuminative displays

The term *alphanumeric* is a contraction of the terms alphabetic and numeric and describes displays of the letters of the alphabet and numbers 0 to 9 with decimal points. Illuminative displays are frequently alphanumeric displays. A commonly used format for such displays involves the use of seven segments to generate the various alphabetic and numeric characters. Figure 9.4 shows the segments and Table 9.1 shows how a 4-bit binary code input can be used to generate the inputs to switch on the various segments. Another format involves a 7 by 5 or 9 by 7 dot matrix (Figure 9.5). The characters are then generated by the excitation of appropriate dots.

Figure 9.4 *Seven-segment display*

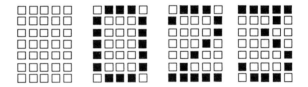

Figure 9.5 *7 by 5 dot matrix display*

Table 9.1 *Seven-segment display*

Binary input				Segments activated							Number displayed
				a	b	c	d	e	f	g	
0	0	0	0	1	1	1	1	1	1	0	0
0	0	0	1	0	1	1	0	0	0	0	1
0	0	1	0	1	1	0	1	1	0	1	2
0	0	1	0	1	1	1	1	0	0	1	3
0	1	0	0	0	1	1	0	0	1	1	4
0	1	0	1	1	0	1	1	0	1	1	5
0	1	1	0	0	0	1	1	1	1	1	6
0	1	1	1	1	1	1	0	0	0	0	7
1	0	0	0	1	1	1	1	1	1	1	8
1	0	0	1	1	1	1	0	0	1	1	9

Commonly used illuminative displays include:

1 *Light emitting diodes (LEDs)*

2 *Liquid crystal displays (LCDs)*

3 *Cathode ray tube displays*

The following sections give more details of these forms of display.

9.2.1 Light emitting diodes

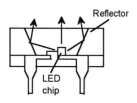

Figure 9.6 *LED*

These diodes when forward biased emit light over a certain band of wavelengths. Figure 9.6 shows the basic form of an LED, the light emitted from the diode being enhanced in one direction by means of reflectors. Commonly used LED materials are gallium arsenide, gallium phosphide and alloys of gallium arsenide with gallium phosphide. Gallium phosphide when doped with nitrogen emits green light with light doping and yellow light with heavy doping. A 60% gallium arsenide–40% gallium phosphide alloy gives red light, a 35%–65% alloy orange light and a 15%–85% alloy yellow light.

Typical maximum currents for LEDs are in the range 10 to 30 mA. A current limiting resistor is generally required in order to limit the current for each LED to below the maximum rated current. A 'turn-on' voltage has to be reached before there is any significant current and so illumination. Figure 9.7 shows how LEDs might be connected to a driver so that when a line is driven low, the turn-on voltage is applied and the LED in that line is switched on. Such an arrangement is known as the *common anode* form of connection since all the LED anodes are connected together. An alternative arrangement is a *common cathode*.

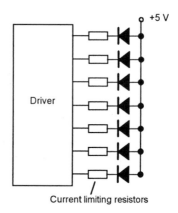

Figure 9.7 *LEDs with common anode connection*

LEDs are available as single light displays, seven and sixteen segment alphanumeric displays, in dot matrix format and bar graph form. In Figure 9.7 the seven LEDs could be the seven segments of a display. Figure 9.8 shows the basic form used for a 5 by 7 dot matrix display. The array consists of five column connectors, each connecting the anodes of seven LEDs. Each row connects to the cathodes of five LEDs. To turn on a particular LED, power is applied to its column and its row is grounded.

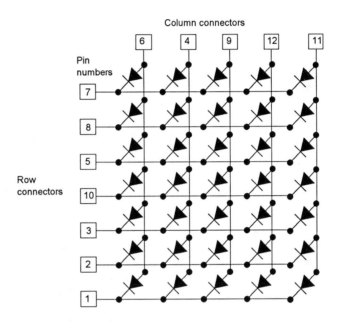

Figure 9.8 *Dot matrix display*

9.2.2 Liquid crystal displays

Unlike light emitting diodes which emit light, liquid crystal displays emit no light of their own but rely upon affecting reflected or transmitted light. In general, liquids have their molecules completely randomly arranged. However, some liquids have rod-shaped molecules which in the liquid state are able to take up defined orientations with respect to each other. Such an orderly packing is characteristic of crystals and hence the term *liquid crystal* is used. If the liquid crystal is sandwiched between sheets of polariser and illuminated, then the first polariser only permits light in one plane of polarisation to be transmitted. Suppose we make this direction such that the molecules are aligned in a direction at 90° to it. The plane of polarisation is then rotated by the molecules so that at the other side of the crystal it is at 90° to its initial direction. If the second polariser is in this direction then the light is transmitted through it. If now an electric field is applied to a liquid crystal then, provided it is above some critical value, the rod-shaped molecules align themselves with the field. If this alignment direction is at 90° to their initial direction, then the effect of such alignment is to cause the plane polarised light not to have its plane of polarisation rotated. It is thus not transmitted through the second polariser but reflected as though at a mirror (Figure 9.9). The result is that the electric field causes the light to be reflected from the liquid crystal, whereas the absence of the field causes it to be transmitted. By using a seven segment form of display, the electric field applied to the appropriate segments can result in them reflecting light and so appearing bright while the other segments which transmit light appear dark. This is the form of display used in battery-operated devices such as watches and calculators. 5 by 7 dot matrix forms are also available.

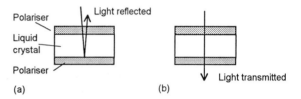

Figure 9.9 *Electric field: (a) applied, (b) not applied*

9.2.3 Cathode ray tube displays

Cathode ray tubes are used in oscilloscopes and visual display units. The *cathode ray tube* (Figure 9.10) consists of an electron gun which produces a focused beam of electrons and a deflection system. In the gun, electrons are produced by heating the cathode, the number of electrons which form the electron beam and determine the brilliance being determined by the potential applied to the modulator. The electrons are accelerated down the tube by the potential difference between the cathode and the anode, and focused into a beam by adjustment of the potentials applied to the focusing plates. The beam is focused so that when it reaches the phosphor-coated screen it forms a luminous spot.

The screen is coated with a phosphor which glows on impact by electrons. The light produced by the phosphor takes a little time to build up and time to decay. The time taken for the light to fall to some specified values of its initial value is termed the *decay* time or persistence. A medium decay phosphor (P1) has a decay time to 0.1% of 95 ms. A medium/short decay phosphor (P31), which is widely used, has a decay time to 0.1% of 32 ms. Colour displays have a screen involving groups of three phosphor dots, one emitting a red colour, one green and one blue.

The beam of electrons in the cathode ray tube is deflected in the Y direction by a potential difference applied between the Y-deflection plates and in the X direction by a potential difference between the X-deflection plates.

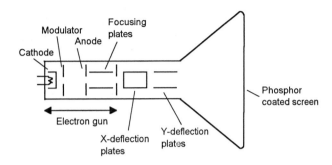

Figure 9.10 *The basic form of the cathode ray tube*

The function of the Y-deflection plates is to produce deflections of the electron beam in the vertical direction. With a cathode ray oscilloscope, a switched attenuator and amplifier enable different deflection factors to be obtained. A general purpose oscilloscope is likely to have deflection factors which vary between 5 mV per scale division to 20 V per scale division. In order that a.c. components can be viewed in the presence of high d.c. voltages, a blocking capacitor can be switched into the input line. When the amplifier is in its a.c. mode its bandwidth typically extends from about 2 Hz to 10 MHz and when in the d.c. mode from d.c. to 10 MHz.

The Y-input impedance is typically about 1 MΩ shunted with about 20 pF capacitance. When an external circuit is connected to the Y input, problems due to loading and interference can distort the input signal. Interference can be reduced by the use of coaxial cable. However, the capacitance of the coaxial cable and any probe attached to it can be enough, particularly at low frequencies, to introduce a relatively low impedance across the input impedance of the oscilloscope and so introduce significant loading. A number of probes exist for connection to the input cable which are designed to increase the input impedance and so avoid this loading problem. A passive voltage probe that is often used is a 10 to 1 attenuator (Figure 9.11) in the form of a 9 MΩ resistor and variable capacitor in the probe tip. This not only reduces the capacitive loading but also the voltage sensitivity. An active voltage probe using an FET can overcome this problem of reduced voltage sensitivity.

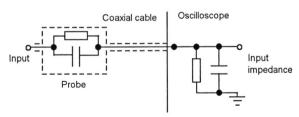

Figure 9.11 *Passive voltage probe*

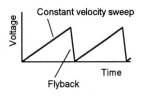

Figure 9.12 *Timebase signal*

The purpose of the X-deflection plates is to deflect the electron beam in the horizontal direction. With a cathode ray oscilloscope they are generally used with an internally generated signal which sweeps the beam from left to right at a constant rate with a very rapid return, i.e. flyback, which is too fast to leave a trace on the screen (Figure 9.12). The constant rate of sweeping across the screen means that the distance moved in the X direction is proportional to the time elapsed and so gives a horizontal time axis, i.e. a *time base*. A general purpose oscilloscope will have a range of time bases from about 1 s per scale division to 0.2 μs per scale division.

For a periodic input signal to give rise to a steady trace on the oscilloscope screen it is necessary to synchronise the time base and the input signal. This can be achieved using a *trigger circuit*. This circuit can be adjusted so that it responds to a particular voltage level and also as to whether the voltage is increasing or decreasing. It thus can be adjusted to respond to particular points on the input waveform and produce pulses which trigger the time base into action and start the sweep from left to right. Since the time base then always starts at the same point on the waveform, successive scans of the input signal are superimposed and so a stationary trace is produced.

Double beam oscilloscopes enable two separate inputs to be observed simultaneously on the screen. This can be achieved by having two independent electron gun assemblies and consequently two electrons beams, each beam with its own Y-deflection plates but a common set of X-deflection plates and so a common time base. Alternatively, a single electron gun can be used and the Y-deflection plates switched from one input to the other. If this switching is done each time the time base is triggered it is termed alternative mode; if more frequently it is termed chopping mode.

The upper frequency limit of the conventional oscilloscope is of the order of 100 MHz. However, higher frequencies can be achieved with a *sampling oscilloscope*. Such an instrument constructs a continuous display from samples taken at different points during a number of cycles of the waveform. The sampling frequency may be one-hundredth of the input frequency and so the upper frequency limit can be extended by a factor of one hundred.

The trace on the conventional cathode ray oscilloscope quickly dies away and does not persist on the screen when the input is removed. With *storage oscilloscopes*, the trace produced remains on the screen after the input has ceased and requires a deliberate action of erasure to remove it. There are a number of techniques used to achieve this. With the *bistable storage tube* (Figure 9.13) there are three electron guns. Two of the guns, called flood electron guns, are on all the time and flood the screen with low velocity electrons. The screen consists of phosphor particles on a dielectric sheet, backed by a conducting layer. The low velocity flood electrons do not have enough energy to cause the phosphor to glow but do cause the phosphor particles to become negatively charged. The writing gun emits high velocity electrons, these having sufficient energy to overcome the repulsive forces due to the negatively charged phosphor particles and eject electrons from them. These ejected electrons are gathered and conducted away by the conducting layer. The result of the phosphor particles loosing electrons is that they become positively charged and emit light. This positive charge and light emission persists, even when the writing gun is no longer emitting electrons. This is because it is still being bombarded by flood electrons and the positive charge means that they acquire sufficient energy to cause the emission of further electrons and light.

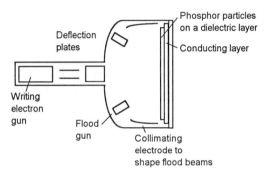

Figure 9.13 *Bistable storage cathode ray tube*

Digital storage oscilloscopes convert analogue input signals into digital ones, the digital signals then being stored in memory. Figure 9.14 shows the systems used in a basic digital oscilloscope. The Y input signal is sampled and the samples then converted to digital form by means of an analogue-to-digital converter. The digital signals can then be stored in the memory. In this form the signals can be analysed and then, using a digital-to-analogue converter, converted back to analogue form for input to the Y-deflection plates of a cathode ray tube. Because of the memory, the signal can be held indefinitely and provide outputs not only for a cathode ray tube but also to computer systems or x-y plotters.

For further reading about cathode ray oscilloscopes see *Oscilloscopes* by I. Hickman (Butterworth-Heinemann, 3rd edn 1990).

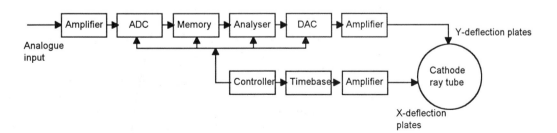

Figure 9.14 *Basic digital storage oscilloscope*

Visual display units (VDUs) are basically just a form of cathode ray tube which is used to display alphanumeric, graphic and pictorial data. With the *raster form* of display, saw-tooth signals are applied to both the X- and the Y-deflection plates. Figure 9.15 illustrates the principle. The Y signal causes the beam to move at a constant rate from top to bottom of the screen before flying back to the top again. The X signal causes the beam to move at a constant rate from left to right of the screen before flying back to the left again. The consequence of both these signals is that the beam pursues a zigzag path down the screen before flying back to the top left corner and then resuming its zigzag path down the screen. During its travel down the screen, the electron beam is switched on or off, with the result that a picture or character can be 'painted' on the screen. The standard monochrome VDU has a 312-line raster display.

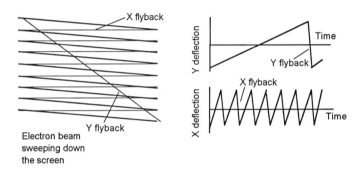

Figure 9.15 *Raster form of display*

Figure 9.16 *An interlaced display*

The raster form of display illustrated in Figure 9.15 is said to be *non-interlaced*, the electron beam just following a single zigzag path down the screen. An *interlaced* display has two beams following a zigzag pattern down the screen (Figure 9.16).

The screen of the visual display unit is coated with a large number of phosphor dots, these dots forming the *pixels*. The term pixel is used for the smallest addressable dot on a display device. Character generation is by the selective illuminations of these pixels. Thus for a 7 by 5 matrix, Figure 9.17 shows how characters are built up by the electron beam moving in its zigzag path down the screen.

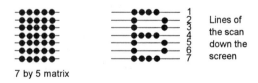

Figure 9.17 *Character build-up by pixels*

The input data to the VDU is usually in digital *ASCII (American Standard Code for Information Interchange)* format so that as the electron beam sweeps across the screen it is subject to on-off signals which end up 'painting' the characters on the screen. The ASCII code is a 7-bit code and so can be used to represent $2^7 = 128$ characters. This enables all the standard keyboard characters to be covered, as well as some control functions such as RETURN which is used to indicate the return from the end of a line to the start of the next line. Table 9.2 gives an abridged list of the code. ASCII is the format used in most computers.

Table 9.2 *ASCII code*

Character	ASCII	Character	ASCII	Character	ASCII
A	100 0001	N	100 1110	0	011 0000
B	100 0010	O	100 1111	1	011 0001
C	100 0011	P	101 0000	2	011 0010
D	100 0100	Q	101 0001	3	011 0011
E	100 0101	R	101 0010	4	011 0100
F	100 0110	S	101 0011	5	011 0101
G	100 0111	T	101 0100	6	011 0110
H	100 1000	U	101 0101	7	011 0111
I	100 1001	V	101 0110	8	011 1000
J	100 1010	W	101 0111	9	011 1001
K	100 1011	X	101 1000		
L	100 1100	Y	101 1001		
M	100 1101	Z	101 1010		

9.3 Graphical recorders

Graphical recorders are devices for the permanent recording on paper of data that is in a graphical form, i.e. presented as a relationship between two variables with one often being time. There are a number of forms of recorders and a number of different ways that data can be recorded on paper. The forms of recorder include:

1 *Direct reading recorders*
These are sometimes referred to as *circular chart recorders* because the paper chart used is circular.

2 *Closed-loop recorders*
These are sometimes referred to as *strip chart recorders* because the paper chart is in the form of a strip.

3 *Galvanometric recorders*
These are based on the use of the meter movement described earlier for the moving coil meter.

The methods used for recording data on paper include:

1 *Pen and ink*
These use disposable fibre-tipped pens.

2 *Impact printing*
This can involve the impact of the pointer against a carbon ribbon pressing against the paper and so marking it.

3 *Thermal writing*
This uses thermally sensitive paper which changes colour when subject to heat. A heated pointer is used to leave a mark on the paper.

4 *Optical writing*
This involves a beam of ultraviolet light falling on paper sensitive to it. The paper develops in daylight or artificial light without the need for special chemicals.

5 *Electrical writing*
The special paper used is coated with a layer of coloured dye and this in turn is coated with a very thin layer of aluminium. A tungsten wire stylus moves across the surface of this paper and when a potential of about 35 V is applied to it, the resulting electrical discharge removes the aluminium and exposes the dye.

9.3.1 Direct reading recorders

With the direct reading type of chart recorder, a pen or stylus is directly moved over a circular chart by the measurement system (Figure 9.18). With a pressure recorder, the displacement of the end of a Bourdon tube or bellows might be used. With a temperature recorder, the displacement of the end of a bimetallic strip might be used. A circular chart is used and rotates at a constant rate, usually one revolution in 12 hours, 24 hours or 7 days.

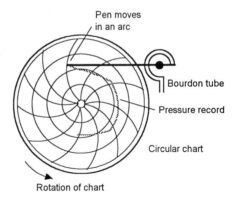

Figure 9.18 *Direct reading recorder*

The pressure or temperature gives displacements out along curved radial lines because the movement of the pen is in the arc of a circle. The chart rotates through equal angles in equal intervals of time. This means that the distance moved by the pen in equal intervals of time will depend on its distance out from the chart centre. These effects make interpolation difficult and there are particular problems in determining values for traces close to the centre of the chart where the radial lines are very close together. Simultaneous recording of up to four separate variables is possible with the use of four different pens. The accuracy is generally of the order of ±0.5% of the full-scale signal deflection.

9.3.2 Closed-loop recorders

A closed-loop recorder is one where there is a feedback loop. The recorder uses a self-balancing potentiometer (Figure 9.19). The output from the measurement system provides an input to a motor. When there is an input to the motor, the motor shaft rotates and moves the pen. The pen continues in motion as long as there is an input to the motor. However, the motion of the pen results in a slider moving along a potentiometer track. This produces a potential difference which is subtracted from the input voltage to the motor by an operational amplifier, the difference signal then being used to operate the motor. The pen thus ends up moving to a position where the result is no difference between the signal from the pen's position and the input to the recorder motor.

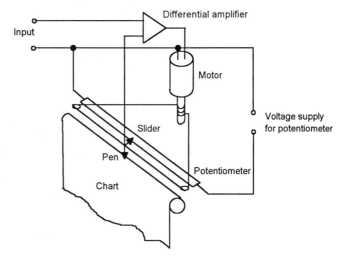

Figure 9.19 *Closed-loop recorder*

Potentiometer recorders are more robust than galvanometric recorders, can be multi-channel and have a linear relationship between deflection and signal input. Typically they have high input resistance, high accuracy (about ±0.1% of full-scale reading), but slow response to changing signals and so a bandwidth from d.c. to only about 1 or 2 Hz. Because of friction there is a minimum current required to get the motor operating, a so-called *dead band*, which is usually about ±0.3% of the range.

9.3.3 Galvanometric recorders

The basis of the galvanometric type of recorder is a moving coil meter movement with the coil hanging from a taut suspension and so the resisting torque provided by the twisting of the suspension. Figure 9.20 shows the basic elements of such a recorder when ultraviolet light is used to provide the beam which leaves a trace on ultraviolet sensitive paper. A number of galvanometric blocks, 6, 12 or 25, are generally mounted side by side in the one magnet so that simultaneous recordings can be made of a number of variables.

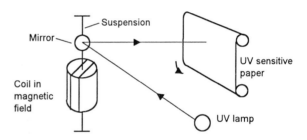

Figure 9.20 *UV galvanometric recorder*

The UV galvanometric recorder operates to the same principles as outlined earlier in this chapter for the meter movement. Thus the differential equation describing the dynamic response of a coil is:

$$\frac{d^2\theta}{dt^2} + \frac{K_c^2}{J(R_t + R_c)}\frac{d\theta}{dt} + \frac{K_s}{J}\theta = \frac{K_c}{J(R_t + R_c)}E_t$$

where θ is the angular deflection.

Figure 9.21(a) shows, in plan view, the initial position of the coil and the reflected beam and Figure 9.21(b) the position of the reflected beam when the coil deflects through an angle θ.

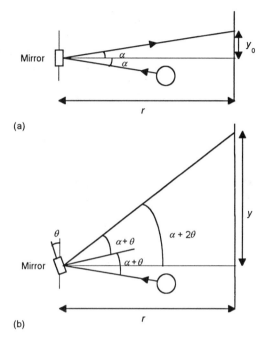

(a)

(b)

Figure 9.21 *(a) Initial position, (b) position when coil deflected*

The initial position y_0 of the spot of light on the paper (Figure 9.21(a)) is given by $y_0 = r \tan \alpha$, where α is the initial angle of incidence at the mirror, and also the angle of reflection, and r the distance of the mirror from the paper. When the coil deflects through an angle θ (Figure 9.21(b)), the angle of incidence and the angle of reflection both become $\alpha + \theta$. Thus we now have $y = r \tan(\alpha + 2\theta)$, the reflected beam having been deflected through twice the angle the mirror rotated through. Thus the deflection of the spot on the paper from its initial position is:

deflection $= y - y_0 = \tan(\alpha + 2\theta) - \tan \alpha$

There is thus a non-linear relationship between the deflection of the coil and the deflection of the point of impact of the beam of light on the paper. However, suppose we only consider small changes in θ.

$$\frac{dy}{d\theta} = 2r \sec^2(\alpha + 2\theta)$$

Thus, for finite changes, we can write for the change in deflection Δy resulting from small changes in deflection $\Delta \theta$, neglecting the 2θ term in relation to the larger α term:

$$\Delta y = 2r \sec^2 \alpha \, \Delta\theta$$

Thus the deflection for small angular changes is proportional to the angular change. Small angles mean angles less than about 10°. Since the steady state voltage sensitivity for angular deflections is (see earlier in this chapter for its derivation):

$$\frac{\Delta\theta}{E_t} = \frac{K_c}{K_s(R_t + R_c)}$$

then the voltage sensitivity of the recorder is:

$$\text{voltage sensitivity} = \frac{\Delta y}{E_t} = \frac{\Delta y}{\Delta\theta}\frac{\Delta\theta}{E_t} = \frac{2rK_c \sec^2\theta}{K_s(R_t + R_c)}$$

The sensitivity depends on the bandwidth required, i.e. the natural angular frequency. With a bandwidth of d.c. to about 50 Hz, the sensitivity is typically about 5 cm/mV, the coil having a resistance of about 80 Ω. A bandwidth up to 5 kHz would typically result in a sensitivity of about 0.0015 cm/mV and a coil resistance of about 40 Ω. The limiting frequency for this type of instrument is about 13 kHz. The accuracy is about ±2% of the full-scale deflection.

9.4 Printers

Printers provide a record of data on paper. The commonly encountered printers are:

1 *Dot matrix printer*
   This is an impact method using movable needles to form the dots needed to print alphanumeric characters on paper.

2 *Ink/bubble jet printer*
   This uses a jet of drops of ink to form alphanumeric characters on a sheet of paper.

3 *Laser printer*
   This uses electrostatic fields to transfer toner to a sheet of paper in the form of the required characters.

9.4.1 Dot matrix printer

The dot matrix printer is an impact printer using movable needles to create dots and so form the alphanumeric characters of the data. Figure 9.22 shows the basic form of one such needle. At rest the needle is held away from the inked ribbon and the sheet of paper by the return spring. When a current passes through the solenoid, it attracts the hammer which then forces the print needle into contact with the inked ribbon and presses it against the paper, leaving a mark. When the current to the solenoid ceases, the print needle is retracted under the action of the return spring. The overrun spring is there to prevent excessive movement of the needle when the hammer moves to force the needle against the inked ribbon. The number of dots used to make up a character is determined by the number of such print needle assemblies used on the print head. Typically it has a vertical row of nine, each of which can be fired separately and so form the characters as the print head tracks across the paper. A more expensive print head might have twenty-four.

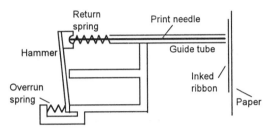

Figure 9.22 *Dot matrix print head mechanism*

9.4.2 Ink/bubble jet printer

The ink jet printer uses conductive ink being forced through a small nozzle to produce a jet of drops of ink. Very small, constant diameter, drops of ink are produced at a constant frequency and so regularly spaced in the jet. With one form a constant stream of ink passes along a tube and is pulsed to form fine drops by a piezoelectric crystal which vibrates at a frequency of about 100 kHz (Figure 9.23). Another form uses a small heater in the print head with vaporised ink in a capillary tube, so producing gas bubbles which push out drops of ink (Figure 9.24). In one form each drop of ink is given a charge as a result of passing through a charging electrode. The charged drops then pass between deflection plates which can deflect the stream of drops in a vertical direction, the amount of deflection depending on the charge on the drops. In another form, a vertical stack of nozzles is used and the jets of each just switched on or off on demand.

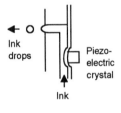

Figure 9.23 *Continuous flow printing*

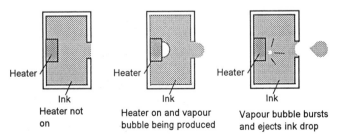

Figure 9.24 *Principle of the on-demand system*

Inkjet printers can give colour prints by the use of three different colour inkjet systems. The fineness of the drops is such that prints can be produced with 600 dots per inch.

9.4.3 Laser printer

The laser printer uses electrostatic fields to transfer toner to the paper. The basic assembly has a photosensitive drum which is coated with a selenium-based material that is light sensitive (Figure 9.25). In the dark the selenium has a high resistance and becomes charged as it passes close to the charging wire. This is a wire which is held at a high voltage and charge leaks off it. A light beam is made to scan along the length of the drum by a small eight-sided mirror which rotates and so reflects the light that it scans across the drum. When light strikes the selenium its resistance drops and it can no longer remain charged. By controlling the brightness of the beam of light, so points on the drum can be discharged or left charged. As the drum passes the toner reservoir, the charged areas attract particles of toner which then stick to its surface. Thus the result is a pattern of toner on the drum with toner on the areas that have not been exposed to light and no toner on the areas exposed to light. The paper is given a charge as it passes another charging wire, the so-called corona wire, with the result that as the paper passes close to the drum it attracts the toner off the drum. A hot fusing roller is then used to melt the toner particles so that, after passing between rollers, they firmly adhere to the paper. General-use laser printers are currently able to produce 600 dots per inch.

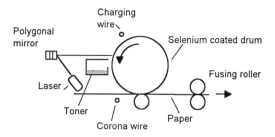

Figure 9.25 *Basic elements of a laser printer*

9.5 Magnetic recording

Magnetic recording involves data being stored in a thin layer of magnetic material as a sequence of regions of different magnetisation. The material used may be in the form of tape or disk, floppy or hard.

Figure 9.26 shows the basic elements of a head used to record data on magnetic tape. The head is made of a ferromagnetic material with a small air gap and forms a magnetic circuit. The material to be magnetised consists of a magnetic coating on a plastic base and this passes just under the air gap. With tapes and floppy disks, the recording head is in contact with the magnetic coating, whereas with a hard disk the head flies on an air cushion a very small distance above it. When a current is passed through a coil wound on the head, magnetic flux is produced in the head and the air gap. However, if the magnetic coating is present it offers a much lower reluctance path than the air gap. As a result, a significant amount of the magnetic flux that would have passed through the air gap is diverted to flow through the magnetic coating. Hence the magnetic coating becomes magnetised. As the tape or disk moves under the head, so the pattern of magnetisation produced in it will reflect the changes in the magnetising current.

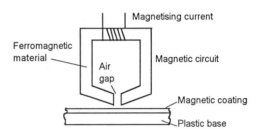

Figure 9.26 *Recording/replay head*

The replay head, often termed the reading head, is the same form as the recording head and often the same head is used for both purposes. When the magnetised magnetic tape or disk passes under the head, it completes the magnetic circuit and produces magnetic flux in the head. As a result of the magnetic flux changing as the magnetism on the coating changes, an e.m.f. is induced in the coil wrapped round the head. This e.m.f. is proportional to the rate of change of the magnetic flux in the circuit and thus a current in an external circuit can be produced which is related to the changing magnetism in the magnetic tape or disk.

9.5.1 Analogue recording

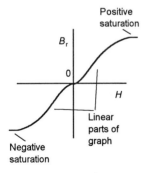

Figure 9.27 *Remanent magnetisation graph*

Ideally the remanent flux density B_r produced in the magnetic tape would be proportional to the magnetising field strength H and so the magnetising current. The remanent flux density is the flux density remaining in the tape when the magnetising current has ceased. However, in practice the relationship tends to be of the form shown in Figure 9.27. At high positive or high negative magnetising fields, the material becomes saturated and the magnetic flux assumes a constant value which does not change as the magnetising field is further increased. Near the zero magnetising field there is non-linearity, the relationship only being reasonably linear at magnetising fields in a region between the zero and the saturation values. This non-linearity means that a current input which fluctuated about the zero gives a distorted variation in remanent flux density in the tape (Figure 9.28(a)).

This distortion can be reduced by adding a steady d.c. current to the signal, i.e. a *d.c. bias*, to shift the signal to the linear part of the graph. An alternative is to add a high frequency a.c. current to the signal, i.e. an *a.c. bias*, to give an amplitude modulated signal (Figure 9.28(b)). The amplitude variation represents the signal and this occurs on the more linear parts of the graph. Biasing can thus produce a more linear response.

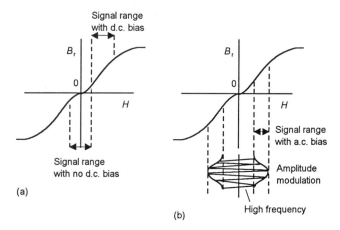

Figure 9.28 *Use of (a) d.c. bias, (b) a.c. bias*

The tape moves at a constant velocity under the recording head and so for a sinusoidal input of frequency f to the record head a sinusoidal variation in magnetisation is produced along the tape. The time interval for one cycle is $1/f$ and so for a tape moving with a constant velocity v, the distance along the tape taken by one cycle is v/f. This distance is called the *recorded wavelength*. The minimum size this recorded wavelength can have is the gap width in the recording head, since the average magnetic flux across the gap is then the average of one cycle and therefore zero. The upper limit to the frequency response of the recorder is thus set by the gap width and the tape velocity. Typically tape velocities range between about 23 mm/s to 1500 mm/s with a gap width of 5 μm, hence the upper frequency limit ranges from about 4.6 kHz to 300 kHz.

For the replay head, a sinusoidal variation of flux on the recording tape will produce a sinusoidal variation of flux ϕ in the core:

$$\phi = \phi_\mathrm{m} \sin \omega t$$

where ϕ_m is the maximum value of the flux in the recorded signal. The output from the recording head is proportional to the rate of change of flux in the head, i.e. $\mathrm{d}\phi/\mathrm{d}t$. Hence:

$$\text{head output} \propto \omega\phi_\mathrm{m} \cos \omega t$$

Thus the head output depends on both the recorded flux and also its frequency. To overcome this problem of the head output being dependent on the flux frequency, the output is amplified using an amplifier which has a gain that varies with frequency in such a way that the frequency effect is cancelled out. This process is called *amplitude equalisation*.

A consequence of the replay head output being proportional to the frequency is that for very low frequencies the output may be very small and become comparable with the noise picked up by the replay head. Thus there is a lower frequency limit of about 100 Hz. This problem can be overcome by *frequency modulation*. With frequency modulation the recording head signal is used to modulate a high frequency carrier signal, the carrier frequency being altered in accordance with the fluctuations of the signal from the recording head. Because the carrier frequency is high the low frequency problem does not occur and so frequency modulation recording can be used right down to the lowest frequency. The upper frequency limit is, however, less than with direct recording, typically being about a third of the carrier frequency. This puts it roughly in the range 2 to 80 kHz. Frequency modulation tends to give a better signal to noise ratio than direct recording. However, accurate control of the tape speed is vital since fluctuations in this can lead to apparent frequency fluctuations.

Pulse duration modulation (PDM) is another technique that is used and involves sampling the input signal and converting each sample into a pulse, the duration of which is proportional to the amplitude of the input signal at that instant (Figure 9.29). The pulses are then recorded on tape and the original signal reconstituted on playback by passing the discontinuous readings through a decoder which turns them back into amplitude signals and then through a suitable filter which smoothes them out to give the original analogue signal.

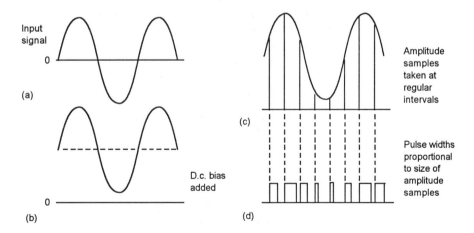

Figure 9.29 *Pulse width modulation*

9.5.2 Digital recording

In the recording of binary signals, the surface of the tape or disk is divided into bit cells. A *bit cell* is the element of the surface where the magnetism is either completely saturated in the positive direction or completely saturated in the negative direction (see Figure 9.27). The surface is magnetised in opposite directions in the two forms of cells, Figure 9.30 illustrating this. This form of recording is termed *saturation recording*. It enables data which has previously been written on a tape to be easily overwritten, all that is necessary is to lay down a new sequence of cells.

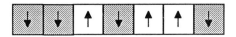

Figure 9.30 *Cells on the magnetic surface, the arrows indicating the directions of magnetisation*

In order to maximise the storage capacity of a tape or disk, it is necessary to have as many bit cells per unit length of track as possible. Thus the cells need to be made as small as possible, consistent with no ambiguity regarding the data stored and the ability to read that data.

One possible method of storing data would seem to be to use magnetisation in one direction to represent a 0 and in the opposite direction to represent 1. The question is then how do we distinguish between successive 1s or successive 0s and recognise them as being more than just a single 0 or single 1? This can be achieved by a *return-to-zero* method of the signal returning to zero magnetism after each 0 or 1 (Figure 9.31). This method does, however, require a regular clock pulse to indicate when each state of magnetism is to be read and also is wasteful of track in that the zero magnetism bits between the 0s and 1s take up space.

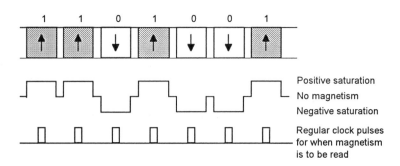

Figure 9.31 *Return-to-zero method*

A *non-return-to-zero* method has the advantage of not wasting track space. With such a method the transition from one state of magnetism to another can be used to indicate the digit. One commonly used version of this method uses no change in flux to represent 0 and a change in flux to represent 1. An external clock is still needed so that the no change in flux points can be determined. Figure 9.32 illustrates this method.

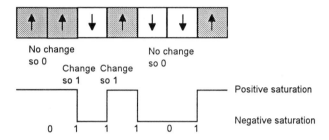

Figure 9.32 *Non-return-to-zero method*

Phase encoding (PE) is another method that can be used and has the advantage of being self-clocking, i.e. no external clock signals have to be supplied. With this method, each half of each cell is magnetised in opposite directions. A digit 0 is then recorded as a half-bit positive saturation followed by a half-bit negative saturation, a 1 digit by a half-bit negative saturation followed by a half-bit positive saturation. The mid-cell transition positive to negative thus indicates a 0 and negative to positive a 1 (Figure 9.33).

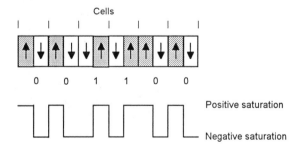

Figure 9.33 *Phase encoding method*

Another method is *frequency modulation* (FM). This is similar to phase modulation but there is always a flux direction reversal at the beginning of each cell. For a 0 bit there is then no additional flux reversal during the cell; for a 1 bit there is an additional reversal during the cell. A 1 is thus a wave of twice the frequency of a 0. Figure 9.34 illustrates this.

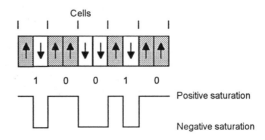

Figure 9.34 *Frequency modulation method*

There are various modifications of the above methods with the aim of improving the storage capacity of a track, making it self-clocking and avoiding interpretation problems.

With computer systems the forms of medium used for the magnetic recording of data are the *floppy disk* and the *hard disk*. Floppy disks are designed to be removable from the computer drive mechanism but hard disks are generally permanently fixed within the drive mechanism. The form of floppy disk currently in use is rigid and 3½ inches in diameter (Figure 9.35). Data is stored along concentric circles termed tracks. The 3½ inch floppy disk has 135 tracks per inch and can store up to 1.44 MB of data. No part of the magnetic surface is exposed when the disk is outside the computer. There is a sliding protective metal cover at the top of the disk and this slides back automatically to reveal the magnetic surface when the disk is loaded into the drive unit. Data on a disk can be prevented from being overwritten by a write protect plastic slider at the bottom of the disk cover. This covers a hole. When the hole is uncovered overwriting is possible, when it is covered it is not. There is another hole at the bottom of the disk, this being used to indicate that the disk is high capacity.

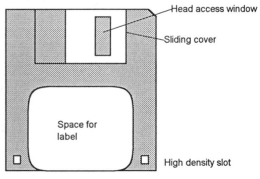

Figure 9.35 *Floppy disk*

Hard disks are sealed units with data stored on the disk surface along concentric circles, the unit generally having more than one such disk and having a magnetic coating on both sides. The disks are rotated at high speed, typically 3600 rpm, 5400 rpm or 7200 rpm, and the tracks accessed by moving the read/write heads. Hard disks can store a large amount of information, storages of the order of 1 GB or more are now common. Figure 9.36 shows the basic form of a hard disk.

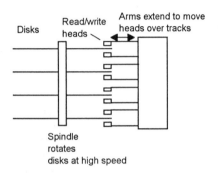

Figure 9.36 *Basic principles of a hard disk*

9.6 Optical disks

Optical disk storage of data involves data being stored by modifying the surface of a disk in such a way that light reflected from it is affected. There are three main types of optical disk:

1   *Compact disk read-only memory (CD-ROM)*
    Information can only be stored on such a disk during its manufacture. Such disks typically can hold more than 600 MB of data. The CD-ROM disk is the same as the audio compact disk. The disk stores its data along a single spiral track by a sequence of pits pressed into the surface. Figure 9.37 shows how such tracks can be read.

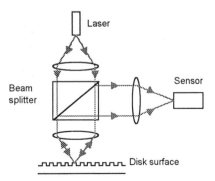

Figure 9.37 *Reading data*

2 *Write once read many (WORM)*
Such a disk can have data written on to it many times. However, data once written is not erased and cannot be written over. Thus the capacity of such a disk continually decreases as more data is written to it. Such a disk is the same basic format as a CD-ROM but the sequence of pits along the spiral track are burnt into the disk surface or bubbles raised from it by heating.

3 *Erasable optical disk*
Such a disk can be used just like a magnetic disk with data being written to it, read and erased or overwritten. One form of such a disk is the *magneto-optical disk (MO)*. Thus uses a heat resistance magnetic layer under protective glass. A magnetic head is used to give the magnetic layer magnetic polarities that reflect light differently.

Problems

1 On what factors does the current sensitivity of a moving coil meter movement depend?

2 A moving coil meter movement has a coil with 100 turns of wire, a cross-sectional coil area of 1.0×10^{-4} m^2, a moment of inertia of 2.4×10^{-5} kg m^2, a resistance of 60 Ω, a magnetic field of flux density 80 T and springs which apply a restoring torque of 1.2×10^{-2} N m per radian of coil rotation. A voltage source with a resistance 240 Ω is connected to the meter. Determine (a) the steady-state voltage sensitivity, (b) the damping ration, (c) the natural frequency.

3 A galvanometric recorder has a moving coil meter movement with a damping factor of 0.6 and a resistance of 60 Ω when a sensor of resistance 160 Ω is connected across it. How is the damping factor and the steady-state voltage sensitivity changed if a resistance of 100 Ω is connected in series with the sensor?

4 A UV galvanometric recorder has a galvanometer coil with a resistance of 60 Ω and a damping factor of 2.8 when there is no external circuit connected to it. A sensor of resistance 120 Ω is to be connected to the galvanometer coil. What additional resistance should be connected in series with the sensor if a damping factor of 0.7 is required?

5 A magnetic-tape recorder has a recording head with a gap width of 5 μm and is used with a tape running at a speed of 95.5 mm/s. What will be the maximum frequency signal that can be recorded?

6 A magnetic-tape recorder can be operated in either the direct or frequency modulated modes. Which mode should be used for recording signals which are (a) slow-varying d.c., (b) alternating at about a frequency of 200 kHz?

7 Explain how, for a magnetic tape/disk recorder, the non-return-to-zero method of recording binary data operates and its advantages when compared with return-to-zero recording.

8 Explain the reasons for the use of bias and amplitude equalisation with magnetic-tape recorders.

10 Force, torque, pressure & strain measurements

This chapter, together with Chapters 11, 12 and 13, is concerned with showing the application of principles outlined in earlier chapters to the making of particular measurements. This chapter is concerned with measurements of force, torque, pressure and strain.

Sensors that are used for the measurement of force, torque or pressure often contain an elastic member that converts the mechanical quantity into a deflection or strain which can then be transformed using another sensor into an electrical signal. Electrical resistance strain gauges are widely used in this capacity. For the sensing of deflections, potentiometers and linear variable differential transformers (LVDTs) are often used.

10.1 Force measurement

The unit of force is the *newton* (N) and is that force which when applied to a mass of one kilogram gives it an acceleration of 1 m/s². The *weight* of a body of mass *m* at rest relative to the surface of the earth is the force exerted on it due to gravity and is equal to *mg*, where *g* is the acceleration the body would experience if allowed to fall freely under just the action of gravity. This acceleration due to gravity is typically about 9.8 m/s² at the surface of the earth. Since weight is a force it has the unit of force, i.e. the newton. However, weight measurements are often specified in terms of the mass that would give rise to the weight force at the local value of the acceleration due to gravity.

The following are examples of commonly used force measurement systems. They have been grouped under the general headings of:

1 *Lever balance methods*
   These methods are based on the principle of moments and are typically used for weights up to 1000 kg and can be very accurate.

2 *Force balance methods*
   These depend on the force causing a displacement which is then monitored by a displacement sensor, the output of this sensor then being used to produce a force which balances out the unknown force. They can be very accurate with high stability and used for dynamic force measurements. The range is typically 0.1 N to 1 kN.

3 *Elastic element methods*
   The force is applied to some member and causes it to change in length, the change in length being monitored by another sensor. The accuracy, range and whether static or dynamic forces can be measured will

depend on the form of the elastic member and how its change in length is monitored.

4   *Pressure methods*
This depends on the measurement of the pressure resulting from the application of the force over some area. Typically such methods are used for forces in the range 5 kN to 5 MN and for static or dynamic forces.

With the exception of the equal arm balance, which can be considered to directly compare masses, calibration of the other methods in terms of the gravitational forces acting on standard masses requires a knowledge of the acceleration due to gravity. An equation that is used for the value of the acceleration due to gravity g at different geographical latitudes ϕ and height h about sea level is:

$$g = 9.780\ 318(1 + 0.005\ 302\ 4 \sin^2 \phi - 0.000\ 005\ 9 \sin^2 2\phi)$$
$$- 3.086 \times 10^{-6}h \ \text{m/s}^2$$

In Britain at sea level a working value of 9.807 m/s² is generally used.

In using standard masses, account has also to be taken of upthrust forces acting on them when they are in air. According to Archimedes' principle the upthrust is equal to the weight of fluid displaced. Thus for a mass of volume V and density ρ in air of density σ, the weight of air displaced is $V\sigma g$. Thus the downward force experienced by the mass is $V\rho g - V\sigma g$.

10.1.1 Lever balance methods

Lever balance methods depend on the *principle of moments*, i.e. at static equilibrium the algebraic sum of the clockwise moments about an axis equals the anticlockwise moments. The *moment of a force* about an axis is the product of the force and the perpendicular distance from its line of action to the axis.

With the *equal arm balance*, a rigid beam rests horizontally across a knife edge. Figure 10.1 shows the principle when applied to the simple analytical balance used for weighing objects. The unknown force is applied to one end of the beam to cause a moment about the axis through the knife edge and a balancing force applied to the other end of the beam so that the net result is that the beam remains horizontal. The unknown force is applied at an equal distance from the pivot axis to that of the balancing force. Thus if we apply the principle of moments we must have the unknown force equal in size to the balancing force. Since the forces usually compared by this method are weights and the acceleration due to gravity is the same at both ends of the beam, then we must have the unknown mass equal to the balancing mass. This type of balance is capable of high accuracy and versions are available which can be used for weights up to 1000 kg.

The *unequal arm balance* has the unknown force balanced by moving the point of application of a constant force to different distances from the pivot

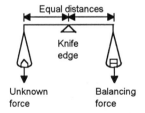

Figure 10.1 *Equal arm balance*

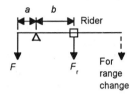

Figure 10.2 *Unequal arm balance*

axis (Figure 10.2). The unknown force F is balanced by moving a rider to different distances from the pivot. Thus if balance is achieved with the rider of weight F_r at a distance b from the pivot when the unknown force is at a distance a:

$$Fa = F_r b$$

Since a is a constant, then F is proportional to the distance b and so a graduated scale along the beam can be used to enable the weight to be directly read from the position of the rider. The range of the balance can be altered by adding masses to the end of the beam. Such a form of balance is widely used in industry. It is bulky but can be very accurate.

10.1.2 Force balance methods

Force balance methods depend on the unknown force causing a displacement which can be monitored by means of a displacement sensor, the output of this sensor then being used to produce a force which then balances out the unknown force. Figure 10.3 shows the principles involved in one form of force balance. The unknown force causes the ferromagnetic core of a linear variable differential transformer (LVDT) (see Section 5.4.2) to become displaced from its central position. This results in an electrical output from the LVDT which increases as the displacement increases as a result of the applied force. After amplification, this is fed back to provide an opposing force as a result of the current passing through a coil in a magnetic field. The opposing force increases as the displacement increases until a point is reached when the opposing force just balances the unknown force. When this happens, the amplified LVDT signal is a measure of the unknown force. Such methods of measuring forces have high stability, high accuracy, a range which is typically from about 0.1 N to 1 kN and can be used for both static and dynamic force measurements.

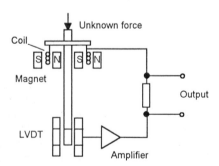

Figure 10.3 *An example of a force balance*

10.1.3 Elastic element methods

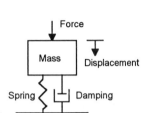

Figure 10.4 *Model of an elastic element*

In general, elastic elements used for the measurement of force can be considered to be represented by the model shown in Figure 10.4. Thus when a force acts on the mass of the element, a displacement is produced which varies with time before attaining a steady-state value. The displacement variation with time is described by a second order differential equation. See Section 3.3 for a derivation of this equation and its solution. Often it is not the displacement itself that is directly monitored but the strain experienced by a strain gauge attached to what is the equivalent of the spring. Since strain is the change in length per unit length it gives an output which is proportional to the displacement and so the same basic second order differential equation still applies.

Various forms of elastic member are used. The simplest is just a spring to give the device referred to as the *spring balance*. The steady-state extension of the spring is used as a measure of the applied force. As the extensions produced are rather small, spring balances have low accuracy. They have ranges within about 0.1 N to 10 kN and are only suitable for static force measurements.

Load cells, i.e. elastic members which transform forces into displacements or strains, can take many forms. *Proving rings* (see Section 5.7.1 and Figure 5.30) are a ring type of load cell. They can be accurate, ±0.2 to ±0.5% and have a range of the order of 2 kN to 2000 kN. They are only used for static force measurements. Other forms of load cell are beams and columns (see Section 5.7.1 and Figures 5.31 and 5.32). These generally employ strain gauges to monitor the strains produced when forces are applied to the elastic member. Generally four strain gauges are used with a load cell, two in tension and two in compression, and connected to form the four arms of a Wheatstone bridge (see Section 7.1.3). This gives temperature compensation. Load cells employing strain gauges have a rapid response and can be used for both static and dynamic force measurements. They typically have an accuracy within the range ±0.01 to ±1.0% and a range within about 5 N to 40 MN depending on the form of the load cell.

10.1.4 Pressure methods

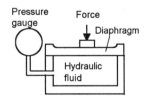

Figure 10.5 *Hydraulic load cell*

The change in pressure of a hydraulic fluid or air that is produced by the application of a force can be used as a measure of the force. Figure 10.5 shows the basic form of a hydraulic load cell. The application of the force to a flexible diaphragm results in a change in the pressure of the fluid, this pressure being registered by some form of pressure gauge. This may be a direct-reading Bourdon tube or a diaphragm sensor giving an electrical output. Hydraulic load cells tend to be used for forces up to 5 MN and have an accuracy of the order of ±0.25 to ±1.0%.

10.2 Torque measurement

Torque measurements are used for rotating shafts to determine the power transmitted and also to monitor against failure as a result of shear stresses.

For a rotating shaft (Figure 10.6) of radius r, the distance travelled by a point on its surface in one revolution is $2\pi r$. Thus if the shaft makes n revolutions per second, the distance travelled by the point on its surface is $2\pi rn$. The torque T applied to the shaft to cause it to rotate is Fr, where F is the tangential force at a radius r. Thus, since the work done is the product of the force and the distance moved in the direction of the force by its point of application:

Figure 10.6 *Torque applied to a shaft*

$$\text{work done per second} = F \times 2\pi rn = \frac{T}{r}2\pi rn = 2\pi nT$$

Since the work done per second is the power transmitted:

$$\text{power transmitted by a rotating shaft} = 2\pi nT = \omega T$$

where ω is the angular velocity of the shaft.

When a shaft is twisted as a result of the application of a torque, shear stresses and strains are produced in the shaft material. Thus for the strip of material on the surface of the circular shaft shown in Figure 10.7, the applied torque results in shear forces being applied to it in the way shown.

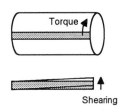

Figure 10.7 *Shearing*

For further discussion of torque applied to shafts the reader is referred to texts such as *Mechanical Science* by W. Bolton (Blackwell Science 1993) or *Mechanics of Materials* by J.M. Gere and S.P. Timoshenko (Chapman and Hall, 3rd ed. 1991).

Basic methods of torque measurement are:

1   *Cradled dynamometer*
    This involves measuring the reaction force in cradled shaft bearings.

2   *Prony brake method*
    This involves measuring the force in a rope wrapped round the shaft.

3   *Strain measurement*
    This involves the measurement of the strain produced in a rotating body as a result of the torque twisting it.

10.2.1 Cradled dynamometer

Any system involved in the transmission of torque through a shaft involves a power source and a power absorber where the power is dissipated. The cradled dynamometer method of measuring torque involves cradling either the power source or the power absorber end of the shaft in bearings and then measuring the reaction force to prevent movement of the cradle and the arm length of this force. Figure 10.8 illustrates this, with (a) being where the power source is cradled and (b) where the power absorber is cradled. The torque is FL.

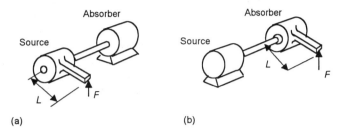

Figure 10.8 *(a) Cradled source, (b) cradled absorber*

10.2.2 Prony brake

A Prony brake is an example of an absorption dynamometer. It consists of a rope wound round the shaft with one end of the rope being attached to a spring balance and the other to a load (Figure 10.9). When the shaft rotates a frictional force develops between the rope and the brake drum and the rope tightens, thus resulting in a force on the spring balance.

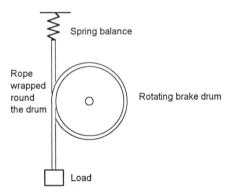

Figure 10.9 *The basic principle of the Prony brake*

If the force indicated by the spring balance is F_s and m is the mass of the load, then the force exerted by the rope on the shaft is $mg - F_s$. This force is tangential to the drum. Thus the torque on the drum is $(mg - F_s)r$, where r is the radius at which the force is applied. The power output of the engine driving the shaft is thus the product of this torque and the angular velocity of the drum.

10.2.3 Strain measurement

The torque T transmitted by a shaft is related to the maximum shear stress τ produced on the shaft surface by the equation:

$$\frac{T}{J} = \frac{\tau}{r}$$

where J is the polar second moment of area of the shaft section and r the radius of the shaft (see texts on mechanical science for a derivation of this equation, e.g. *Mechanical Science* by W. Bolton (Blackwell Science 1993) or *Mechanics of Materials* by J.M. Gere and S.P. Timoshenko (Chapman and Hall, 3rd ed. 1991)). For a solid circular shaft $J = \pi D^4/32$, where D is the shaft diameter. Thus the maximum shear stress is:

$$\tau = \frac{16T}{\pi D^3}$$

For a circular shaft subject to pure torsion, the direction of the maximum stresses resulting from this shear are at 45° to the shaft axis (Figure 10.10) and are:

$$\sigma_1 = -\sigma_2 = \tau = \frac{16T}{\pi D^3}$$

These stresses at right angles to each other will give rise to strains in these directions of:

$$\varepsilon_1 = \frac{16T}{\pi D^3}\left(\frac{1+v}{E}\right)$$

and

$$\varepsilon_2 = -\frac{16T}{\pi D^3}\left(\frac{1+v}{E}\right)$$

where E is Young's modulus for the material and v Poisson's ratio. These strains can be measured by the use of resistance strain gauges aligned in these directions.

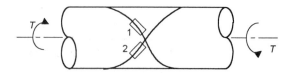

Figure 10.10 *Strain gauges in the directions of maximum strain*

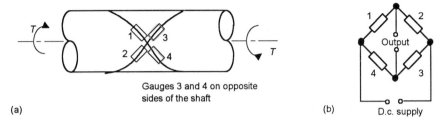

Gauges 3 and 4 on opposite
sides of the shaft

(a)

(b)

D.c. supply

Figure 10.11 *The strain gauges and the Wheatstone bridge*

In order to compensate for temperature changes, four resistance strain gauges are used. These are attached to the shaft in the positions indicated in Figure 10.11(a) and connected into a Wheatstone bridge in the way shown in Figure 10.11(b). Gauges 1 and 3 are in tension and 2 and 4 in compression, the magnitude of the strain being the same for each gauge. Thus the fractional changes in resistance of each of the strain gauges is:

$$\frac{\Delta R_1}{R_1} = -\frac{\Delta R_2}{R_2} = \frac{\Delta R_3}{R_3} = -\frac{\Delta R_4}{R_4} = G\frac{16T}{\pi D^3}\left(\frac{1+v}{E}\right)$$

where G is the gauge factor of the strain gauges. If the strain gauges give an initially balanced bridge, then the out-of-balance voltage V_0 is given by (see Section 7.1.3):

$$V_0 = \frac{V_s R_1 R_4}{(R_1 + R_2)(R_3 + R_4)}\left(\frac{\Delta R_1}{R_1} - \frac{\Delta R_2}{R_2} - \frac{\Delta R_3}{R_3} + \frac{\Delta R_4}{R_4}\right)$$

and thus:

$$V_0 = \frac{16TGV_s}{\pi D^3}\left(\frac{1+v}{E}\right)$$

The output voltage V_0 from the bridge is thus proportional to the torque T acting on the shaft.

When the torque is measured for a rotating shaft, it is necessary to obtain the output signal from the bridge connection of the strain gauges which are on the rotating shaft and provide the d.c. supply voltage for the bridge. One method of doing this is via slip rings. A problem with such an arrangement is the noise generated by variations in contact resistance between the rings and the brushes. An alternative method is to mount the entire bridge, together with its voltage supply, on the rotating shaft and use the bridge output to modulate a radio signal which is then transmitted to nearby, stationary, equipment for display or recording.

10.3 Pressure measurement

The unit of pressure is the pascal (Pa) with 1 Pa being 1 N/m². The term *absolute pressure* is used for the pressure measured relative to zero pressure, the term *gauge pressure* for the pressure measured relative to the atmospheric pressure. At the surface of the earth, the atmospheric pressure is generally about 100 kPa. This is sometimes referred to as a pressure of 1 bar.

The following are some of the commonly used methods for the measurement of pressures. Calibration of pressure gauges in the region 20 Pa to 2000 kPa is generally by means of the dead-weight tester, though U-tube manometers can be used for 20 Pa to 140 kPa. The *dead-weight tester* has the basic form shown in Figure 10.12. Pressure is produced in a fluid by winding in a piston. The pressure is determined by adding weights to the platform so that it remains at a constant height. If the total mass of the platform and its weights is M and the cross-sectional area of the platform piston is A, then the pressure is Mg/A.

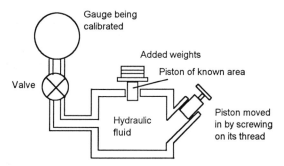

Figure 10.12 *Dead-weight tester*

10.3.1 Manometers

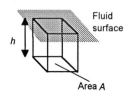

Figure 10.13 *Pressure at a depth h*

The gauge pressure at some depth h in a fluid at rest due to the weight of fluid above it is the weight of that fluid above a certain area divided by the area considered (Figure 10.13). The weight of that volume of fluid is $HA\rho g$, where ρ is the density of the fluid. Thus the pressure p is:

$$p = \frac{hA\rho g}{A} = h\rho g$$

The basic *U-tube manometer* consists of a U-tube containing a liquid (Figure 10.14). A pressure difference between the gases above the liquid in the two limbs produces a difference h in vertical heights of the liquid in the two limbs. For a liquid at rest, the pressure at the base of each limb must be the same. Thus we must have:

$$P_1 + h_1\rho g = P_2 + h_2\rho g$$

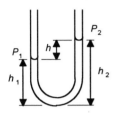

Figure 10.14 *U-tube manometer*

where ρ is the density of the manometric liquid. Hence the pressure difference between the gases above the two limbs is:

$$P_1 - P_2 = (h_2 - h_1)\rho g = h\rho g$$

where h is the vertical difference in heights of the liquid in the two limbs. If one of the limbs is open to the atmosphere then the pressure difference is the gauge pressure.

Water, alcohol and mercury are commonly used manometric liquids. U-tube manometers are simple and cheap and can be used for pressure differences in the range 20 Pa to 140 kPa. Errors can arise due to the height measured not being truly vertical, the effects of temperature on the density of the liquid, and incorrect values of the acceleration due to gravity being used. The accuracy is also affected by difficulties in obtaining an accurate reading of the level of the manometric liquid in a tube due to the meniscus. The accuracy is thus typically about ±1%.

The correction that has to be applied for the effect of temperature on the density of the manometric liquid is derived as follows. A mass m of liquid at 0°C has a volume V_0 and a density ρ_0 which are related by $m = \rho_0 V_0$. At a temperature θ the same mass of liquid has a volume V_θ and density ρ_θ, where $m = \rho_\theta V_\theta$. The volume at temperature θ is related to the volume at 0°C by $V_\theta = V_0(1 + \gamma\theta)$, where γ is the coefficient of cubical expansion of the liquid. Hence:

$$\rho_\theta = \frac{\rho_0 V_0}{V_\theta} = \frac{\rho_0}{1 + \gamma\theta}$$

Thus the pressure when measured by a U-tube manometer at temperature θ, when the manometer liquid density at 0°C is known, is given by:

$$P = h\rho_\theta g = \frac{H\rho_0 g}{1 + \gamma\theta}$$

An *industrial form of the U-tube manometer* or *cistern* manometer is shown in Figure 10.15. It has one of the limbs with a much greater cross-sectional area than the other. A difference in pressure between the two limbs causes a difference in liquid level with liquid flowing from one limb to the other. For such an arrangement:

pressure difference = $P_1 - P_2 = H\rho g$

But $H = h + d$, where h and d are the changes in level in each limb from the level that existed when there was no pressure difference. Thus:

pressure difference = $(h + d)\rho g$

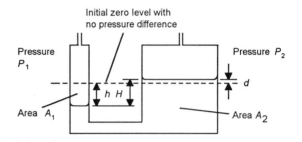

Figure 10.15 *The cistern manometer*

The volume of liquid leaving one limb must equal the volume entering the other. Hence we must have $A_1h = A_2d$, where A_1 and A_2 are the cross-sectional areas of the two limbs. Hence:

$$\text{pressure difference} = \left(\frac{A_2 d}{A_1} + d\right)\rho g = \left(\frac{A_2}{A_1} + 1\right)d\rho g$$

$$= \text{a constant} \times d\rho g$$

Thus the movement of the liquid level d in the wide tube from its initial level is proportional to the pressure difference. This form of manometer thus only requires the level of liquid in one limb to be measured from a fixed point. Usually this change in level is determined by using a float and lever system to move a pointer across a scale.

The *inclined tube manometer* (Figure 10.16) is a U-tube manometer with one limb having a larger cross-section than the other and the narrower limb being inclined at some angle θ to the horizontal. It is generally used for the measurement of small pressure differences and gives greater accuracy than the conventional U-tube manometer.

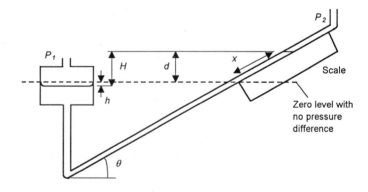

Figure 10.16 *The inclined tube manometer*

The vertical displacement d of the liquid level in the inclined limb is related to the movement x of the liquid along the tube by $d = x \sin \theta$. The displacement x is the measured quantity, thus:

$$\text{pressure difference} = P_1 - P_2 = \left(\frac{A_2}{A_1} + 1\right) d\rho g = \left(\frac{A_2}{A_1} + 1\right) \rho g x \sin \theta$$

Since A_2 is much greater than A_1, the equation approximates to:

$$\text{pressure difference} = \rho g x \sin \theta$$

Thus for a particular angle and manometer liquid, the movement x of the liquid along the tube is proportional to the pressure difference. Since this movement is greater than would occur with the conventional U-tube manometer, greater accuracy is possible.

10.3.2 Diaphragms

With diaphragm pressure gauges (see Section 5.7.2), a difference in pressure between two sides of a diaphragm results in it bowing out to one side or the other. Diaphragms may be flat, corrugated or dished, the form determining the amount of displacement produced and hence the pressure range which can be measured. It also determines the degree of non-linearity. If the fluid for which the pressure is required is admitted to one side of the diaphragm and the other side is open to the atmosphere, the diaphragm gauge gives the gauge pressure. If fluids at different pressures are admitted to the two sides of the diaphragm, the diaphragm gauge gives the pressure difference.

There are a number of methods used to detect and give a measure of the deformation of the diaphragm, the following indicating some of the more commonly encountered forms:

1 *Reluctance diaphragm gauge*
   Figure 10.17(a) shows the basic form generally used. The displacement of the central part of the diaphragm increases the reluctance of the coil on one side of the diaphragm and decreases it on the other (see Section 5.4.1). With the two coils connected in opposite arms of an a.c. bridge, the out-of-balance voltage is related to the pressure difference causing the diaphragm displacement. The range is generally about 1 Pa to 100 MPa with an accuracy of about ±0.1% and a bandwidth up to 1 kHz.

2 *Capacitance diaphragm gauge*
   Figure 10.17(b) shows two basic forms of capacitance diaphragm gauge. With one, the displacement of the diaphragm relative to a fixed plate changes the capacitance between the diaphragm and the fixed plate (see Section 5.3.1). The capacitor can form part of the tuning circuit of a frequency modulated oscillator and so give an electrical

output related to the pressure difference across the diaphragm. With the other, the diaphragm is between two fixed plates and its movement thus increases the capacitance with respect to one fixed plate and decreases it with respect to the other (see Section 5.3.1). Such a gauge is generally used with the two capacitors in opposite arms of an a.c. bridge, the out-of-balance voltage then being related to the pressure difference across the diaphragm. The range of capacitance diaphragm gauges is generally about 1 kPa to 200 kPa with an accuracy of about ±0.1% and a bandwidth up to 1 kHz.

3   *Strain gauge diaphragm gauge*
There are a number of ways strain gauges (see Section 5.2.3) can be used to monitor the displacement of a diaphragm (Figure 10.17(c)). One way involves them being attached to a cantilever which is bent when the central part of the diaphragm is displaced. Another way involves them being directly stuck on the diaphragm (see Section 5.7.2), or a silicon sheet used for the diaphragm with the strain gauges being formed in the diaphragm material by the introduction of doping material into it (see Section 6.3.2). Whatever the form used, four strain gauges are generally used with two of the gauges being subject to tension and two to compression. These form the arms of a Wheatstone bridge (see Section 7.1.3) and the out-of-balance voltage is then taken as a measure of the pressure difference across the diaphragm. Typically metal wire strain gauge instruments are used over the range 100 kPa to 100 MPa, with the integrated semiconductor gauge instrument over the range 0 to about 100 kPa. The accuracy is about ±0.1% with a bandwidth up to 1 kHz.

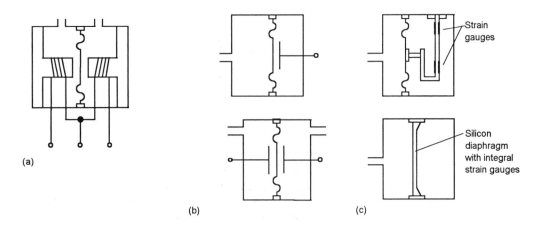

(a)

(b)   (c)

Figure 10.17 *Some basic forms of diaphragm pressure gauges: (a) reluctance, (b) capacitance, (c) strain gauge*

4 *Force-balance diaphragm gauge*

There are a number of versions of this form of instrument. They are all, however, based on the applied pressure causing a displacement of a diaphragm. The displacement is monitored by some form of sensor which produces a signal which, by means of a feedback loop, actuates a response to cancel out the displacement. A common form of such an instrument is the *pneumatic differential pressure cell* (Figure 10.18). A difference in pressure across the diaphragm results in a displacement of the diaphragm and hence the end of force beam. The force beam is pivoted and its movement results in a change in the gap between the flapper and the nozzle. This gap determines the rate at which air escapes from the nozzle and so the pressure in the nozzle part of the system. The pressure change resulting from movement of the force beam is communicated to bellows, the expansion or contraction of which results in forces which act on the force beam and cause it to cancel out the diaphragm displacement. The pneumatic pressure in the flapper-nozzle system is a measure of the pressure difference across the diaphragm. Other forms of the cell use reluctance, LVDT or capacitive means of detecting the displacement of the diaphragm. Typically such cells have a range of about 0 to 100 kPa with accuracies of about ±0.2% and response times of about 1 s.

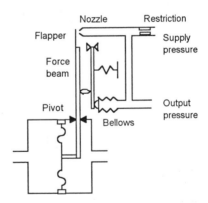

Figure 10.18 *Pneumatic differential pressure cell*

5 *Piezoelectric diaphragm gauge*

This consists essentially of a diaphragm which presses against a piezoelectric crystal (see Section 5.8). Movement of the diaphragm causes the crystal to be compressed and a potential difference is consequentially produced across its faces. Such a gauge can only be used for dynamic pressures and typically for pressures in the range 200 kPa to 100 MPa with a bandwidth of 5 Hz to 500 kHz.

10.3.3 Bourdon tubes

The Bourdon tube may be in the form of a 'C', a flat spiral, a helical spiral, or twisted (see Section 5.7.4). In all forms, an increase in the pressure in the tube causes the tube to straighten out to an extent which depends on the pressure. This displacement may be monitored in a variety of ways, e.g. to directly move a pointer across a scale, via gearing to move a pointer across a scale, to move the slider of a potentiometer, to move the core of an LVDT. Figure 10.19 shows some forms. Bourdon tube instruments typically operate in the range 10 kPa to 100 MPa, the range depending on the form of the tube and on the material from which it is made. C-shaped tubes made from brass or phosphor bronze have a pressure range from about 35 kPa to 100 MPa. Spiral and helical tubes are more expensive and have a greater sensitivity but as a consequence a lower maximum pressure that can be measured, typically about 50 MPa. Bourdon tubes are robust with an accuracy of about ±1% of full-scale reading. The main sources of error are hysteresis, changes in sensitivity due to temperature changes and frictional effects with linkages and pointers.

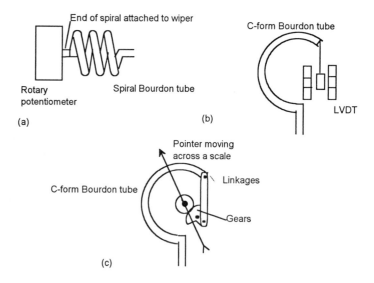

Figure 10.19 *Forms of Bourdon tube instruments*

10.3.4 Measurement of low pressures (vacuum)

The term *vacuum* tends to be used for pressures less than the atmospheric pressure, namely 1.013×10^5 Pa. A unit that is often used for such pressures is the torr, this being the pressure equivalent to that given by a column of mercury of height 1 mm. It is thus a pressure of 7.50×10^{-3} Pa.

In torr units the atmospheric pressure is about 760 torr since it will support a column of mercury with a height of about 760 mm.

With careful design the Bourdon tube can be used for pressures down to about 100 Pa (1 torr), the U-tube manometer for pressures down to about 10 Pa (0.1 torr) and diaphragm gauges down to about 1 Pa (0.001 torr). For lower pressures, other instruments need to be used. Gauges commonly used include:

1   *McLeod gauge*
    This gauge depends on an application of Boyle's law, namely that for a fixed mass of an ideal gas the pressure is inversely proportional to the volume. Thus by compressing a known volume of a gas into a known smaller volume the pressure is increased by the ratio of these volumes and can be made large enough to be measured. Figure 10.20 shows the form of the McLeod gauge based on this.

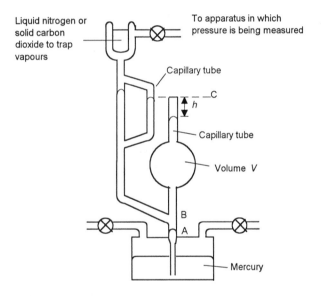

Figure 10.20 *McLeod gauge*

The McLeod gauge initially has the mercury level at A so that the gas in the bulb is at the pressure to be measured. The mercury level is then raised by allowing air to enter the mercury reservoir. When the mercury passes B the sample of the gas in the bulb is trapped. The mercury level is then raised until it reaches C, the level in the open capillary tube corresponding to the top of the capillary tube containing the trapped volume of gas. The difference in height h between the mercury in the two capillary tubes is then a measure of the gas pressure. V is the volume of gas trapped and reduced into a volume hA, where A is the cross-sectional area of the capillary tube, by increasing the pressure by $h\rho g$, where ρ is the density of the mercury. If p is the pressure being measured, then Boyle's law gives:

$$pV = (p + h\rho g)hA$$

Hence:

$$p = \frac{Ah^2\rho g}{V - Ah}$$

Since AH is much smaller than V, the equation approximates to:

$$p = \frac{Ah^2\rho g}{V}$$

The gauge can be used for pressures in the range 5×10^{-4} Pa to atmospheric pressure with an accuracy of about ± 5 to $\pm 10\%$.

2 *Thermal conductivity gauges*
The temperature of a wire or resistance element which is heated as a result of a current passing through it depends on the rate at which heat is conducted away by the gas surrounding the element and hence the gas pressure. The *Pirani gauge* consists of a platinum or tungsten wire in a glass or metal tube. When a current passes through the wire, its electrical resistance depends on its temperature and consequently on the pressure in the tube. Since the resistance is also affected by changes in the ambient temperature a dummy gauge is used, this being a sealed gauge for which the pressure does not change. The active and the dummy gauges are connected in adjacent arms of a Wheatstone bridge and the out-of-balance voltage becomes a measure of the gas pressure in the active gauge (Figure 10.21). The Pirani gauge is used for pressures in the range 10^{-2} to 10^3 Pa with an accuracy of about $\pm 10\%$. A version of this gauge is available using a thermistor instead of the wire. Because thermistors give much larger changes in resistance and can be very small, the gauge is more sensitive and responds more quickly than the metal wire gauge.

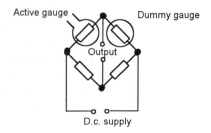

Figure 10.21 *Pirani gauge*

3 *Ionisation gauges*

When a gas is bombarded with electrons ionisation can occur and as a consequence a current flows between two electrodes in the gas. The amount of ionisation produced depends on the number of gas molecules present and hence the gas pressure. The *Penning gauge* (Figure 10.22) is a commonly used form of ionisation gauge. It consists of two parallel flat cathodes about 20 mm apart in a glass or metal envelope. Midway between them is a wire anode. The gauge is placed between the poles of a magnet, either a permanent magnet or an electromagnet, which produces a magnetic field at right angles to the cathode plates. A high potential difference is connected between the anode and the cathodes. Electrons emitted from the cathodes move towards the anode in helical paths, rather than straight lines, as a consequence of the magnetic field. This increased path length increases the probability of collisions between the electrons and the gas molecules. The current between the electrodes which arises from this ionisation is taken as a measure of the pressure. The gauge is used for pressures in the region 10^{-5} Pa to 1 Pa with an accuracy of about ±10 to $\pm20\%$. A problem with the gauge is that it shows a hysteresis effect, the current depending on whether the pressure is increasing or decreasing.

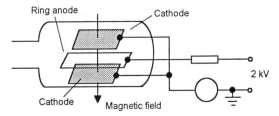

Figure 10.22 *Principle of the Penning gauge*

Another form of ionisation gauge (Figure 10.23) has the electrons emitted by a heated filament. Surrounding the filament is a grid which is positive with respect to the cathode and collects the electrons after they have passed through the gas. When the electrons collide with gas molecules they ionise them with the result that there is an increase in the number of electrons but there is also the production of positive ions. Outside the grid is another electrode for the collection of the positive ions, this electrode being negative with respect to the filament and the grid. The pressure is proportional to the ratio of the positive ions current and the electron current. If the electron current is kept constant, the positive ion current is directly proportional to the pressure. The gauge is used for pressures in the range 10^{-6} to 100 Pa with an accuracy of about ±10 to $\pm30\%$.

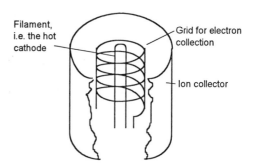

Figure 10.23 *Hot cathode ionisation gauge*

The lowest pressure measurable with the hot cathode gauge is determined by the X-rays produced as a result of the electrons striking the grid. A modification of the design which utilises this X-ray production is known as the *Bayard-Alpert gauge*. Such a gauge can be used for the measurement of pressures from about 1 Pa down to 10^{-8} Pa with an accuracy of about ±10 to ±30%.

10.4 Strain measurement

Tensile and compressive *stress* is defined as the force acting at right angles to the cross-sectional area of a material divided by that area and results in an increase or decrease in length (Figure 10.24(a)). *Strain* is defined as the change in length per unit length. The unit of stress is the pascal (Pa) and strain, being a ratio, has no units. For many materials, for stresses less than those needed to permanently deform the material, the stress σ is proportional to the strain ε (Figure 10.24(b)):

$$\sigma = E\varepsilon$$

where E is the *modulus of elasticity* or *Young's modulus*. When a rod is stretched in the longitudinal direction there is a reduction in the cross-sectional area (Figure 10.24(c)). The change in the width divided by the width is the transverse strain. The ratio of the transverse strain to the longitudinal strain is termed *Poisson's ratio*.

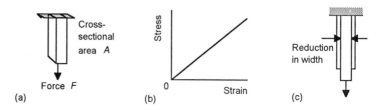

Figure 10.24 *(a) Tensile stress = F/A, (b) stress proportional to strain, (c) transverse strain resulting from longitudinal strain*

10.4.1 The electrical resistance strain gauge

The electrical resistance strain gauge (see Section 5.2.3) is very widely used for the measurement of strain. It is stuck to the surface of the material being tested so that when the surface is subject to strain the gauge is acted on by the same strain. The strain results in a change ΔR in the resistance R of the strain gauge, the fractional change in resistance being proportional to the strain ε:

$$\frac{\Delta R}{R} = G\varepsilon$$

where G is the *gauge factor*.

The gauge factor is supplied by the strain gauge manufacturer from a calibration made of a number of gauges taken from the same production batch. Such a calibration can involve the strain gauges being attached to a tensile test piece and the strain determined using an extensometer. Another method is to use four point bending of a bar to which the gauges have been stuck (Figure 10.25). With this method, a load is applied so that equal forces are applied at equal distances from each end of the bar, the bar resting symmetrically on two knife edges. The applied forces cause the bar to bend into the arc of a circle. The radius of curvature R of the upper surface of the bar is determined from a measurement of the deflection of the bar midpoint using a dial gauge indicator. The strain acting on the strain gauge is then y/R, where y is the distance of the strain gauge element of the bar from the neutral axis of the bar. For a rectangular cross-section bar the neutral axis is in the central axis of the bar and thus y is half the bar thickness plus the distance of the strain gauge wires/foil from the surface of the bar.

Since changes in temperature can also produce resistance changes for the strain gauges, dummy strain gauges might be used with a Wheatstone bridge (see Section 7.1.3). These are gauges not subject to the strain but respond only to temperature changes. An alternative is to use four active strain gauges as the four arms of a Wheatstone bridge (see Section 7.1.3), two of them being subject to tensile strains and two to compressive strains.

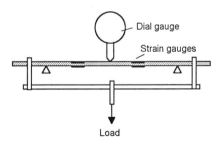

Figure 10.25 *Strain gauge calibration*

Strain gauges measure the strain in the direction of the length of the gauge wires or foil elements. If there is a uniaxial stress σ and the gauge is aligned along this direction, then the strain indicated by the strain gauge is:

$$\text{strain} = \frac{\sigma}{E}$$

If the strain gauge had been attached in any other direction then the strain indicated by the gauge would have been lower. Consider, however, the strain that would be indicated if a strain gauge was attached so that it was at right angles to a uniaxial stress. Despite there being no stress in this direction, the strain gauge would indicate a strain of:

strain at right angles to uniaxial stress $= -v \times$ longitudinal strain

where v is Poisson's ratio.

If the stress acting on a surface is biaxial (Figure 10.26), involving principal stresses mutually at right angles to each other, then at least two strain gauges are needed. If they are aligned along the directions of these stresses and the strains indicated by the gauges are ε_x and ε_y then the strains indicated by each of these gauges will be the result of the stress in their direction and the stress in the perpendicular direction. Thus:

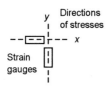

Figure 10.26 *Biaxial stress*

$$\varepsilon_x = \frac{\sigma_x}{E} - v\frac{\sigma_y}{E}$$

$$\varepsilon_y = \frac{\sigma_y}{E} - v\frac{\sigma_x}{E}$$

Hence:

$$\text{stress in } x \text{ direction} = \frac{E(\varepsilon_x + v\varepsilon_y)}{1 - v^2}$$

$$\text{stress in } y \text{ direction} = \frac{E(\varepsilon_y + v\varepsilon_x)}{1 - v^2}$$

Often the directions of the principal stresses are not known and so the strain gauges cannot be guaranteed to be aligned along the stress directions. The principal strains ε_x and ε_y have then to be determined from other strain measurements before the above equations can be used. In such a situation three strain gauges have to be used in an arrangement termed a *rosette*. One form of rosette has the three strain gauges aligned in the way shown in Figure 10.27, a 45° rosette. With such an arrangement the values of the principal strains ε_x and ε_y, i.e. the strains along the directions of the principal stresses, are related to the strains ε_1, ε_2 and ε_3 by the following simultaneous equations:

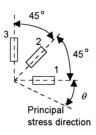

Figure 10.27 *45° rosette*

$$\varepsilon_1 = \left(\frac{\varepsilon_x + \varepsilon_y}{2}\right) + \left(\frac{\varepsilon_x - \varepsilon_y}{2}\right)\cos 2\theta$$

$$\varepsilon_2 = \left(\frac{\varepsilon_x + \varepsilon_y}{2}\right) + \left(\frac{\varepsilon_x - \varepsilon_y}{2}\right)\cos 2(\theta + 45^\circ)$$

$$= \left(\frac{\varepsilon_x + \varepsilon_y}{2}\right) - \left(\frac{\varepsilon_x - \varepsilon_y}{2}\right)\sin 2\theta$$

$$\varepsilon_3 = \left(\frac{\varepsilon_x + \varepsilon_y}{2}\right) + \left(\frac{\varepsilon_x - \varepsilon_y}{2}\right)\cos 2(\theta + 90^\circ)$$

$$= \left(\frac{\varepsilon_x + \varepsilon_y}{2}\right) - \left(\frac{\varepsilon_x - \varepsilon_y}{2}\right)\cos 2\theta$$

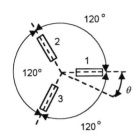

Figure 10.28 *120° rosette*

Another form of rosette has the gauges at 120° angles (Figure 10.28). Then the equations are:

$$\varepsilon_1 = \left(\frac{\varepsilon_x + \varepsilon_y}{2}\right) + \left(\frac{\varepsilon_x - \varepsilon_y}{2}\right)\cos 2\theta$$

$$\varepsilon_2 = \left(\frac{\varepsilon_x + \varepsilon_y}{2}\right) + \left(\frac{\varepsilon_x - \varepsilon_y}{2}\right)\cos 2(\theta + 120^\circ)$$

$$\varepsilon_3 = \left(\frac{\varepsilon_x + \varepsilon_y}{2}\right) + \left(\frac{\varepsilon_x - \varepsilon_y}{2}\right)\cos 2(\theta + 240^\circ)$$

Figure 10.29 *120° rosette*

Another form of the 120° strain gauge rosette has the strain gauges arranged as shown in Figure 10.29.

As an illustration of the use of a strain gauge rosette, consider a 45° rosette for which the strains indicated by the strain gauges are:

$$\varepsilon_1 = 500 \times 10^{-6}, \ \varepsilon_2 = 350 \times 10^{-6}, \ \varepsilon_3 = -300 \times 10^{-6}$$

Using the equations given above for the 45° rosette:

$$\varepsilon_1 = 500 \times 10^{-6} = \left(\frac{\varepsilon_x + \varepsilon_y}{2}\right) + \left(\frac{\varepsilon_x - \varepsilon_y}{2}\right)\cos 2\theta$$

$$\varepsilon_2 = 350 \times 10^{-6} = \left(\frac{\varepsilon_x + \varepsilon_y}{2}\right) - \left(\frac{\varepsilon_x - \varepsilon_y}{2}\right)\sin 2\theta$$

$$\varepsilon_3 = -300 \times 10^{-6} = \left(\frac{\varepsilon_x + \varepsilon_y}{2}\right) - \left(\frac{\varepsilon_x - \varepsilon_y}{2}\right)\cos 2\theta$$

Adding the first and last of the three equations gives:

$$200 \times 10^{-6} = \varepsilon_x + \varepsilon_y$$

Substituting this value into the first and second equations gives:

$$500 \times 10^{-6} = 100 \times 10^{-6} + \tfrac{1}{2}(\varepsilon_x - \varepsilon_y)\cos 2\theta$$

$$350 \times 10^{-6} = 100 \times 10^{-6} - \tfrac{1}{2}(\varepsilon_x - \varepsilon_y)\sin 2\theta$$

Hence, dividing the second equation by the first gives:

$$\tan 2\theta = -\frac{250 \times 10^{-6}}{400 \times 10^{-6}}$$

and so $\theta = 164°$. This value can be substituted into the equations to give $\varepsilon_x = 572 \times 10^{-6}$ and $\varepsilon_y = -372 \times 10^{-6}$.

For further discussion of stress and strain analysis the reader is referred to texts such as *Mechanical Science* by W. Bolton (Blackwell Science 1993) or *Mechanics of Materials* by J.M. Gere and S.P. Timoshenko (Chapman and Hall, 3rd ed. 1991).

10.4.2 Whole surface strain measurement

There are a number of methods that can be used to give the strains over an entire surface. Three methods that are commonly used are:

1 *Brittle lacquers*
  The surface of the test piece is coated with a special lacquer. When the lacquer dries it becomes brittle and when the test piece is stressed, cracks appear where the strain is above a certain threshold. The cracks form at right angles to the maximum principal stresses and thus the pattern of cracks shows the location of the highest stresses. The method is of most use for the determination of the directions of the principal stresses. The accuracy in relation to their magnitude is very low.

2 *Moiré fringes*
  A line pattern is etched on the surface of the test piece. A transparent grating with the same line pattern is then held over the surface and the surface pattern viewed through it (Figure 10.30). A Moiré fringe pattern is seen, the pattern changing as the test piece is subject to strain and the spacing of the line pattern on the test piece changes.

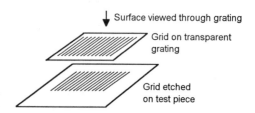

Figure 10.30 *Using Moiré fringes*

3 *Photoelasticity*
  Photoelasticity is a technique for the determination of stresses in models made of certain materials, e.g. Perspex. Such materials have the property called *birefringence*. Light waves are transverse waves with sinusoidally oscillating electric and magnetic fields that are at right

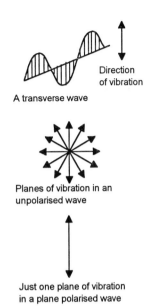

A transverse wave

Planes of vibration in an
unpolarised wave

Just one plane of vibration
in a plane polarised wave

Figure 10.31 *Polarised light*

angles to the direction of propagation. You can think of the situation as being rather like a cork on a water surface as a water wave passes the cork bobs up and down in a direction at right angles (more or less) to the direction of propagation of the wave. In general a light source emits waves containing vibrations in all perpendicular directions (Figure 10.31). A light wave is said to be *plane polarised* if the transverse vibrations are restricted to just one direction. If plane polarised light is passed through a birefringent material, then when that material is stressed the ray of light splits into two plane polarised waves. These have their planes of vibration at right angles to each other, the planes being along the directions of the principal stresses. These two waves travel through the material at different velocities, the velocity of each being proportional to the size of the stress in its plane of vibration. Thus if σ_1 and σ_2 are the principal stresses:

$$\text{velocity for wave 1, } v_1 \propto \sigma_1$$

$$\text{velocity for wave 2, } v_2 \propto \sigma_2$$

Thus on emerging from the material, the two waves will be longer in step and show a phase difference:

$$\text{phase difference} \propto (v_1 - v_2)t \propto (\sigma_1 - \sigma_2)t$$

where t is the thickness of the material. If the two waves are now passed through a polariser (Figure 10.32) then the components of the two that lie in the direction of transmission of the polariser will be combined. If the two waves had become completely out of step, i.e. a phase difference of $180°$, then mutual extinction would occur and zero light intensity would occur and an interference fringe produced joining all such positions on the material. The first out-of-step fringe is said to have the order number 1. The stresses may, however, be sufficient for the waves to get out of step by more than this and higher order fringes are then produced:

$$\text{principal stress difference} = \frac{nf}{t}$$

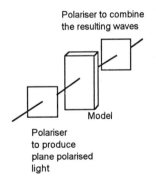

Polariser to combine
the resulting waves

Model

Polariser
to produce
plane polarised
light

Figure 10.32 *Photoelasticity*

where n is the order number and f a constant for the material termed the material fringe value. Thus by determining the order number of the fringe at a point the difference in principal stress values can be obtained. One way of doing this is to count the number of fringes occurring between the point concerned and a zero stress position for the model. The above represents just a very sketchy and superficial discussion of photoelasticity, for more details the reader is referred to texts such as *Experimental Stress Analysis* by G.S. Holister (Cambridge University Press 1967).

Problems

1 A plastic material with a density of 1200 kg/m³ is weighed on an equal arm balance and balance conditions obtained with brass weights totalling 140 g. The balance is used in a room where the ambient temperature is 20°C and the pressure 1.01×10^5 Pa. The density of the brass weights is 8490 kg/m³. What is the percentage error if air buoyancy is neglected? (Hint: the density of the air can be obtained by the use of $PV = mRT$, with $R = 287$ J/kg K.)

2 A load cell is being designed for the measurement of forces in the range 5 to 30 N. It is in the form of a cantilever made of a material with a tensile modulus of elasticity of 200 GPa and has breadth b of 10 mm and depth d of 4 mm. Two strain gauges are on its upper surface 50 mm from the free end and two on the lower surface 50 mm from the free end. The strain gauges have a gauge factor of 2.0 and are connected as the arms of a Wheatstone bridge. If the Wheatstone bridge has a supply voltage of 4.0 V, what will be the minimum out-of-balance voltage? (Hint: the stress on the surface of a cantilever a distance L from the free end when a force F is applied at that end is $6FL/bd^2$.)

3 A load cell is in the form of a cantilever made of a material with a tensile modulus of elasticity of 200 GPa and has breadth b of 10 mm and depth d of 2 mm. Two strain gauges are on its upper surface 40 mm from the free end and two on the lower surface 40 mm from the free end. The strain gauges have a gauge factor of 2.0 and are connected as the arms of a Wheatstone bridge. The Wheatstone bridge has a supply voltage of 9.0 V and the out-of-balance voltage feeds an amplifier with a voltage gain of 200. Determine the output from the amplifier per newton when the force is applied at the free end of the cantilever. (Hint: the stress on the surface of a cantilever a distance L from the free end when a force F is applied at that end is $6FL/bd^2$.)

4 The torque developed by a shaft is to be measured by the use of two resistance strain gauges, of resistance 120 Ω and gauge factor 2.1, which are attached on the surface of the shaft at 45° to its axis, one gauge detecting tensile strain and the other compressive strain. The strain gauges are connected in adjacent arms of a Wheatstone bridge, the other two arms being resistances of 500 Ω. The bridge has a supply voltage of 6.0 V. What will be the out-of-balance output from the bridge when the strain on the gauges is $+400 \times 10^{-6}$ and -400×10^{-6}?

5 An inclined tube manometer has the indicating tube at 30° to the horizontal. What will be the change in level in the inclined tube that is produced by a pressure difference between the two limbs of 1000 Pa if the manometric liquid has a relative density of 13.56? Take the acceleration due to gravity to be 9.81 m/s².

6 A McLeod gauge has a bulb with a volume of 100 cm³ and a capillary of diameter 1 mm. Determine the pressure that will be indicated by a height difference of 30 mm. Take the manometric fluid, mercury, to

have a relative density of 13.6 and the acceleration due to gravity to be 9.81 m/s^2.

7   A 120° strain gauge rosette gave the following results:

> Gauge 1 a strain of +200 × 10^{-6}
> Gauge 2, at 120° to gauge 1, a strain of +530 × 10^{-6}
> Gauge 3, at 240° to gauge 1, a strain of −400 × 10^{-6}

Determine the principal strains and their directions.

8   A 45° strain gauge rosette gave the following results:

> Gauge 1 a strain of +160 × 10^{-6}
> Gauge 2, at 45° to gauge 1, a strain of +40 × 10^{-6}
> Gauge 3, at 90° to gauge 1, a strain of −320 × 10^{-6}

Determine the principal strains and their directions.

9   Select forms of instruments that might be suitable for the following measurements:
(a) A static force of about 100 N with high accuracy.
(b) A slowly changing force of about 1 MN with an accuracy of ±1%.
(c) The power transmitted by a rotating shaft.
(d) A pressure difference of about 1 MPa which varies with time.
(e) A pressure difference of about 10 MPa with a frequency of 1 kHz.
(f) Air pressure of the order of 10^{-3} Pa.
(g) A biaxial strain at a point on the surface of a shaft when the directions of the principal stresses are not known.

11 Position and motion measurement

This chapter is about the measurements concerned with detecting the presence or proximity of objects and the measurement of displacement, velocity and acceleration. The situations in which such measurements are required can be very varied.

For example, there are many control systems where the need is for devices which indicate the presence or otherwise of some item, e.g. an automatic door where the presence of a person approaching it needs to be detected, a device for counting items moving along a conveyor belt where the presence of an item is required to produce a pulse which can be counted, robot 'hands' which need to determine when they have touched and gripped an object. With machine tools there is generally a need to monitor the displacement of the tool and compare it with the displacement required. Thus there are often situations where feedback is required which is a measure of the presence or displacement of an item.

Measurements of displacement, velocity and acceleration are often associated with shocks or vibrations resulting from the application of forces to items. Thus measurements may be required for short-duration acceleration produced by impact on a body, or the instantaneous values of the displacements, velocities and accelerations occurring during linear or rotational vibrations.

11.1 Proximity detection

Proximity detection is generally concerned with sensors which, with their associated signal conditioning, give an output which is essentially an on-off signal. They are connected to the input ports of controllers such a programmable logic controllers (PLCs) (see Section 8.5) to give an input to the system to indicate the presence or otherwise of an item so that control actions can be initiated.

The sensors used in such situations can be classified as contact or non-contact. The following outlines some of the commonly used forms of such sensors.

11.1.1 Contact sensors

A *mechanical switch* is a very commonly used proximity sensor, giving an on-off signal as a result of a mechanical input. Such switches can take a number of forms. They can be supplied with normally open or normally closed contacts, generally being configured as either by the choice of the relevant contacts. A normally open switch has contacts which are open in the absence of a mechanical input and the mechanical input closes the switch, thus giving an off to on signal. A normally closed switch has its contacts closed in the absence of a mechanical input and the input is used to open the switch, thus giving an on to off signal. The term *limit switch* is used for a switch which is used to detect the presence or passage of a moving part. Such switches can be actuated in a number of ways, Figure 11.1 showing some of them. The lever (Figure 11.1(a)) or roller (Figure 11.1(b)) might be used to actuate a switch as a result of a workpiece moving into position for a machine tool. The cam (Figure 11.1(c)) can be rotated at a constant rate and so can be used to switch the switch on or off for particular time intervals.

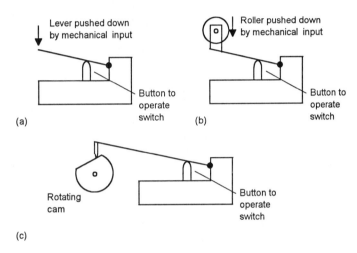

(a)

(b)

(c)

Figure 11.1 *Limit switches actuated by: (a) lever, (b) roller, (c) cam*

11.1.2 Non-contact sensors

Switches that are used to detect the presence of an item without making contact with it include the eddy current proximity switch, the inductive proximity switch, the reed switch, the capacitive proximity switch and photoelectric switches.

The *eddy current proximity switch* has a coil which is energised by a constant alternating current and produces a constant alternating magnetic field. When a metallic object is close to it, eddy currents are induced in it (Figure 11.2). The magnetic field due to these eddy currents induces an e.m.f. back in the coil with the result that the voltage amplitude needed to maintain the constant current coil changes. The voltage amplitude is thus a measure of the proximity of metallic objects. The voltage can be used to activate an electronic switch circuit and so give an on-off device. The range over which such objects can be detected is typically about 0.5 to 20 mm.

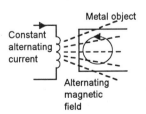

Figure 11.2 *Eddy current proximity switch*

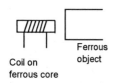

Figure 11.3 *Inductive proximity switch*

The *inductive proximity switch* consists of a coil wound round a ferrous metal core (Figure 11.3). When one end of this core is placed near to a ferrous metal object there is a change in the amount of metal associated with the coil and so a change in its inductance. This change in inductance can be monitored using a resonant circuit, the presence of the ferrous metal object thus changing the current in the circuit. This current can be used to activate an electronic switch circuit and so give an on-off device. The range over which such an object can be detected is typically about 2 to 15 mm.

The *reed switch* (Figure 11.4) consists of two overlapping, but not touching, strips of a springy ferromagnetic material sealed in a glass or plastic envelope. When a magnet or current carrying coil is brought close to the switch, the strips become magnetised and attract each other. The contacts then close. Typically a magnet closes the contacts when it is about 1 mm from the switch.

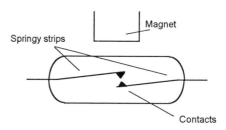

Figure 11.4 *Reed switch*

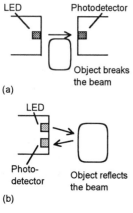

Figure 11.5 *Photoelectric sensors*

The *capacitive proximity switch* can be used with both metallic and non-metallic objects. The capacitance of a pair of parallel plates separated by some medium depends on their separation and the medium. The sensor of the capacitive proximity switch is just one of the plates of the capacitor. If an earthed metal object is brought near to it then it can be considered to act as the other plate of the capacitor and so the capacitance changes. If a non-metallic object is brought near, it can be considered to be altering the dielectric between the capacitor plate and the earth and so also alters the capacitance. The change in capacitance can be used to activate an electronic switch circuit and so give an on-off device. Capacitive proximity switches can be used to detect objects when they are typically between 4 and 60 mm from the sensor head.

Photoelectric switch devices can either operate as *transmissive types* where the object being detected breaks a beam of light, usually infrared radiation, and stops it reaching the detector (Figure 11.5(a)) or *reflective types* where the object being detected reflects a beam of light onto the detector (Figure 11.5(b)). In both types the radiation emitter might be a phototransistor, often a pair of transistors known as a *Darlington pair* being used. The Darlington pair increases the sensitivity. Another possibility is the photodiode. For both the transistor and the diodes, depending on the circuit used, the output can be made to switch high or low when light strikes the transistor/diode. Another possibility is a photoconductive cell. The resistance of such a cell, often cadmium sulphide, depends on the intensity of the light falling on it.

Another form of light-reflection sensor is the *laser distance sensor*. This is used to determine when machine tools are in the right position relative to the workpiece. Thus it might be used with a robot welding machine to enable the welding torch to be located in the right position relative to the workpiece edges so that edge welding can take place. The sensor works on the principle of the diffuse reflection of the laser beam from the workpiece (Figure 11.6). The reflection is detected by an array of photodetectors and when the distance of the assembly from the workpiece is just right the reflection falls on a particular detector and so when the response occurs from this detector the assembly is at the correct distance from the workpiece.

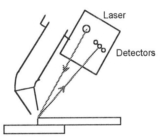

Figure 11.6 *Laser distance sensor*

11.2 Linear displacement measurement

Linear displacements can be measured in a number of ways. These include:

1 *Mechanical methods*
This includes steel rules, micrometer screw gauges and vernier callipers and instruments that often involve levers and/or gears to magnify the movement of a plunger in contact with the object whose displacement is being monitored.

2 *Pneumatic methods*
These work on the principle that the rate at which air leaves a jet depends on the proximity of that jet to the surface of the object whose displacement is being monitored.

3 *Electrical methods*
There are a number of methods that are used to give an electrical output related to a displacement. These include the linear potentiometer, capacitive systems, variable reluctance systems, LVDTs and linear inductosyns.

4 *Optical methods*
These include position sensitive photocells, the use of Moiré fringes and interferometers.

11.2.1 Mechanical methods

The measurement of length in the range of a micrometre to about a metre can be carried out using steel rules, micrometer screw gauges and vernier callipers. *Steel rules* are available in lengths up to about one metre and have an accuracy of about ±1 to ±0.3 mm. External *micrometer screw gauges* can be used for the measurement of shorter lengths. They depend for their accuracy on the accuracy of the screw thread responsible for the movement of the anvil, the amount of rotation that is needed to close the anvils of the instrument on the object being determined. Modern versions include the use of an electrical sensor to convert the movement of the screw into a digital display. The range of such instruments lies within 0 to 600 mm and they have an accuracy of about ±0.002 mm. *Vernier callipers* are essentially just steel rules with jaws to enable the sides of the object being measured to be located. They have ranges within 0 to 1 m and an accuracy of about ±0.02 to ±0.06 mm.

Other instruments used to measure lengths tend to involve levers and/or gears to magnify the movement of a plunger in contact with the object whose displacement is being monitored. Figure 11.7 shows the basic form of the *dial gauge*. This uses a system of gears to convert the linear displacement of its plunger into a highly magnified rotation of a pointer over a scale. With scale divisions of 0.01 mm, they are able to indicate small gradual changes of the order of ±0.025 mm to within 0.003 mm and have an accuracy of about ±0.005 to ±0.020 mm, the higher figure being for when the pointer has made many revolutions round the scale and so large displacements are involved.

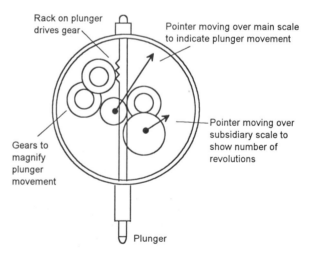

Figure 11.7 *Dial gauge*

11.2.2 Pneumatic methods

Figure 11.8(a) shows the form of a *pneumatic comparator*. This works on the principle that the rate of flow of air through an opening depends on the movement of the needle into the opening and hence the displacement of the plunger responsible for the needle movement. The flow rate of the air can be measured using a rotameter flow gauge (see Chapter 12). Other forms can be used for other forms of displacement. For example, the internal measurements of diameters (Figure 11.8(b)) can be measured by the proximity of the cylinder from which the air escapes to the internal walls of the hole determining the rate at which air escapes. Similarly the method can be used to give a measure of bore straightness (Figure 11.8(c)) or measurements of hole taper (Figure 11.8(d)).

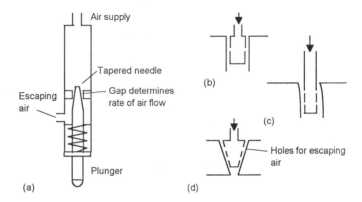

Figure 11.8 *Pneumatic comparator form for (a) linear displacement, (b) hole diameter, (c) bore straightness, (d) hole taper*

11.2.3 Electrical methods

There are a number of methods that are used to give an electrical output related to a displacement. These include the linear potentiometer, capacitive systems, variable reluctance systems, LVDTs and linear inductosyns.

The *linear potentiometer* (see Section 5.2.1) uses the displacement to move the sliding contact over a linear potentiometer and so produce a voltage output related to the displacement. The range of displacements that can be measured depend on the track used but tend to be within the range 1 mm to 1 m. The resolution of a wirewound track is limited by the diameter of the wire used and typically ranges from about 0.5 mm for a finely wound one to 1.5 mm for a coarsely wound one. Conductive plastic tracks have no such resolution problems. Errors due to non-linearity of track tend to range for wirewound tracks from less than 0.1% to about 1% and for conductive plastics to be about 0.05%. The conductive plastic is more sensitive to temperature changes and thus such changes affect its accuracy. Loading can also introduce non-linearity (see Section 5.2.1).

Capacitive systems can involve the displacement varying the separation of capacitor plates, varying the area of overlap of capacitor plates or varying the amount of dielectric between capacitor plates (see Section 5.3.1). Simple systems involving just a single capacitor have the problem that air is generally the dielectric, or at least part of it, and variations in the relative permittivity of air as a result of changes in pressure and humidity can thus affect the accuracy possible. An alternative which overcomes this is the push-pull form of capacitor system which involves two capacitors. Figure 11.9 shows two such forms of capacitive system. In (a) the displacement causes a central plate to move in such a way that it moves closer to one fixed plate and so increases the capacitance between it and the central plate and further away from the other fixed plate and so decreases the capacitance between it and the central plate. This form of system is particularly useful for small displacements. In (b) the movement of the central plate is along the axis between two pairs of fixed parallel plates. Because the central plate is earthed we essentially have for each pair of fixed plates two capacitors in series: fixed plate to central earthed plate to other fixed plate. Thus the capacitances between a pair of fixed plates will depend on the amount of the central plate between them, this determining the areas of overlap. Hence the displacement of the central plate affects the capacitances of the two sets of fixed electrodes. This form of capacitance system can be used for large displacements. With both forms, the two capacitors can be incorporated in an a.c. bridge (c) so that the out-of-balance voltage becomes a measure of the displacement. Capacitive systems have high output impedance and can have high accuracy, of the order of ±0.01%.

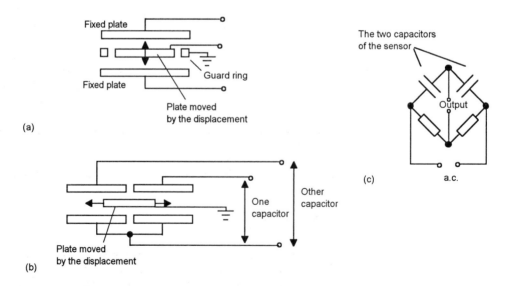

Figure 11.9 *(a) and (b) capacitive displacement sensors, (c) a.c. bridge*

Variable reluctance systems (see Section 5.4.1) can take a number of forms. One form (Figure 11.10) involves a central ferromagnetic plate being moved so that it reduces the reluctance in one magnetic circuit and increases it in the other. It is a push-pull form of system. The two coils are connected in adjacent arms of an a.c. bridge and the out-of-balance signal becomes a measure of the displacement. Such comparators have short ranges, typically 0 to 10 mm, poor linearity and an accuracy of about ±0.5%.

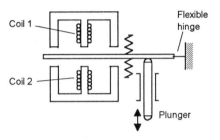

Figure 11.10 *Variable reluctance system*

The *linear variable differential transformer* (LVDT) (see Section 5.4.2) is widely used for the measurement of displacement. LVDTs are available with ranges which cover small displacements of the order of 0 to 0.2 μm or less to larger displacements of the order of 0 to 500 mm. They are robust, linear, have high reliability, high sensitivity and an accuracy of about ±0.5%.

The *linear inductosyn* (Figure 11.11) consists of a track along which a slider moves, the slider being attached to the object for which the displacement is required. The track, which may be several metres long, has mounted on it a fine metal wire formed into a single, continuous, rectangular waveform. Typically the pitch of the waveform is about 2 mm. The slider sits on top of the track and is much shorter, typically about 50 or 100 mm, and has two separate wires formed into the rectangular waveform. The waveforms are, however, displaced from each other by one-quarter of the waveform pitch. This means that when one of the slider wires is aligned with the track wire, the other is out of step. When an alternating current $V \sin \omega t$ flows through the track wire, e.m.f.s are induced in the slider wires. For one slider wire, the induced e.m.f. v_1 can be represented by:

$$v_1 = kV \sin \omega t \sin\left(\frac{2\pi x}{p}\right)$$

where x is the displacement, p the pitch, V the maximum value of the track alternating voltage and k a constant. The other slider wire will always be a quarter of cycle out of phase with the above wire and so its induced e.m.f. v_2 can be represented by:

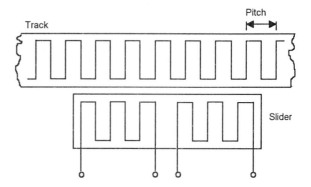

Figure 11.11 *Linear inductosyn*

$$v_2 = kV \sin \omega t \sin\left(\frac{2\pi x}{p} + 90°\right)$$

The sum of these two voltages is an alternating voltage with an amplitude which is cyclic, repeating itself every time the displacement x changes by the pitch p.

$$v_1 + v_2 = kV \sin \omega t \sin\left(\frac{2\pi x}{p}\right) + kV \sin \omega t \sin\left(\frac{2\pi x}{p} + 90°\right)$$

$$= kV \sin \omega t \left[\sin\left(\frac{2\pi x}{p}\right) + \sin\left(\frac{2\pi x}{p} + 90°\right) \right]$$

$$= kV \sin \omega t \left[2 \sin\left(\frac{2\pi x}{p} + 45°\right) \cos(-45°) \right]$$

$$= \left[KV \sin\left(\frac{2\pi x}{p} + 45°\right) \right] \sin \omega t$$

Because of this repetition, the inductosyn cannot unambiguously interpret displacements greater than the pitch. For this reason, an inductosyn with a fine pitch is generally used in conjunction with one of a coarser pitch. The accuracy of a fine pitch inductosyn can be about ±2.5 μm.

11.2.4 Optical methods

These include position sensitive photocells, the use of Moiré fringes and interferometers.

With the one form of *position sensitive photocell* (Figure 11.12), termed a *split cell*, a defined beam of light falls on two identical photocells which are side by side. When the beam of light is central then equal segments of each photocell are illuminated. The output from each cell is connected to a differential amplifier and so there is no output from the amplifier. However, a displacement of the light beam results in more light falling on one cell than the other. The consequence of this is that there is more output from one cell than the other and so the differential amplifier gives an output. The output is proportional to the displacement of the beam across the cells until the point is reached when all the light falls on just one cell. Four photocells placed in two perpendicular directions in a plane can be used to sense two axes of motion.

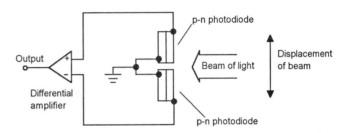

Figure 11.12 *Split cell*

An *array of diodes* (see Section 6.9) is often used in motion/distance determining instruments. For example, the auto-focus camera uses such a device to determine distances of objects from the camera and hence set the focusing. Other instruments might use the arrangement for pattern recognition. The output from a diode depends on the intensity of light falling on it. Generally each diode has its own charge storage capacitor so that its output can be stored until it is selected, a multiplexer being used to select such outputs for processing by a microprocessor.

Moiré fringes are produced when light passes through two gratings which have rulings inclined at a slight angle to each other. Movement of one grating relative to the other causes the fringes to move. Figure 11.13(a) illustrates this. Figure 11.13(b) shows a transmission form of instrument using Moiré fringes and Figure 11.13(c) a reflection form. With both, a long grating is fixed to the object being displaced. With the transmission form, light passes through the long grating and then a smaller fixed grating, the transmitted light being detected by a photocell. With the reflection form, light is reflected from the long grating through a smaller fixed grating and onto a photocell.

Coarse grating instruments might have 10 to 40 lines per millimetre, fine gratings as many as 400 per millimetre. Movement of the long grating relative to the fixed short grating results in fringes moving across the view of the photocell and thus the output of the cell is a sequence of pulses which can be counted. The displacement is thus proportional to the number of pulses counted. Displacements as small as 1 μm can be detected by this means. Such methods have high reliability and are often used for the control of machine tools.

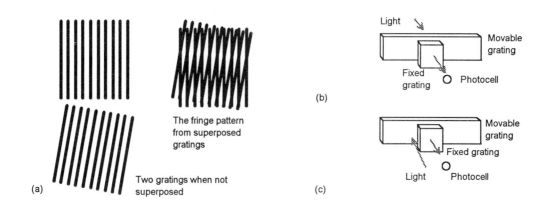

Figure 11.13 *(a) Moiré fringes, (b) transmission form of instrument, (c) reflection form of instrument*

Laser interferometers can be used for highly accurate measurements of distances. Figure 11.14 shows one form. A helium-neon laser is used since it emits light at two distinct frequencies, both about 5×10^{14} Hz but differing by 2×10^6 Hz. They are plane polarised at right angles to each other. The light of both frequencies from the laser is split at the first beam splitter so that a reference beam is provided for detection by photodetector A. There the two frequencies are combined to produce a beat frequency fluctuation in output of 2×10^6 Hz. The beam transmitted by the first beam splitter then passes to the second beam splitter where use is made of the different polarisations of the two frequencies to separate them, light of one frequency being reflected by a fixed cube corner back to the beam splitter and from it to photodetector B. The light of the other frequency is transmitted through the second beam splitter to a movable cube corner. This reflects it back to the second beam splitter and through it to photo-detector B. The two frequencies are then combined to produce a beat frequency. With the movable cube corner stationary, the beat frequency produced by detector B will be the same as that produced by detector A. However, when it is moving there is a Doppler shift in frequency, the shift depending on the velocity and being about 3.3 MHz per m/s. The difference in the two frequencies given by detector B is now no longer the same as that given by detector A. The outputs from the two detectors are subtracted and the number of cycles counted, the count being proportional to the distance moved by the movable cube corner. A velocity of 0.01 m/s would give a Doppler shift of about 33 000 Hz and thus if this velocity was maintained for 1 s, i.e. a displacement of 0.01 m, then 33 000 cycles would be counted. Since a count of 1 can be resolved, this means a resolution of 0.01/33 000 or 3.0×10^{-7} m.

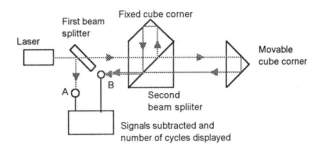

Figure 11.14 *Laser interferometer*

11.3 Angular displacement measurement

The following are some of the methods commonly used for the measurement of angular displacement:

1 *Circular and helical potentiometers*
These give a change in resistance related to the angle through which the slider has rotated.

2 *Synchos and resolvers*
These are electromechanical devices with alternating currents in stator windings being used to induce an alternating current in a rotor winding.

3 *Encoders*
Encoders give a digital output related to the angular position of a shaft.

11.3.1 Circular and helical potentiometers

Circular and helical potentiometers (see Section 5.2.1) give an output of a change in resistance for an angular rotation of the shaft moving the wiper over the resistance track. Circular potentiometers have a track which limits the angular displacement possible to less than 360° and typically have an accuracy of about ±1%. Helical potentiometers have many turns and so can be used for rotations greater than 360°. They have an accuracy which can be as good as ±0.002%. In any use of potentiometers, consideration must be given to any non-linearity introduced by loading.

11.3.2 Synchros and resolvers

The term *synchro* is given to a range of electromechanical devices which are used for the measurement of angular displacements, angle transmission from one machine to another and computation of the rectangular components of vectors. They are small single phase rotary transformers. A synchro element consists of three stator windings at 120° spacing around a cylindrical stator case (Figure 11.15). A concentric inner rotor coil is free to rotate within these stator coils. An alternating current input to the rotor coil induces outputs in each of the secondary coils. The input to the rotor is generally by means of slip rings. The relationship between the outputs from the three stator coils depends on the angular position of the rotor and thus can be used as a measure of the angular position.

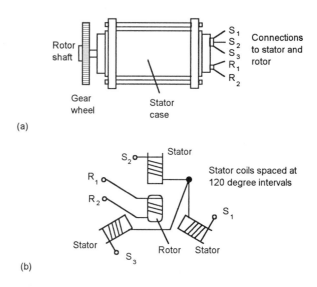

(a)

(b)

Figure 11.15 *Synchro: (a) external appearance, (b) internal form of arrangement of the elements*

Synchro elements are frequently used in pairs (Figure 11.16). One acts as a transmitter and the other as a receiver. When the rotors of the two elements are at the same angular positions with reference to their stator coils then the e.m.f.s in corresponding stator coils are identical and no potential difference exists between the transmitter and the receiver. However, if the rotors are not in the same angular positions, the e.m.f.s induced in corresponding stator coils are not the same. Currents then pass between the coils and produce magnetic fields which in turn produce forces on the rotors which cause them to become aligned at the same angle. Thus movement of the rotor of one synchro results in movement of the other rotor to the same angular position. This means that the transmitting rotor can be mechanically coupled to a shaft and the receiving rotor some distance away. The arrangement might thus be used for remote indications of angular data, the receiver being used to move a pointer across a scale, or as part of a control system.

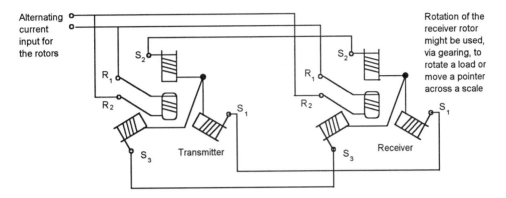

Figure 11.16 *A synchronous transmission link*

Sometimes there is a requirement for the resolution of a vector into its components or the generation of cosine and sine functions, i.e. a function generator. This can be achieved by the use of resolvers. *Resolvers* are a form of synchro with two stator windings at right angles to each other and generally two rotor windings at right angles to each other (Figure 11.17). If one of the stator windings is excited with an alternating signal of constant amplitude and the other stator is short circuited, alternating currents are induced in the rotor. If the rotor is rotated through an angle θ from the position where it gave zero output, then the output from one rotor is proportional to $\sin \theta$ and the other to $\cos \theta$.

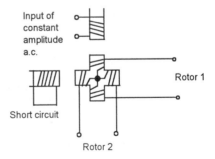

Figure 11.17 *Resolver*

11.3.3 Encoders

The term *encoder* is used for a device that provides a digital output as a result of angular or linear displacement. An *increment encoder* gives an output related to changes in angular or linear displacement from some datum, while an *absolute encoder* gives the actual angular or linear position.

Figure 11.18 shows the basic form of an incremental encoder for the measurement of angular displacement. A beam of light, from perhaps a light-emitting diode (LED), passes through slots in a disc and is detected by a light sensor, e.g. a photodiode or phototransistor. When the disc rotates, the light beam is alternately transmitted and stopped and so a pulsed output is produced from the light sensor. The number of pulses produced is proportional to the angle through which the disc has rotated, the resolution being proportional to the number of slots on a disc. With 60 slots equally spaced, a movement from one slot to the next is a rotation of 360°/60 = 6°. By using offset slots it is possible to have over a thousand slots for one revolution and so a resolution of the order of 0.36°.

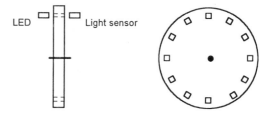

Figure 11.18 *Basic form of an incremental encoder*

The absolute encoder differs from the incremental encoder in having a pattern of slots with each angular position being uniquely defined by the slot pattern. Figure 11.19 shows the form of such an encoder using three sets of slots and so giving a three-bit output. Typical encoders tend to have up to 10 or 12 tracks. The number of bits in the binary output is equal to the number of tracks. Thus with 3 tracks there will be 3 bits and so the number of positions that can be detected is $2^3 = 8$, i.e. a resolution of 360/8 = 45°. With 10 tracks there will be 10 bits and the number of positions that can be detected is $2^{10} = 1024$ and the angular resolution is 360/1024 = 0.35°.

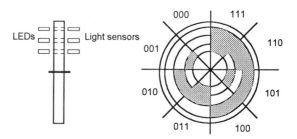

Figure 11.19 *A 3-bit absolute encoder*

The code shown in Figure 11.19 is conventional binary system. However, many encoders use different code patterns to avoid errors resulting from small misalignments. To illustrate this, Figure 11.20(a) shows the binary pattern used in Figure 11.19 and Figure 11.20(b) the Gray code pattern which can aid in avoiding such errors. With the normal binary code consecutive numbers can give changes in more than one of the binary digits, e.g. 011 to 100 requires all three bits to change, and so more than one exposed or blocked hole in the encoder. There may be differences in the transition times of the circuits used for the different bits and so we might, for example, have 011 momentarily giving 111 before 100 is reached. This occurrence of 111 is only momentary but could end up giving an erroneous result. The Gray code is designed so that consecutive numbers only give a change in one binary digit and so only one exposed or blocked hole in the encoder. Since only one change of bit occurs there can be no intermediate erroneous result. Table 11.1 shows how the Gray code compares with the binary code for the decimal numbers 0 to 9. Conversion from binary to Gray code and vice versa can be accomplished by using standard logic gates (Figure 11.21).

Table 11.1 *The Gray code*

Decimal	Binary	Gray code	Decimal	Binary	Gray code
0	0000	0000	5	0101	0111
1	0001	0001	6	0110	0101
2	0010	0011	7	0111	0100
3	0011	0010	8	1000	1100
4	0100	0110	9	1001	1001

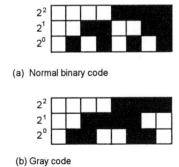

(a) Normal binary code

(b) Gray code

Figure 11.20 *Code patterns*

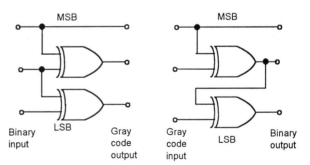

Figure 11.21 *Conversions from binary to Gray code and vice versa using XOR logic gates*

11.4 Measurement of velocity

This section is a brief consideration of methods that are used for the measurement of linear velocities and angular velocities. One general method that can be used is to use a linear or angular displacement sensor, e.g. a potentiometer, and then electrically differentiate the signals from the sensor to give the velocity, or alternatively use an accelerometer and integrate the output.

11.4.1 Measurement of linear velocity

The average velocity over some distance can be determined by generating a signal when the moving object passes the start point and another signal when it passes the end point of the measured distance. The time interval between these two signals is measured and the average velocity then computed. The sensors used to detect the passage of the moving object might be photoelectric sensors (Figure 11.22), or proximity sensors of the form described earlier in this chapter. The time measurement might be achieved by displaying the two pulses on a cathode ray oscilloscope and using the calibrated time base to determine the time between the pulses. Another possibility is to use an electronic timer which is triggered to start by the first pulse and stopped by the second.

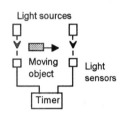

Figure 11.22 *Average velocity measurement*

Electromagnetic induction can be used to provide a linear velocity sensor. The e.m.f. generated by a conductor moving in a magnetic field or a magnetic field moving past a fixed conductor is BLv, where B is the flux density at right angles to the direction of motion, L the length of the conductor and v the velocity. Thus the induced e.m.f. is proportional to the velocity. Figure 11.23 shows the basic form of a sensor based on this principle. The moving object is used to move a permanent magnet core axially in a pair of coils. The two coils are connected in series in such a way that the induced e.m.f.s due to each coil add up.

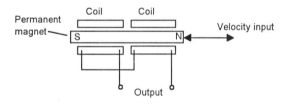

Figure 11.23 *Linear velocity sensor*

11.4.2 Measurement of angular velocity

Various forms of *tachometer generators* are used for the measurement of the angular velocities of rotating shafts. They are essentially just d.c. or a.c. generators. The *d.c. tachometric generator* (Figure 11.24) consists of a rotor which rotates in the field of a permanent magnet or electromagnet stator, the output from the rotor coil being extracted, via slip rings, as a measure of the angular velocity.

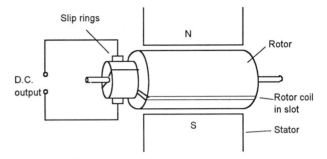

Figure 11.24 *The d.c. tachometric generator principle*

The output of the d.c. tachometric generator is approximately proportional to the angular velocity. The generator gives a relatively high voltage output, typically 5 V or more per 1000 revs per minute, and so has a high sensitivity. Non-linearity is about 0.1%.

The *a.c. tachometric generator* (Figure 11.25) is generally a two-phase squirrel-cage induction motor. One of the stator windings is supplied with an a.c. voltage and the measurement signal taken from the other stator winding. The moving conductors of the squirrel cage have voltages induced in them by passing through the alternating magnetic field produced by the input voltage. These then produce a magnetic field which sweeps past the output coil and produces an alternating voltage in it. When the rotor is stationary, the output voltage is zero. When the rotor is rotating, the amplitude of the output voltage is proportional to the angular velocity. The range of measurement is typically 0 to 4000 revs per minute with an accuracy of ±0.05% of the full-scale reading.

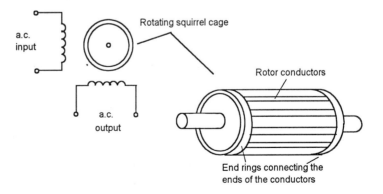

Figure 11.25 *The a.c. tachometric generator principle*

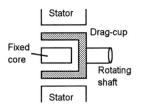

Figure 11.26 *Drag-cup*

Another form of a.c. tachogenerator has a *drag-cup rotor* (Figure 11.26) instead of the squirrel cage form. The rotor is in the form of a cup. The principle of operation is the same as for the squirrel cage. The input alternating voltage in one stator coil induces e.m.f.s in the rotor. When the rotor is rotating these give rise to an alternating voltage output from the other stator coil. This gives a greater accuracy than the squirrel cage form.

The *eddy current drag-cup tachometer* (Figure 11.27) has a permanent magnet which is rotated by the rotating shaft. The rotating magnet induces voltages in the cup. These produce eddy currents which interact with the magnetic field to produce a torque on the cup. This torque is proportional to angular velocity of the magnet. The torque causes the cup to rotate and, in doing so, twist a spring and so produce an opposing torque. The cup thus rotates to an angle where the torque on the cup due to the rotating magnet is balanced by the torque produced by the spring. Thus, in the steady state, the angle through which the cup is twisted is proportional to the angular velocity. The accuracy is typically about ±0.5%.

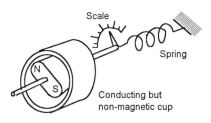

Figure 11.27 *Drag-cup tachometer*

The angular velocity of a rotating shaft can be measured using the *variable reluctance tachogenerator* (see Section 5.5.1). The rotating shaft has a toothed wheel of ferromagnetic material attached to it and as the wheel rotates, so the magnetic flux linked by a pickup coil wound round a permanent magnet changes each time a tooth passes. The maximum value of the induced e.m.f. in the coil is a measure of the angular velocity, alternatively the number voltage pulses produced per second can be counted and used as a measure of the angular velocity.

Another possibility is an *incremental encoder* wheel (see Section 11.3.3) which rotates with the shaft. The number of pulses produced per second is a measure of the angular velocity.

11.5 Accelerometers

Accelerometers are used to measure absolute acceleration. There are a number of forms of accelerometer; they all, however, can be considered to have the elements shown in Figure 11.28, such types of instruments being termed *seismic*. The accelerometer is rigidly fastened to the body undergoing the acceleration and the relative displacement measured of a mass connected by a spring to the accelerating body. Damping is present.

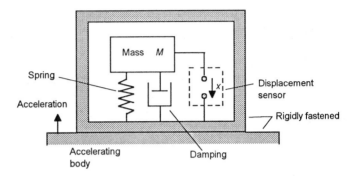

Figure 11.28 *Basic structure of an accelerometer*

Consider the forces acting on the mass M. If the casing is given a constant acceleration a then the mass experiences a force of ma in the opposite direction. The situation is rather like a car accelerating, the car driver experiences a force in the opposite direction to the acceleration. This force on the mass is opposed by the force exerted by the spring. This is proportional to the extension/contraction x_s of the spring and thus can be expressed as kx_s, where k is the force constant for the spring and x_s the steady-state value of the displacement. Thus at equilibrium when the mass is at rest relative to the casing, we have:

$$ma = kx_s$$

The steady-state sensitivity is thus:

$$\text{steady-state sensitivity} = \frac{x_s}{a} = \frac{m}{k}$$

If, however, we have a sudden change in acceleration to a at time $t = 0$, then the steady-state conditions cannot be considered to prevail and the mass can no longer be assumed to be at rest relative to the casing. The mass is acted on by a force ma. Opposing this force is that arising from the spring kx, where x is the displacement of the mass at some instant of time. Also opposing the motion is the damping. The damping may be taken as being proportional to the velocity and so expressed as $c\,dx/dt$, where c is a constant. Thus the net force acting on the mass:

$$\text{net force} = ma - kx - c\frac{dx}{dt}$$

This force causes the mass to accelerate with an acceleration d^2x/dt^2, this being different to the acceleration a. Thus:

$$ma - kx - c\frac{dx}{dt} = m\frac{d^2x}{dt^2}$$

This equation can be written as:

$$m\frac{d^2x}{dt^2} + c\frac{dx}{dt} + kx = ma$$

This second order differential equation (see Section 3.3) can be written in terms of the natural angular frequency ω_n, i.e. the frequency with which the system will oscillate with in the absence of damping, and the damping constant ζ where $\omega_n = \sqrt{(k/m)}$ and $\zeta = c/2\sqrt{(km)}$:

$$\frac{1}{\omega_n^2}\frac{d^2x}{dt^2} + \frac{2\zeta}{\omega_n}\frac{dx}{dt} + x = \frac{m}{k}a$$

Multiplying by ω_n^2 and taking the Laplace transform gives:

$$s^2X(s) + 2\zeta\omega_n sX(s) + \omega_n^2 X(s) = \frac{m}{k}\omega_n^2 A(s) = A(s)$$

The transfer function $G(s)$ for the system, output displacement input acceleration (see Section 3.4.2), is thus:

$$G(s) = \frac{X(s)}{A(s)} = \frac{1}{s^2 + 2\zeta\omega_n s + \omega_n^2}$$

A damping factor ζ less than 1 describes an oscillating system and when greater than 1 a system which sluggishly climbs to the steady-state value. A damping factor of about 0.7 is often employed to minimise the response time without giving too great an overshoot. See Section 3.3 for further discussion of second order systems.

11.5.1 Vibrations

If the accelerometer system is designed mainly as a vibration sensor then the input is sinusoidal, or can be considered to be the sum of a number of sinusoidal waveforms. We are then concerned with the response of the system to such an input. A vibration can, in general, be described by an equation of the form $y = Y \sin \omega t$, where y is the displacement at time t, ω the angular frequency and Y the maximum value of the displacement from the equilibrium position. The variation of the velocity v with time is obtained by differentiating the equation. Thus:

$$v = \frac{dy}{dt} = -\omega Y \cos \omega t$$

The acceleration a is obtained by a further differentiation. Thus:

$$a = \frac{dv}{dt} = -\omega^2 Y \sin \omega t = -\omega^2 y$$

The acceleration is proportional to the displacement from the equilibrium position, being greatest at the maximum displacement. The maximum value of the acceleration is thus $\omega^2 Y$. The maximum acceleration depends on the square of the frequency. Thus a frequency of 1000 Hz produces, for the same maximum displacement, an acceleration which is 10 000 times that given by a frequency of 10 Hz.

The differential equation describing the behaviour of the accelerometer can thus be written as:

$$\frac{1}{\omega_n^2}\frac{d^2x}{dt^2} + \frac{2\zeta}{\omega_n}\frac{dx}{dt} + x = \frac{m}{k}a = \frac{\omega^2}{\omega_n^2}y$$

Hence, multiplying by ω_n^2 and taking the Laplace transform:

$$s^2 X(s) + 2\zeta\omega_n X(s) + \omega_n^2 X(s) = \omega^2 Y(s)$$

The transfer function relating the input displacement y and the output displacement x is thus:

$$G(s) = \frac{X(s)}{Y(s)} = \frac{\omega^2}{s^2 + 2\zeta\omega_n s + \omega_n^2}$$

We can use the transfer function with such an input or use the frequency response function $G(j\omega)$. Then (see Section 3.5 and 3.5.2):

$$G(j\omega) = \frac{\omega^2}{(j\omega)^2 + 2\zeta\omega_n(j\omega) + \omega_n^2}$$

and the amplitude ratio of the output to input displacement is:

$$\text{amplitude ratio } r = \frac{\left(\dfrac{\omega^2}{\omega_n^2}\right)}{\sqrt{\left[\left(1 - \dfrac{\omega^2}{\omega_n^2}\right)^2 + 4\xi^2\dfrac{\omega^2}{\omega_n^2}\right]}}$$

and

$$\text{phase angle } \phi = -\tan^{-1}\left[\frac{2\xi\dfrac{\omega}{\omega_n}}{1 - \dfrac{\omega^2}{\omega_n^2}}\right]$$

Thus if the input is a displacement $y = Y \sin \omega t$ then the output is a displacement $x = rY \sin(\omega t + \phi)$.

If the applied angular frequency ω is much greater than the natural frequency ω_n, i.e. $\omega/\omega_n \gg 1$, then the amplitude ratio tends to the value 1. The amplitude of the mass thus becomes the same as that of the casing and so the measured displacement is a true indication of the movement of the casing. This form of instrument is often called a *vibrometer*. Thus to measure displacement amplitudes with a seismic sensor, it must have a natural frequency below the frequencies that are to be applied to it. Figure 11.29 shows how the amplitude ratio varies with the angular frequency and damping ratio.

When ω_n is much smaller than ω then the amplitude ratio tends to the value $(\omega/\omega_n)^2$. But, for an input of $y = Y \sin \omega t$ the maximum acceleration a is $-\omega^2 Y$. Thus the amplitude ratio is proportional to the maximum acceleration and therefore the output displacement is proportional to the input acceleration. Figure 11.30 shows how the ratio of the output displacement to maximum acceleration varies with angular frequency. The instrument under such conditions is referred to as an *accelerometer* rather than a vibrometer.

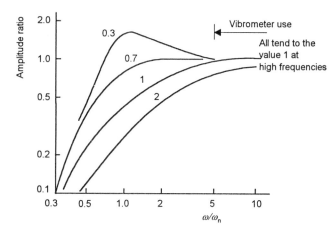

Figure 11.29 *Amplitude ratio at different damping ratios*

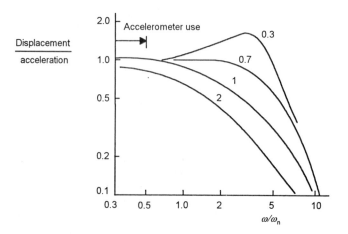

Figure 11.30 *Seismic instrument used as an accelerometer*

11.5.2 Displacement sensors used with accelerometers

The acceleration is obtained from an accelerometer by measurement of the displacement of the mass within the accelerometer. An important part of the specification of a displacement sensor is its sensitivity to accelerations at right angles to the sensing axis, this being termed the *cross-sensitivity*. Accelerations are generally specified in terms of the acceleration due to gravity, i.e. what multiple of that acceleration they are. Displacement sensors that are used include:

1 *Potentiometers*
The movement of the mass causes the slider of the potentiometer to move across the track. Potentiometers can only be used for slowly varying accelerations or low frequency vibrations, accelerations typically in the range 0 to $50g$ where g is the acceleration due to gravity. The natural frequency is low, of the order of 10 to 90 Hz. The cross-sensitivity is typically about ±1% of the full-scale output and the accuracy about ±1%.

2 *Strain gauges*
Strain gauges are often attached to the spring element, this then being in the form of a cantilever. Another possibility is to use unbonded strain gauges, i.e. just a wire or wires attached between the mass and the case. They typically have a natural frequency in the region 20 Hz to 800 Hz. Strain gauges give a cross-sensitivity of about ±2% of the full-scale output and an accuracy of ±1%. They are typically used for accelerations in the range 0 to $200g$.

3 *Capacitive sensors*
These involve the displacement changing the separation of capacitor plates. One form has the seismic element in the form of a disk with spiral elements attached to it. The disk is sandwiched between the capacitor plates. The movement of air through holes in the disk provides the damping, the spiral elements provides the spring force. Such an instrument has typically a cross-sensitivity of ±1% of the full-scale output and is used for accelerations up to $1000g$, having a natural frequency of 3000 Hz.

4 *Piezoelectric sensor*
These are very widely used for shock and vibration measurements. Because of the basic characteristics of such sensors (see Section 5.8), they are not, however, suitable for constant acceleration measurements. They typically have a cross-sensitivity of ±2 to 4% of the full-scale output and an accuracy of ±1%. Instruments are available in a wide range of specifications, with sensitivities ranging from about 0.003 pC/g to 1000 pC/g. A sensor for use with low frequency vibrations might have a natural frequency of 7 kHz while a shock accelerometer might have a natural frequency of 250 kHz. Figure 11.31 shows the typical form of two piezoelectric accelerometers.

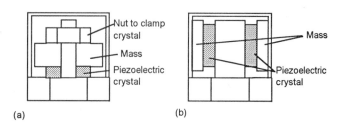

Figure 11.31 *Basic forms of piezoelectric accelerometers: (a) single-ended compression, (b) shear. Note that the piezoelectric crystal as well as being the sensor also provides the spring force and the damping*

11.5.3 Shock

Shock measurements involve the measurement of high accelerations which last for a very short time. A simple example is the shock occurring when an object falls and hits the ground. As an illustration, consider an object falling freely from a height of 3 m onto a concrete floor. The velocity v with which the object hits the ground is given by:

$$v = \sqrt{2gh} = \sqrt{2 \times 9.8 \times 3} = 7.7 \text{ m/s}$$

When it hits the ground it quickly decelerates and comes to rest. Suppose this deceleration takes 2 ms. The average acceleration during that time is thus:

$$\text{average acceleration} = \frac{7.7 - 0}{0.002} = 3850 \text{ m/s}^2$$

Expressed in terms of the acceleration due to gravity, this is $393g$.

Problems

1 A linear potentiometer is to be used for the measurement of linear displacement. The potentiometer selected is wirewound with a resistance of 10 kΩ and a total of 1000 turns. What will be the resolution of the potentiometer and the loading error when at its mid-range displacement it is connected to an output display of resistance 10 kΩ.

2 A rotary potentiometer is to be used for the measurement of angular displacement. The potentiometer has a resistance of 4 kΩ and a maximum angular rotation of 320° and is required to give an output of 10 mV per degree of rotation. Neglecting any consideration of loading, what voltage is required for the supply?

3 A measurement of angular displacement to a resolution of 1 minute of arc is required. An absolute encoder is to be used. Determine the number of tracks required.

4 An absolute encoder has 7 tracks. What will be the maximum decimal number that will be indicated by one revolution through 360°?

5 Explain why Gray coding is often used with encoders rather than normal binary coding.

6 An inductive pickup is being used with a toothed wheel attached to a rotating shaft in order to measure the angular velocity of that shaft. The toothed wheel has 120 teeth. The output from the pickup is displayed on a frequency meter which counts the number of cycles occurring over a period during which its gate is open. If the gating period is set to 10 ms and the count is 0110 on a four-digit display, what is the angular velocity?

7 Suggest a method that could be used to provide the signal for the speedometer of a car.

8 A vibrating object is vibrating with simple harmonic motion. What will be the maximum acceleration in terms of the acceleration due to gravity of the body when it has a frequency of 10 Hz and an amplitude of 1 mm?

9 A vibrometer has a natural frequency of 20 Hz and a damping ratio of 0.7. What will be the amplitude ratio between the amplitude of seismic instrument mass and the input vibration if the input has a frequency of (a) 30 Hz, (b) 1 kHz? Explain the significance of the results in relation to the measurements made.

10 A vibrometer has a natural frequency of 1 Hz. What is the lower frequency limit of the instrument for 2% error if the damping ratio is effectively zero?

11 An accelerometer gives an output of 10 mV/g. Design a signal conditioning system that could be used to give a velocity signal of 0.25 V per m/s.

12 A seismic instrument with a natural frequency of 3 Hz and damping factor 0.5 is used to monitor the vibration of an object which is vibrating with a displacement x in millimetres given by:

$$x = 0.10 \sin 4\pi t + 0.05 \sin 8\pi t$$

Determine the displacement variation with time indicated by the instrument.

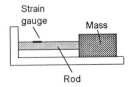

Figure 11.32 *Problem 13*

13 An accelerometer for the measurement of shock is of the form shown in Figure 11.32, consisting of a strain gauge of resistance 120 Ω and gauge factor 2.0 attached to the surface of a rod. The rod has a cross-sectional area of 2×10^{-4} m and is made of a material with a Young's modulus of 1 GPa, the mass is 50 g. Determine the change in gauge resistance per g of acceleration.

12 Flow measurement

Fluid flow measurements are carried out in a wide variety of industries and for a wide range of applications. For example, such measurements are made in industrial process control where fluid flow along pipes is controlled, in the measurement of flow rate of petroleum and natural gas through pipes, in the determination of the amount of petrol you have to pay for when you fill up your car, in the measurement of water along channels or in rivers, etc. Measurements may be of the volume flow rate, the mass flow rate, the flow rate in open channels or the velocity at a point in a fluid. There are a wide variety of methods involved in such flow measurements and this chapter can but touch on the more widely used methods.

12.1 Basic principles of fluid flow

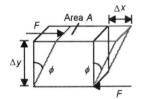

Figure 12.1 *Shear*

The term *fluid* is can refer to both liquids and gases. *Viscosity* is a measurement of a fluid's resistance to flow. Thus a low viscosity fluid will flow more easily than a high viscosity one. In this context of ease of flow, you can think of a high viscosity fluid in terms of treacle or a thick oil and a low viscosity fluid as water.

Consider a shear force F applied to a rectangular block along a face of area A (Figure 12.1). The shear stress τ is the force per unit area, i.e. F/A. The effect of the shear force is to deform the block. With a solid, the angle ϕ through which the block is sheared is constant, not changing with time, and is proportional to the shear stress. The angle is termed the *shear strain*. However, with a fluid the angle ϕ changes with time and flow occurs. This is what basically distinguishes a fluid from a solid (note that solids can show a phenomenon called creep where they do change very slowly with time). In a fluid termed a *Newtonian fluid*, the rate of change of the shear strain with time is proportional to the shear stress. The constant of proportionality, termed the *dynamic viscosity* η, is thus:

$$\text{dynamic viscosity } \eta = \frac{\text{shear stress}}{\text{shear strain rate}}$$

The unit of the dynamic viscosity is Pa s.

If Δv is the velocity of the top surface of the rectangular block relative to its base, then $\Delta v = \Delta x/\Delta t$. But, for small angles of deformation, ϕ is approximately $\Delta x/\Delta y$. Thus $\phi = \Delta v \, \Delta t/\Delta y$ and the strain rate is $\Delta v/\Delta y$. Hence:

$$\text{dynamic viscosity } \eta = \frac{\tau}{(\Delta v/\Delta y)}$$

$\Delta v/\Delta y$ is the velocity gradient. Thus, writing the velocity gradient in its differential form, for a Newtonian fluid we have:

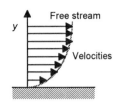

Figure 12.2 *Velocity gradient*

shear stress $\tau = \eta \dfrac{dv}{dy}$

When a fluid flows over a solid surface as a result of a shear force, there is a velocity gradient (Figure 12.2). As the distance away from the solid surface increases then the velocity increases until well away from the surface the fluid assumes its free stream velocity. The layers of fluid between the free stream and the boundary in which there is the velocity gradient are termed the *boundary layers*.

Viscosities vary with temperature. The term *kinematic viscosity* is used for the dynamic viscosity divided by the fluid density. It has the unit m²/s.

Not all fluids are Newtonian fluids. A non-Newtonian fluid is characterised by a shear stress which is not proportional to the velocity gradient. We can consider that the viscosity depends on the rate of shearing, i.e. the velocity gradient. Paints are designed to be non-Newtonian in that they have a viscosity which is low at high rates of shearing and lower at low rates of shearing. This means that when you apply the paint by a brush and give high rates of shearing it has a low viscosity and flows readily onto the surface being painted, but when you stop painting the viscosity increases and the paint does not flow off the surface.

12.1.1 Liquids and gases

Liquids and gases are fluids. However, they differ in that liquids are difficult to compress and effectively might be regarded as incompressible. This means they have densities which are independent of pressure. However, because liquids expand when the temperature rises, they have densities which depend on temperature. Gases are easy to compress and have a density which depends on both the pressure and temperature. For an *ideal gas* we have:

$$PV = mRT$$

where P is the pressure, V the volume and T the temperature on the Kelvin scale of a mass m of the gas. R is a constant for the gas. Since the density ρ is m/V, then:

$$P = \rho RT$$

The heat required to raise the temperature of a gas depends on whether the gas is allowed to expand and do work. A gas is thus specified as having two specific heats, the specific heat at constant pressure and the specific heat at constant volume. If the change in volume of a gas is carried out in such a way that no heat enters or leaves the system, the change being said to be *adiabatic*, then the corresponding relationship between the pressure and volume of the gas is:

PV^γ = a constant

where γ is the ratio specific heat at constant pressure/specific heat at constant volume.

12.1.2 Laminar and turbulent flow

A fluid can flow along a pipe in either of two different ways, depending on whether the value of its Reynold's number is less than or more than some critical value. *Reynold's number* is a dimensionless number which is defined as being:

$$\text{Reynold's number} = \frac{\rho v d}{\eta}$$

with ρ being the fluid density, η the dynamic viscosity, v the mean velocity of the fluid and d the diameter of the pipe.

When the Reynold's number is below about 2000, the flow of the fluid is *orderly*, the terms *laminar* and *streamline* also being used. With such a form of flow, every particle in the fluid moves in straight lines down the tube parallel to the walls. When the Reynold's number is above about 2000, the flow is *turbulent*. The motion of particles of the fluid down the pipe is now chaotic, each particle following a very irregular path. Figure 12.3 shows the type of velocity (in a direction parallel to the axis of the tube) profile that occurs with both these types of flow. With orderly flow the velocity profile is parabolic and the mean fluid velocity along the pipe is half the maximum flow velocity occurring along the central axis of the tube. With turbulent flow the velocity profile is very much flatter. The mean fluid velocity along the pipe is then about 1.2 times the maximum flow velocity, the actual value depending on the value of the Reynold's number. The profile can, however, be considerably distorted by the presence of a bend in the pipe or a valve or flow meter.

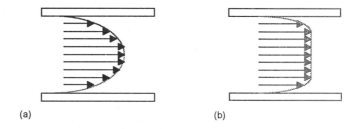

(a) (b)

Figure 12.3 *Velocity profiles: (a) orderly flow, (b) turbulent flow*

12.1.3 Volume and mass flow rate

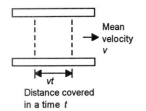

Mean velocity v

vt
Distance covered in a time *t*

Figure 12.4 *Volume flow*

Consider a pipe of cross-sectional area A through which a fluid flows with a mean velocity v (Figure 12.4). In a time t a particle of fluid will have moved a distance of vt down the pipe. Thus the volume of fluid moving down the pipe in a time t is Avt. Hence the volume flow rate Q is given by:

$$Q = Av$$

If ρ is the fluid density, then the volume Avt has a mass of ρAvt and so the rate of mass flow along the tube is:

$$\text{mass flow rate} = \rho AV$$

Consider a tube through which there is steady flow (Figure 12.5). The mass of fluid entering the tube in some time t must equal the mass of fluid leaving it in that time t. Thus:

mass flow rate in = mass flow rate out

We thus have:

$$\rho_1 A_1 v_1 = \rho_2 A_2 v_2$$

If the fluid is incompressible, then $\rho_1 = \rho_2$ and so:

$$A_1 v_1 = A_2 v_2 = \text{the volume flow rate}$$

This equation is called the *equation of continuity*.

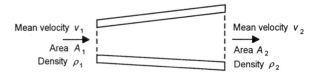

Mean velocity v_1
Area A_1
Density ρ_1

Mean velocity v_2
Area A_2
Density ρ_2

Figure 12.5 *Steady flow through a pipe*

12.1.4 Steady flow energy equation

Consider the application of the principle of the conservation of energy to the system shown in Figure 12.6 with fluid entering it and leaving it.

1 Fluid of mass m at a height z above some datum line will have a potential energy mgz. Thus the potential energy per unit mass of fluid entering the system is gz_1 and the potential energy per unit mass of fluid leaving the system is gz_2.

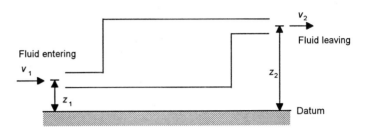

Figure 12.6 *Fluid flow through a system*

2   The kinetic energy of a mass m of fluid in motion with a velocity v is $\frac{1}{2}mv^2$. Thus the kinetic energy per unit mass of fluid entering the system is $\frac{1}{2}v_1^2$ and leaving the system $\frac{1}{2}v_2^2$.

3   The volume of fluid entering the system has to displace the volume ahead of it in order to progress. Consider the element of fluid shown in Figure 12.7. If the pressure exerted is P and this acts over an area A, then the force exerted is PA. If, as a result of this force, a mass m of fluid is moved, then because such a mass has a volume m/ρ, where ρ is the fluid density, the work done is:

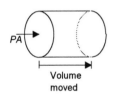

Figure 12.7 *Displacement energy*

$$\text{work done} = \text{force} \times \text{distance} = \frac{mP}{\rho}$$

This is termed the *displacement* or *pressure* energy. Thus the displacement energy per unit mass for the fluid entering at a pressure P_1 and density ρ_1 is P_1/ρ_1 and for the fluid leaving at pressure P_2 and density ρ_2 it is P_2/ρ_2.

If we assume that no external work is done on the system or that no heat enters or leaves it, then applying the principle of the conservation of energy we must have:

$$gz_1 + \tfrac{1}{2}v_1^2 + \frac{P_1}{\rho_1} = gz_2 + \tfrac{1}{2}v_2^2 + \frac{P_2}{\rho_2}$$

For an incompressible fluid $\rho_1 = \rho_2$ and so:

$$gz_1 + \tfrac{1}{2}v_1^2 + \frac{P_1}{\rho} = gz_2 + \tfrac{1}{2}v_2^2 + \frac{P_2}{\rho}$$

This is known as *Bernoulli's equation.*

For a compressible fluid the density depends on the pressure and the displacement energy term has to be modified. For an ideal gas the change in volume arising from a change in pressure is likely to take place so quickly that the energy to change the volume cannot be taken from the surroundings and so the internal energy of the gas supplies it and the temperature drops. The result is an adiabatic change. As a consequence a factor called the *expansibility* factor is introduced into the equation. This involves the ratio of the specific heats γ of the gas (see Section 12.2).

12.2 Measurement of volume flow rate

Instruments commonly used for the measurement of the volume flow rate can be grouped under the following headings:

1 *Differential pressure flow meters*
   A constriction is placed in a pipe and the differential pressure developed across the constriction measured.

2 *Mechanical flow meters*
   Targets, vanes or turbines in the path of the flow are acted on by forces.

3 *Positive displacement meters*
   The fluid is broken up into known volume packets and the number of packets counted.

4 *Vortex flow meters*
   These depend on measurements of the vortices shed by a body as a result of flow over it.

12.2.1 Differential pressure flow meters

Consider a fluid flowing through a constriction in a pipe (Figure 12.8). In order that the same mass and hence volume of fluid can flow per second through both pipes we must have the fluid moving with a higher velocity through the constriction. We have (see Section 12.1.3 for the derivation of the equation):

$$A_1 v_1 = A_2 v_2 = \text{the volume flow rate } Q$$

Figure 12.8 *Flow through a constriction*

and thus a decrease in cross-sectional area from A_1 to A_2 must mean that v_2 is greater than v_1. Thus there is a gain in kinetic energy. For a horizontal pipe with energy conserved, this gain in energy can only occur from a drop in the displacement energy. But the displacement energy per unit mass is the pressure divided by the density. Thus there is a pressure drop at the constriction from P_1 to P_2. It is the pressure difference which has produced the force causing the fluid to accelerate and so increase its velocity.

For an incompressible fluid we can apply Bernoulli's equation:

$$gz_1 + \tfrac{1}{2}v_1^2 + \frac{P_1}{\rho} = gz_2 + \tfrac{1}{2}v_2^2 + \frac{P_2}{\rho}$$

and since we assume a horizontal pipe and $z_1 = z_2$:

$$\tfrac{1}{2}v_1^2 + \frac{P_1}{\rho} = \tfrac{1}{2}v_2^2 + \frac{P_2}{\rho}$$

This can be rearranged to give:

$$\frac{v_2^2 - v_1^2}{2} = \frac{P_1 - P_2}{\rho}$$

Thus, using $A_1 v_1 = A_2 v_2 = Q$:

$$Q = \frac{A_2}{\sqrt{1 - \left(\frac{A_2}{A_1}\right)^2}} \sqrt{\frac{2(P_2 - P_1)}{\rho}}$$

Thus a measurement of the pressure difference $(P_2 - P_1)$ enables the volume flow rate Q to be determined. Note that the relationship between the pressure and the volume rate of flow is non-linear. The equation is often written in the form:

$$Q = EA_2 \sqrt{\frac{2(P_2 - P_1)}{\rho}}$$

where:

$$E = \frac{1}{\sqrt{1 - \left(\frac{A_2}{A_1}\right)^2}}$$

and is termed the *velocity of approach factor*.

In practice this equation is only an approximation. This is because some energy losses occur as a result of friction and they have not been allowed for in the above equation. A correction factor C, termed the *discharge coefficient*, is thus inserted into the equation:

$$Q = \frac{CA_2}{\sqrt{1 - \left(\frac{A_2}{A_1}\right)^2}} \sqrt{\frac{2(P_2 - P_1)}{\rho}} = CEA_2 \sqrt{\frac{2(P_2 - P_1)}{\rho}}$$

C is a function of the pipe size, the Reynold's number for the flow and the form of instrument used. Tables and equations are available from standards organisations to enable the C values to be determined for particular configurations.

The equations have also to be modified if a gas is involved since it is compressible. The factor involves the ratio of the specific heats of the gas and can be represented by the *expansibility factor ε*:

$$Q = \varepsilon CEA_2 \sqrt{\frac{2(P_2 - P_1)}{\rho}}$$

where:

$$\varepsilon = \left[\left(\frac{P_2}{P_1} \right)^{2/\gamma} \frac{\gamma}{\gamma - 1} \frac{1 - (P_2/P_1)^{(\gamma-1)/\gamma}}{1 - (P_2/P_1)} \frac{1 - (A_2/A_1)^2}{1 - (A_2/A_1)^2 (P_2/P_1)^{2/\gamma}} \right]^{1/2}$$

The expansibility factor has the value 1 for an incompressible fluid. As a general rule, the factor is only significant when the fractional change in pressure, i.e. $(P_2 - P_1)/P_1$, is greater than about 0.1.

There are a number of forms of differential pressure devices based on Bernoulli's equation and involving constant size constrictions: the venturi tube, nozzles, Dall tube and the orifice plate. In addition there are other devices involving variable size constrictions, e.g. the rotameter and the gate meter. The following are discussions of the characteristics of the above devices.

The *venturi tube* (Figure 12.9) has a gradual tapering of the pipe from the full diameter to the constricted diameter. The constricted diameter should not be less than $0.224D$ and not more than $0.742D$, where D is the diameter of the full diameter tube. The inlet taper of the tube should be $10.5° \pm 1°$ and the exit taper between $5°$ and $15°$. For this arrangement the discharge coefficient is about 0.99 and the pressure loss occurring as a result of the presence of the venturi tube is about 10 to 15%, a comparatively low value. The pressure difference between the flow prior to the constriction and the constriction can be measured with a simple U-tube manometer or a differential diaphragm pressure cell. The instrument can be used with liquids containing particles, dilute slurries. It is simple in operation, is capable of accuracy of about $\pm 0.5\%$, has a long-term reliability, but is comparatively expensive and has a non-linear relationship between pressure and the volume rate of flow relationship, the basic relationship being:

$$Q = CEA_2 \sqrt{\frac{2(P_2 - P_1)}{\rho}}$$

and hence the volume flow rate Q is proportional to the square root of the pressure difference.

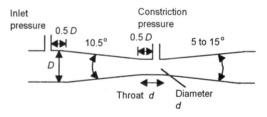

Figure 12.9 *Venturi tube*

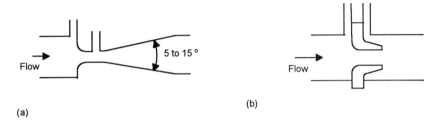

(a)

(b)

Figure 12.10 *(a) Venturi nozzle, (b) flow nozzle*

A cheaper form of venturi is provided by the *nozzle flow meter* (Figure 12.10). Two types of nozzle are used, the venturi nozzle and the flow nozzle. The venturi nozzle (Figure 12.10(a)) is effectively a venturi tube with an inlet which is considerable shortened. The flow nozzle (Figure 12.10(b)) is even shorter. Nozzles have a discharge coefficient value of about 0.96 and produce pressure losses of the order of 40 to 60%. Nozzles are cheaper than venturi tubes, give similar pressure differences, and have an accuracy of about ±0.5%. They have a non-linear relationship between the pressure and the volume rate of flow.

The *Dall tube* (Figure 12.11(a)) is another variation of the venturi tube. It gives a higher differential pressure and a lower pressure drop. The Dall tube is only about two pipe diameters long. An even shorter form, the Dall orifice, is only about 0.3 pipe diameters long. The Dall tube has a discharge coefficient of about 0.66 and is often used where space does not permit the use of a venturi tube.

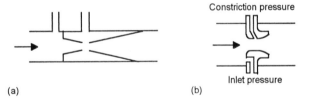

(a)

(b)

Figure 12.11 *(a) Dall tube, (b) Dall orifice*

The *orifice plate* (Figure 12.12) is simply a disc with a hole. The effect of introducing it is to constrict the flow to the orifice opening and the flow channel to an even narrower region downstream of the orifice. The narrowest section of the flow is not through the orifice but downstream of it and is referred to as the *vena contracta*. Because the diameter of the vena contracta cannot be measured, calculations of the flow rate are based on the diameter of the orifice and this results in the low value of about 0.6 for the discharge coefficient used. The pressure difference is measured between a point equal to the diameter of the tube upstream of the orifice and a point equal to half the diameter downstream.

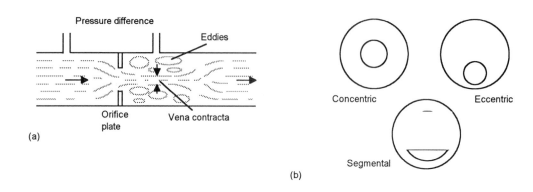

Figure 12.12 *The orifice plate: (a) flow pattern, (b) various forms of orifice*

There are a number of forms of orifice plate, the most widely used being one with a central circular hole. Other forms include the eccentric with an off-centre circular hole, for use where condensed liquids are present in a gas flow or undissolved gases in a liquid flow, and the segmental with just a segment of the central circular hole, for use where particles are present in the liquid flow. The orifice plate has a non-linear relationship between the pressure difference and the volume rate of flow. The orifice plate is simple, reliable, produces a greater pressure difference than the venturi tube and is cheaper but less accurate, about ±1.5%. It also produces a greater pressure drop. Problems of silting and clogging can occur if particles are present in liquids.

The rotameter and the gate meter are examples of *variable area flow meters*. These involve maintaining a constant pressure difference between the main flow and that at the constriction by changing the area of the constriction. The *rotameter* (Figure 12.13) has a float in a tapered vertical tube with the fluid flow pushing the float upwards. The fluid has to flow through the constriction which is the gap between the float and the walls of the tube and so there is a pressure drop at that point. Since the gap between the float and the tube walls increases as the float moves upwards, the pressure drop decreases. The float moves up the tube until the fluid pressure is just sufficient to balance the weight of the float. The greater the flow rate the greater the pressure difference for a particular gap and so the higher up the tube the float has to move to increase the gap and so get to a height where the pressure balances the weight of the float. A scale alongside the tube can thus be calibrated to read directly the flow rate corresponding to a particular height of the float. The rotameter is cheap, reliable, has an accuracy of about ±1% and can be used to measure float rates from about 30×10^{-6} m³/s to 1 m³/s.

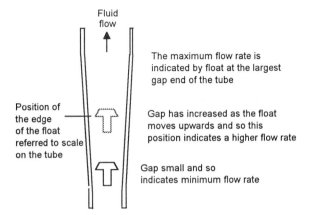

Figure 12.13 *Rotameter*

The *gate meter* (Figure 12.14) is essentially an orifice meter which has the orifice area adjusted manually or automatically to maintain a constant pressure difference. Thus in the following equation A_2 is adjusted.

$$Q = \frac{CA_2}{\sqrt{1 - \left(\frac{A_2}{A_1}\right)^2}} \sqrt{\frac{2(P_2 - P_1)}{\rho}}$$

Because of the $\sqrt{[1 - (A_2/A_1)]}$ term, the relationship between Q and A_2 is non-linear. However, a linear relationship can be produced between the vertical displacement of the gate and the volume rate of flow if the width of the opening decreases towards the top in an appropriate manner, i.e.

$$\text{displacement} \propto \frac{A_2}{\sqrt{1 - \left(\frac{A_2}{A_1}\right)^2}}$$

Figure 12.14 shows the basic form of such an aperture.

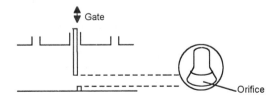

Figure 12.14 *Gate meter*

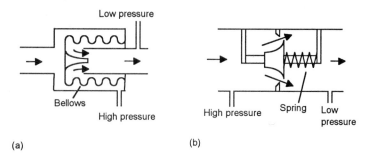

Figure 12.15 *Variable orifice meter: (a) low flow rate, (b) high flow rate*

Two other forms of variable orifice meters are shown in Figure 12.15. In both forms the fluid causes a cone to move axially and so change the orifice size. In one case the cone moves against the force exerted by a bellows and in the other that exerted by a spring. Such is the movement that the differential pressure is directly proportional to the volume flow rate. The bellows form is used for low flow rates up to about 0.05 m^3/s while the spring form is used for high flow rates in the region 1 to 3 m^3/s. They can be used with pressures up to 2×10^7 Pa and temperatures of 500°C, often being used with steam.

12.2.2 Mechanical flow meters

Mechanical flow meters involve the moving fluid impinging on a target or vane and the resulting force or motion measured. With the *target flow meter* the fluid impinges on a target, a disk, and it experiences a force. Consider a fluid of density ρ moving with a velocity v and impinging on a target of area A at right angles to the flow (Figure 12.16). The mass of fluid per second hitting the target is $\rho A v$. The fluid is brought to rest on hitting the target. This mass of fluid has a momentum $(\rho A v)v$ and thus the rate of change of momentum per second of the impinging fluid is $\rho A v^2$. The force F acting on the target will be the rate of change of momentum and so:

$$F = \rho A v^2$$

Figure 12.17 shows the basic form of the target flow meter, the fluid impinging on the target being brought to rest while the remaining fluid flows through the annular space between the target and the walls of the tube. If d is the diameter of the target, then:

$$F = \rho(\tfrac{1}{4}\pi d^2)v^2$$

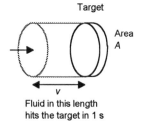

Target

Area A

v

Fluid in this length hits the target in 1 s

Figure 12.16 *Fluid hitting target in 1 s*

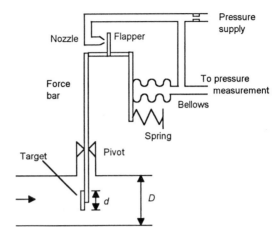

Figure 12.17 *Target flow meter*

If the pipe diameter is D then the annular area through which the fluid flows past the target is $\frac{1}{4}\pi(D^2 - d^2)$. The quantity of fluid flowing per second through this annular area is the product of the area and the velocity and is thus:

$$Q = \frac{1}{4}\pi(D^2 - d^2)v$$

Hence:

$$Q = \frac{\pi(D^2 - d^2)}{4}\sqrt{\frac{4F}{\rho\pi d^2}} = \sqrt{\frac{\pi}{4}}\left(\frac{D^2 - d^2}{d}\right)\sqrt{\frac{F}{\rho}}$$

Thus we can write:

$$Q = C\left(\frac{D^2 - d^2}{d}\right)\sqrt{\frac{F}{\rho}}$$

where C is a constant.

This force then, via a force bar, causes movement of the flapper of a flapper–nozzle arrangement and so affects the rate at which air escapes from the nozzle. The resulting change in pneumatic pressure is a measure of the force. This change in pneumatic pressure is used, by means of bellows, to restore the force bar and target to their undeflected position. The force on the target is thus balanced through the force bar by the pneumatic pressure in the bellows. Target flow meters have a range up to about 0.03 m³/s with an accuracy of ±0.5% and can be used for both liquids and gases. They can also be used with viscous and dirty fluids.

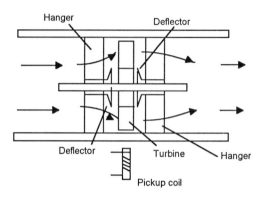

Figure 12.18 *Turbine flow meter*

The *turbine flow meter* (Figure 12.18) consists of a multi-bladed rotor that is supported centrally in the pipe along which the flow occurs. The rotor rotates as a result of the fluid impinging on its blades, the angular rate of rotation being proportional to the flow rate. The rate of revolution of the rotor can be determined using a magnetic pickup. This could be a variable reluctance form of pickup with the blades made of a ferromagnetic material and every time they pass the pickup coil a change in reluctance is produced. The meter is expensive, offers some resistance to the fluid flow, is easily damaged by particles in the fluid, has good repeatability, an accuracy of about ±0.3% and a range up to about 1 m³/s. Turbine meters can be used for both liquids and gases.

12.2.3 Positive displacement meters

Positive displacement meters provide a direct measurement of the volume passing through a meter. If the volume delivered over a particular time is measured then the volume rate of flow can be established. Such meters have mechanical elements which divide the fluid into known volume packets and then count the number of packets to give a total volume. They are widely used for water meters, gas meters and fuel pump meters to determine the volume delivered.

There are many forms of positive displacement meters. Among the main forms for liquids are the rotary piston, the reciprocating piston, the nutating disc, the rotating impeller and the sliding vane. For gases there are the diaphragm meter, the liquid sealed drum and the rotating vane.

The *rotary piston meter* (Figure 12.19) consists of a cylindrical working chamber in which an offset hollow cylindrical piston rotates. Fluid is trapped by the rotating piston and swept round and out through the outlet. The number of rotations of the piston driveshaft is a measure of the volume that has passed through the meter. The meter is widely used for metering domestic water supplies. The accuracy is about ±1%.

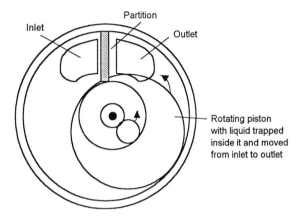

Figure 12.19 *Rotary piston meter*

The *reciprocating piston meter* (Figure 12.20) has a piston which is driven by the incoming fluid to fill up a chamber as the piston is displaced to its maximum position. In the figure this means the piston moving from right to left. When this position is reached, the slide valve reverses the side of the piston to which the fluid is admitted and so causes the piston to reverse its path, in the figure this is from left to right. This drives the fluid already in the chamber out of the outlet. A ratchet attached to the piston rod is used to drive a counter. Each count represents a particular volume of liquid. The meter is capable of high accuracy, typically about ±0.1%, and can operate over a very wide range.

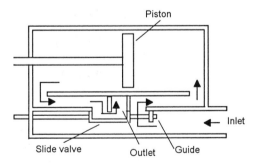

Figure 12.20 *Reciprocating piston meter*

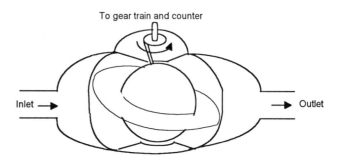

Figure 12.21 *Nutating disc meter*

The *nutating disc meter* (Figure 12.21) has a disc which is eccentrically mounted and is caused to 'wobble' or 'nutate' about a vertical axis by the incoming fluid, both the top and bottom of the disc remaining in contact with the chamber. Thus the disc first moves upwards at its left edge and so fills the chamber with fluid. Then the incoming fluid is diverted to enter the chamber at the other end and forces the right side of the disc upwards. This then results in the fluid already trapped in the chamber being expelled through the outlet. This movement of the disc up and down causes the spindle, protruding from the sphere on which the disc is mounted, to move in a circular path and drive a geared counter. The meter has an accuracy of about ±1%.

Figure 12.22 shows a version of the *rotating impeller meter*. The meter has two fluted rotors. The fluid flow causes them to rotate. Each time they rotate a volume of the liquid is trapped and moved from the inlet to the outlet. Thus measured volumes are moved between inlet and outlet. A counter is driven by a rotating rotor and the count is thus a measure of the volume moved. This form of rotating impeller meter is often used for oil flows up to about 1 m³/s and a pressure of 80 bar.

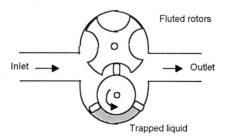

Figure 12.22 *Rotating impeller meter*

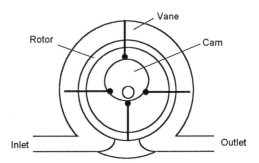

Figure 12.23 *Sliding vane meter*

The *sliding vane meter* (Figure 12.23) consists of a cylindrical rotor from which four retractable vanes protrude. The fluid flow against the vanes causes the rotor to rotate. As the rotor rotates, the trapped fluid between the vanes is swept round from the inlet to the outlet. The number of revolutions of the rotor is thus a measure of the volume of fluid that has passed through the meter. Accuracy is high, about ±0.1%, with the meter being widely used for measurements with oil and fuel. It can be damaged by dirt particles and so filtration of the inflowing liquid is usual.

A very commonly used meter for gases is the *diaphragm meter*. Figure 12.24 shows one form. It has four chambers A, B, C and D which fill and empty with gas. Two of the chambers, B and D, are enclosed by flexible diaphragms which expand and contract as they are filled or emptied of gas. The other two chambers, A and C, are the annular space around the bellows. The stages of operation are shown in Figure 12.24, being controlled by the double slide valve. In (a), A is emptying, B is filling, C is empty and D has just filled up. In (b), A is empty, B is full, C is filling and D is emptying. In (c), A is filling, B is emptying, C is full and D empty. In (d), A is full, B is empty, C is emptying and D is filling. A mechanical linkage allows the number of such cycles to be counted and hence give a measure of the volume of gas moved from the inlet to the outlet. Typically an accuracy of about ±1% is achieved for flow rates up to about 0.1 m³/s. The pressures and temperatures must, however, be close to the ambient. The meter is widely used for the measurement of the volume of gas delivered to domestic premises.

Another form of meter used for the measurement of the volume rate of flow of gases is the *liquid sealed drum* (Figure 12.25). The drum rotates within the outer casing and has four thin metal sheets balanced about a central spindle. Gas enters the drum by an inlet close to the drum centre and is trapped between the shaped sheet and the liquid until the sheet emerges from the liquid. Then the trapped volume is discharged to the outlet. The speed of rotation of the spindle must be kept low enough to avoid any significant disturbance of the liquid level. This meter is capable of an accuracy of about ±0.25%.

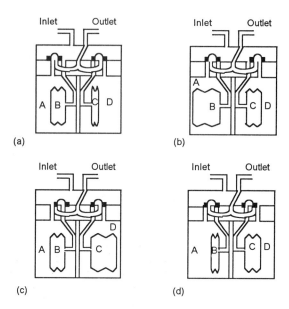

Figure 12.24 *Diaphragm meter*

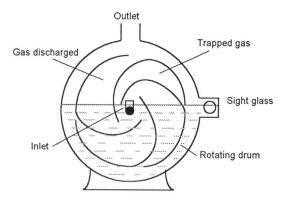

Figure 12.25 *Liquid sealed drum*

The *rotating lobe meter* (Figure 12.26), a form of rotating impeller meter (see Figure 12.22), has two rotors. The fluid flow causes them to rotate and each time they rotate they trap a volume of gas and move it from the inlet to the outlet. The number of revolutions of a rotor is a measure of the volume of gas transferred and is indicated by a counter driven by a rotor. Such a meter typically has an accuracy of ±1% and a range of about 0.003 m^3/s to 3 m^3/s.

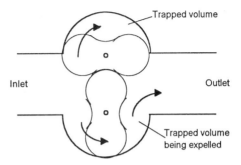

Figure 12.26 *Rotating lobe meter*

Figure 12.27 shows the principle of a *rotating vane meter*. Such a form can be used for low rates of flow of liquids or gases, though with gases the vanes need to have a larger area in order to be readily deflected by the change of momentum of the gas impacting on them. The rotation of the rotor is monitored by a suitable pickup or gearing to the shaft of the rotor. The form used for gases is often referred to as an *anemometer*.

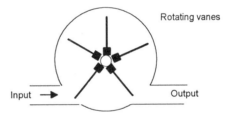

Figure 12.27 *Rotating vane meter*

12.2.4 Vortex flow meters

When a fluid flow encounters a body, the layers of fluid close to the surfaces of the body are slowed down. With a streamlined body, these boundary layers follow the contours of the body until virtually meeting at the rear of the object. This results in very little wake being produced. With a non-steamlined body, a so-called *bluff body*, the boundary layers detach from the body much earlier and a large wake is produced. When the boundary layer leaves the body surface it rolls up into vortices. These are produced alternately from the top and bottom surfaces of the body (Figure 12.28). The result is two parallel rows of vortices moving downstream with the distance between successive vortices in each row being the same, a vortex in one row occurring half way between those in the other row.

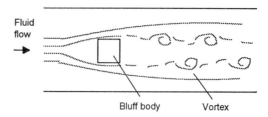

Figure 12.28 *Vortex shedding*

The number of vortices produced per second f from each surface of a bluff body is given by the equation:

$$f = \frac{Sv_b}{d}$$

where v_b is the mean fluid velocity at the bluff body, d the width of the body and S a virtually constant quantity termed the *Strouhal number*. At the body the fluid flows through an area of $\frac{1}{4}\pi D^2 - Dd$, where D is the diameter of the tube and the bluff body is assumed to be a strip of width d extending right across the diameter. The velocity v_b at this point is related to the velocity v some distance from the body and the volume rate of flow Q by:

$$Q = \tfrac{1}{4}\pi D^2 v = \left(\tfrac{1}{4}\pi D^2 - Dd\right)v_b$$

Hence:

$$f = \frac{S\tfrac{1}{4}\pi D^2 v}{\tfrac{1}{4}\pi D^2 - Dd} = \frac{Sv}{1 - \dfrac{4D}{\pi d}}$$

$$= \frac{4SQ}{\pi D^2\left(1 - \dfrac{4D}{\pi d}\right)} = \frac{4SQ}{\pi D^2 d\left(1 - \dfrac{4d}{\pi D}\right)}$$

This is generally written as:

$$f = \frac{4SQ}{\pi D^3}\frac{1}{\dfrac{d}{D}\left(1 - \dfrac{4}{\pi}\dfrac{d}{D}\right)}$$

since it is the ratio d/D which is significant, rather than d. However, the equation needs modification to take account of the form of the bluff body. Thus a coefficient k is introduced, having different values for different-shaped bodies. Hence:

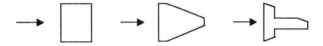

Figure 12.29 *Bluff body shapes*

$$f = \frac{4SQ}{\pi D^3} \frac{1}{\frac{d}{D}\left(1 - \frac{4}{\pi}k\frac{d}{D}\right)}$$

For a circular cross-section body k has the value 1.1, for a rectangle 1.5. Figure 12.29 shows some of the bluff body shapes used. Thus for a particular bluff body, the frequency of the vortices is proportional to the flow rate.

A number of methods are used for the measurement of the frequency. For example, a thermistor might be located behind the face of the bluff body (Figure 12.30(a)). The thermistor, heated as a result of a current passing through it, senses vortices due to the cooling effect caused by their breaking away. Another method has the vortices passing through a beam of ultrasonic waves (Figure 12.30(b)). The resulting amplitude changes to that wave can be monitored. Another method uses a piezoelectric crystal mounted in the bluff body. Flexible diaphragms react to the pressure disturbances produced by the vortices and are detected by the crystal.

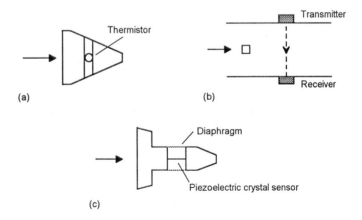

Figure 12.30 *Detection systems: (a) thermistor, (b) ultrasonic, (c) piezoelectric crystal*

Vortex flow meters are used for both liquids and gases, having an output which is independent of density, temperature or pressure, and having an accuracy of about ±1%. They are used at pressures up to about 10 MPa and temperatures of 200°C.

Another meter that depends on oscillations in a fluid is the *swirl meter*. With this meter the fluid is made to swirl or spin by passing through curved blades (Figure 12.31). The oscillations of this swirling fluid are detected by a temperature sensor. This can be a thermistor which is heated as a result of carrying an electrical current. The heat lost, and hence its temperature and consequently resistance, is determined by whether it is a swirl or not. The result is that the resistance of the thermistor oscillates with the same frequency as the swirl. This frequency is proportional to the volume rate of flow. Swirl meters are used with liquids in the range 6×10^{-4} to 2 m³/s, with gases 10^{-3} to 3 m³/s. Accuracy is about ±1%.

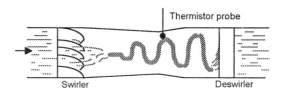

Figure 12.31 *Swirl meter*

12.3 Measurement of mass flow rate

Measurements of the mass flow rate fall into two categories:

1 *True mass flow measurements* in which the quantity measured is a direct measure of the mass flow rate. These methods include the Coriolis mass flow meter and the thermal mass flow meter.

2 *Inferential mass flow measurements* based on measuring the volume flow rate and the fluid density as separate measurements and then computing from those results the mass flow rate. In the case of a pure liquid, since the density depends only on the temperature, if the temperature is reasonably constant then the density may be assumed to be constant and thus a determination of just the volume flow rate gives a measure of the mass flow rate. Where gases and non-homogeneous liquids are concerned, both the volume rate and density need to be measured. One method that is used is a combination of a turbine volume rate flow meter with a vibrating element for the measurement of the density.

12.3.1 True mass flow measurements

A body of mass M moving with constant linear velocity v and subject to an angular velocity ω experiences an inertial force at right angles to the

Figure 12.32 *Coriolis force*

direction of motion, this being known as a *Coriolis force* (Figure 12.32). The Coriolis force is:

Coriolis force = $2M\omega v$

The *Coriolis flow meter* consists basically of a C-shaped pipe (Figure 12.33) through which the fluid flows. The pipe, and fluid in the pipe, is given an angular acceleration by being set into vibration, this being done by means of a magnet mounted in a coil on the end of a tuning fork-like leaf spring. Oscillations of the spring then set the C-tube into oscillation. The result is an angular velocity that alternates in direction. At some instant the Coriolis force acting on the fluid in the upper limb is in one direction and in the lower limb in the opposite direction, this being because the velocity of the fluid is in opposite directions in the upper and lower limbs. The resulting Coriolis forces on the fluid in the two limbs are thus in opposite directions and cause the limbs of the C to become displaced. When the direction of the angular velocity is reversed then the forces reverse in direction and the limbs become displaced in the opposite direction. These displacements are proportional to the mass flow rate of fluid through the tube. The displacements are monitored by means of optical sensors, their outputs being a pulse with a width proportional to the mass flow rate. The flow meter can be used for liquids or gases and has an accuracy of ±0.5%. It is unaffected by changes in temperature or pressure.

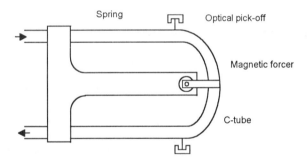

Figure 12.33 *Coriolis flow meter*

The *thermal mass flow meter* (Figure 12.34) consists of two temperature sensors mounted with one upstream and the other downstream of a heater. The difference in temperature between the sensors depends on the rate of mass flow. The two sensors are resistance elements and are mounted in adjacent arms of a Wheatstone bridge. The out-of-balance voltage from the bridge is then a measure of the temperature difference and hence the rate of mass flow. Such a meter is used for gas flows in the range 2.5×10^{-10} to 5×10^{-3} kg/s with an accuracy of ±1%.

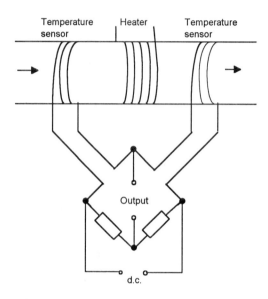

Figure 12.34 *Thermal mass flow meter*

12.3.2 Inferential mass flow measurements

One method used to make an inferential measurement of mass flow involves using a turbine flow meter (see Section 12.2.2 and Figure 12.18) to give a signal related to the volume flow rate and a vibrating element density sensor to give an output related to the fluid density. The two signals are then combined, using a computer, to give an output of the mass flow rate.

Figure 12.35 shows the basic form of the *vibrating tube density method* for the measurement of fluids flowing through the tube. The tube element is fixed at each end to heavy masses, so effectively clamping the ends. The tube is then set in oscillation by means of magnetic forces supplied by alternating current in the drive coil, the coil being located at the midpoint of the clamped length. The arrangement is rather like a string which is clamped at both ends and plucked in the middle. The amplitude of the oscillation of the tube at the midpoint is monitored by a pickup coil. The output from this coil is used as the feedback loop to the amplifier driving the drive coil. The result is that the tube is maintained in oscillation at its natural frequency. This frequency depends on the total mass of the tube and its contents. Thus, since the tube has a constant volume, the frequency is related to the density of the fluid in the tube. The relationship between the frequency f and the density ρ of the fluid is of the form:

$$\rho = \frac{A}{f^2} + \frac{B}{f} + C$$

where A, B and C are constants for the tube.

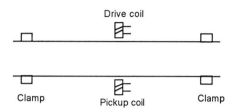

Figure 12.35 *Vibrating tube density method*

12.4 Measurement of velocity

The measurements so far considered in this chapter involve the measurement of the rate of volume or mass flow. There are, however, situations where the velocity at a point in a flowing fluid is required. The following are techniques commonly used:

1 *Pitot and Annubar tubes*
   These are based on using Bernoulli's principle and measuring the pressure difference between a point in the fluid at full flow and a point where it has been brought to rest.

2 *Hot wire anemometer*
   This is based on the resistance of a current-carrying wire depending on the rate at which heat is conducted away from it and hence the velocity of the fluid.

12.4.1 Pitot and Annubar tubes

The *Pitot tube* consists essentially of just a small tube inserted into the fluid with an opening pointing directly upstream (Figure 12.36). The fluid impinging on the open end of the tube is brought to rest and the pressure difference measured between this point and the pressure in the fluid at full flow.

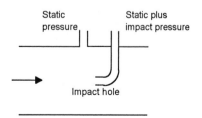

Figure 12.36 *Pitot tube*

The difference in pressure between where the fluid is in full flow and the point where it is stopped is due to the kinetic energy of the fluid being transformed to potential energy. The displacement energy, i.e. pressure energy, per unit mass for the fluid at the impact point is P_1/ρ, where P_1 is the pressure at that point and ρ the density. In full flow at a point just prior to the impact where the fluid has a pressure P_s and a velocity v, the fluid has a displacement energy per unit mass P_s/ρ and kinetic energy per unit mass of $\frac{1}{2}v^2$. Thus, applying the conservation of energy (or Bernoulli's principle):

$$\frac{P_1}{\rho} = \frac{P_s}{\rho} + \frac{1}{2}v^2$$

Hence:

$$v = \sqrt{\frac{2(P_1 - P_s)}{\rho}}$$

The velocity is thus proportional to the square root of the pressure difference. A correction factor C is introduced to correct for not all the fluid incident on the end of the tube being brought to rest, a proportional slipping around it. Then:

$$v = C\sqrt{\frac{2(P_1 - P_s)}{\rho}}$$

The above are the relationships for an incompressible fluid such as a liquid. For a compressible fluid, such as a gas, the relationship needs modification. The equation then becomes:

$$\frac{\gamma}{\gamma - 1}\frac{P_1}{\rho_1} = \frac{\gamma}{\gamma - 1}\frac{P_s}{\rho_s} + \frac{1}{2}v^2$$

where γ is the ratio of the specific heats, ρ_1 the density at the impact point and ρ_s the density at the static point. For adiabatic changes we can use $PV^\gamma = $ a constant and so:

$$\frac{P_1}{\rho_1^\gamma} = \frac{P_s}{\rho_s^\gamma}$$

Hence:

$$v = \sqrt{\left[2\frac{\rho}{\gamma - 1}\frac{P_s}{\rho_s}\left\{ \left(\frac{P_1}{P_s}\right)^{(\gamma-1)/\gamma} - 1 \right\} \right]}$$

For air, the difference in densities between the static and impact holes is negligible for speeds less than 100 m/s and so the incompressible equation can be used.

The pressure difference is often measured with a diaphragm pressure gauge. Accuracies are typically of the order of ±5%.

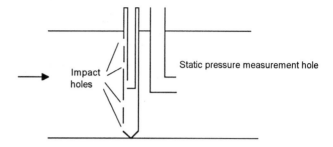

Figure 12.37 *Annubar tube*

The *Annubar tube* (Figure 12.37) is a form of Pitot tube. It has four impact holes in a bar which extends across the width of the tube through which the fluid is flowing. The spacing of the holes is such that each responds to the pressure of equal annular segments of the flow. The average of the pressure from these four holes is then indicated by an inner tube. The static pressure is obtained from a tube facing downstream. This method gives an accuracy of $\pm1\%$ or better.

12.4.2 Hot wire anemometer

The *hot wire anemometer* consists of a small resistance wire element mounted in the fluid flow (Figure 12.38). An electrical current through the wire causes its temperature to rise to a value which is determined by the rate at which it loses heat. This depends on the velocity of the fluid.

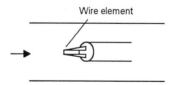

Figure 12.38 *Hot wire anemometer*

At equilibrium, the rate at which the current i through the resistance R dissipates energy is equal to the rate at which heat is conducted away from it. Thus:

$$i^2R = hA(T_s - T_f)$$

where h is the heat transfer coefficient, A the effective area of the resistance element, T_s its temperature and T_f the temperature of the fluid. The heat transfer coefficient depends on the fluid velocity and is given by:

$$h = C_0 + C_1 \sqrt{v}$$

where C_0 and C_1 are constants. Thus:

$$i^2 R = A(C_0 + C_1 \sqrt{v})(T_s - T_f)$$

Usually the resistance, and hence the temperature of the element, is kept constant by changing the current. The current then becomes a measure of the fluid velocity with:

$$i^2 = C_2 + C_3 \sqrt{v}$$

where C_2 and C_3 are constants.

The hot wire anemometer is used for gas velocities from 0.1 to 500 m/s at temperatures up to 750°C and for liquids from 0.01 to 5 m/s. A sensor made using a thin film wrapped round a cylinder, rather than the wire form, can be used with liquids from 0.01 to 25 m/s. Accuracies are about ±1%.

12.5 Open channel measurements

The measurement of the flow rate in open channels is a requirement that is often associated with the water industry, flood control and irrigation. Two methods that are commonly used are:

1   *The weir*
   The weir is a dam over which liquid is allowed to flow, the depth of the liquid over the sill being a measure of the rate of flow.

2   *The hydraulic flume*
   This is generally an open channel version of the venturi meter.

12.5.1 Weirs

A *weir* is a dam across the channel so that the liquid has to flow over its sill. The flow rate over a weir is a function of the weir geometry and of the weir head H, i.e. the vertical distance between the weir crest and the liquid surface in the undisturbed region upstream of the weir. Figure 12.39 shows the type of flow that occurs over a weir. Gravity pulls the liquid surface down from the height H to a lower height as it passes over the weir. Thus we have a change in height of the centre of pressure as the liquid passes over the weir. The discharge equation for liquid flowing over such a weir thus can be obtained by applying Bernoulli's principle.

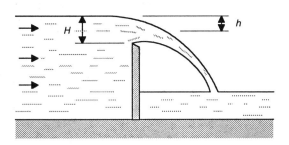

Figure 12.39 *Liquid flow over a weir*

Consider the liquid prior to the weir as having a surface at a height H above the crest of the weir. If we consider the liquid over the weir at a depth h below the original water surface, the liquid has been reduced to a height $(H - h)$. If the velocity of the flow prior to the weir was v_1 and in the flow over the weir at a depth h is v_2, then applying Bernoulli's equation gives:

$$gH + \tfrac{1}{2}v_1^2 = g(H - h) + \tfrac{1}{2}v_2^2$$

Thus:

$$v_2 = \sqrt{2g\left(h + \frac{v_1^2}{2g}\right)}$$

When the velocity v_1 is small then the velocity approximates to:

$$v_2 = \sqrt{2gh}$$

The liquid flowing over the weir is at a range of depths so the velocity will vary with depth. The flow rate δQ through a segment of the cross-sectional area δA through which the liquid flows with a velocity v_2 is $v_2 \delta A$. Thus the total flow rate Q over the weir is given by:

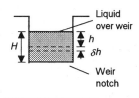

Liquid over weir

Weir notch

$$Q = \int_0^A v_2 \, \mathrm{d}A$$

where A is the total area.

For a *rectangular notch weir* (Figure 12.40), the cross-sectional width of the flow is constant and is the width L of the weir. Thus we have $A = LH$ and $\mathrm{d}A = L \, \mathrm{d}h$. Hence we can write the integral as:

Figure 12.40 *Rectangular notch weir*

$$Q = \int_0^H \sqrt{2gh}\, L \, \mathrm{d}h = \tfrac{2}{3}\sqrt{2g}\, LH^{3/2}$$

The actual flow rate over the weir is less than the flow rate indicated by the above equation. This is because the area of the stream is not LH but less than this because the stream contracts at both the top and the bottom as it flows over the weir (see Figure 12.39). In addition there are frictional losses. These effects are taken account of by the introduction of a *weir discharge coefficient C* to give:

$$Q = \tfrac{2}{3} C \sqrt{2g} \, LH^{3/2}$$

The value of C depends mainly on the relative head of the weir, i.e. the ratio of H to the weir height P above the bed. and the form of the sill. An empirical equation which is sometimes used is:

$$C = 0.60 + 0.075\frac{H}{P}$$

Thus for the height ratio of 1 we have a value of C of about 0.68 and for a ratio of 2 a value of 0.75.

A *triangular notch* (Figure 12.41) is sometimes used, then L is no longer constant. If the angle of the triangular notch is θ, we have for the width of the liquid at a depth h:

$$L = 2(H - h) \tan(\theta/2)$$

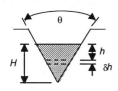

Figure 12.41 *Triangular notch weir*

Hence:

$$Q = \int_0^H \sqrt{2gh} \, 2(H - h) \tan(\theta/2) \, dh = \tfrac{8}{15} \sqrt{2g} \, \tan(\theta/2)H^{5/2}$$

As before, we can include the weir discharge coefficient to obtain:

$$Q = \tfrac{8}{15} C \sqrt{2g} \, \tan(\theta/2)H^{5/2}$$

With H greater than about 2 cm, the discharge coefficient has a value of about 0.58.

12.5.2 Hydraulic flume

A *hydraulic flume* can take a number of forms. One form is to shape the sides of the flow to produce a venturi (Figure 12.42). The depth h_1 of the water prior to the venturi is measured and the depth h_2 in the venturi constriction is measured. Then the rate of flow Q is given by:

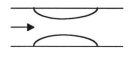

Figure 12.42 *Venturi flum*

$$Q = CL_2 h_2 \sqrt{\frac{2g(h_1 - h_2)}{1 - \dfrac{L_2 h_2}{L_1 h_2}}}$$

where L_1 is the width of the channel prior to the constriction, L_2 the width at the constriction and C the discharge coefficient.

Problems

1 What will be the approximate velocity of water in a pipe of diameter 50 mm when the flow changes from orderly to turbulent? Take the density of water to be 1000 kg/m³ and the viscosity 0.0012 kg/m s.

2 A venturi meter is used to measure the flow of oil through a horizontal pipe. The pipe has a diameter of 120 mm and the venturi constriction has a diameter of 40 mm. The pressure difference between the constriction and the pipe flows is measured and found to be 800 Pa. If the oil has a density of 820 kg/m³, what is the flow rate of the oil? Assume there are no losses.

3 A venturi meter is used to measure the flow of water through a horizontal pipe. The pipe has a diameter of 140 mm and the venturi constriction has a diameter of 30 mm. The pressure difference between the constriction and the pipe flows is measured and found to be 2 kPa. The density of water is 1000 kg/m³. What is the mean velocity of the water entering the meter? Assume there are no losses.

4 An orifice meter is used to measure the flow of air through a 50 mm diameter pipe. The orifice has a diameter of 40 mm. The pressure drop is measured using a water manometer and found to be 64 mm. If the density of air can be taken to be constant at 1.2 kg/m³ and the discharge coefficient has a value of 0.63, what is the volume rate of flow through the pipe?

5 A turbine flow meter used to measure the flow rate of a fluid through a pipe has an angular velocity ω in rad/s related to the volume rate of flow Q in m³/s by the equation $\omega = 50000Q$. The flux Φ in mWb linked by a pickup coil placed outside the tube is related to the angle θ between the turbine blade assembly and coil axis by:

$$\Phi = 4.00 + 0.90 \cos 4\theta$$

Determine the amplitude and frequency of the pickup coil output when the flow rate is 0.001 m³/s.

6 Show that the volume rate of flow Q past the float in a rotameter can be described by:

$$Q = Ca\sqrt{\frac{2gV_f(\rho_f - \rho)}{A\rho}}$$

where C is the discharge coefficient, a the minimum cross-sectional area of the float, A the maximum cross-sectional area of the float, V the volume of the float, ρ_f the density of the float and ρ the density of the fluid. Note: at equilibrium, the downwards forces acting on the float are the weight and the downward pressure force, the upwards forces are the Archimedes' upthrust and upward pressure force; also Bernoulli's equation can be applied.

7 A rotameter is used to measure the volume rate of flow of a gas with a density 0.8 kg/m^3 and the reading on the rotameter scale indicates 2 dm^3/s. However, the rotameter scale had been calibrated for a gas of density 1.2 kg/m^3. What is the real flow rate?

8 A vortex flow meter is used to measure the rate of flow of a fluid through a pipe of diameter 50 mm. The bluff element used for the vortex shedding has a diameter of 10 mm. If the sensor indicates a vortex frequency of 20 Hz, what is the flow rate if the Strouhal number is 0.4?

9 A Pitot tube indicates a pressure difference of 740 kPa between the impact and static pressures for a flow of air. What is the air velocity at the tube if the static air pressure is 95 kPa and the temperature 20°C? Take the gas constant R to have the value 287 J/kg K.

10 A Pitot tube indicates on a water manometer a pressure difference of 100 mm between the impact and static pressures for the flow of water along a pipe. What is the velocity of the water at the impact point? Assume that all the water impacting on the Pitot tube opening is brought to rest.

11 A rectangular notch weir has a height which is 0.60 m above the bed of the channel and a channel width of 1.3 m. The flow over the weir has a head of 0.20 m. What is the rate of flow of water over the weir? Note: use the equation given in this chapter to determine the discharge coefficient.

12 A 60° triangular notch weir has a head of 0.40 m. What is the flow rate of water over the weir? Take the discharge coefficient to be 0.58.

13 Temperature measurement

This chapter is a consideration of temperature scales, how they are defined, and commonly used methods of measuring temperatures.

13.1 Temperature

Temperature can vaguely be defined as the 'degree of hotness' of an object. Thus if you put your hand in contact with one object you might say that it is hot and perhaps another object is cold. You can then say that one is at a higher temperature than the other.

The fundamental law that all temperature measurements rely on is the *zeroth law of thermodynamics*. This states that:

> When two bodies are each in thermal equilibrium with a third body, they are in thermal equilibrium with each other.

Thus if object A does not change its temperature when brought into contact with object B and if object C does not change its temperature when brought into contact with object B, then object C will not change its temperature when brought into contact with object A. Bodies in thermal equilibrium are at the same temperature. Thus if we make B the thermometer, then if it does not change its temperature when brought into contact with object A, and also does not change its temperature when brought into contact with object C, then A and C are at the same temperature.

13.1.1 Specifying a temperature scale

Using the zeroth law we can determine when two objects are the same temperature but cannot say how the temperature of one object compares with that of another. However, before considering temperature consider how we can determine lengths. If we define a rod as being one metre long and then make a second of the same length, we can then define a length of two metres as being the two rods laid end to end. Thus when we define a unit of length we are able to use that definition to specify other lengths by combining rods specified by our units of length. In a similar manner we can do this for mass or time. We cannot, however, do this for temperature. If we define a condition as being at, say, 0 degrees, then adding two such quantities still only gives a temperature of 0 degrees. The combination of two bodies at the same temperature results in no change in the temperature. We need other methods than just a specification of a single entity in order to specify other temperatures. We need to establish a *scale*.

Temperature is an abstract quantity that has to be defined in terms of the behaviour of materials when the temperature changes. To measure and compare temperatures it is necessary to have temperature scales that have been defined in terms of the behaviour of materials. Thermometers are just devices that, when brought into contact with an object, indicate its temperature in terms of some observable property. Thus the mercury-in-glass thermometer indicates temperature in terms of the length of a liquid column, a resistance thermometer indicates temperature in terms of the resistance of a coil and a gas thermometer the temperature in terms of the pressure of a gas at constant volume or the volume of a gas at constant pressure.

A temperature scale requires three key aspects to be defined. These are:

1   Fixed reference points for establishing known temperatures.

2   The definition of the size of the degree. This is effectively done by specifying values for the reference points.

3   A means of interpolating between these fixed points.

Over the years a number of scales have been developed.

13.1.2 Temperature scales

In 1701 Newton proposed that the freezing point of water be taken as the lower reference point and given the value 0. The body temperature of a 'healthy male' was taken as specifying the upper fixed point and given the value 12.

In 1715 Fahrenheit introduced the *Fahrenheit scale* with the fixed points of the freezing point of water defined as being 32 degrees Fahrenheit (°F) and the boiling point of water as 212 degrees Fahrenheit (°F). Thus the interval between the freezing point of water and its boiling point was defined as being divided into 180 degrees. The freezing point of water was set as 32°F because the lowest temperature that had been obtained at that time using ice and water with added salts was specified as 0°F.

Celsius in 1742 divided the same interval into 100 divisions and Linnaeuas later set the freezing point of water on this scale as 0 degrees. This scale was originally known as the *Centigrade scale*, the word centigrade reflecting the fact that it was divided into 100 intervals, but in 1948 this was changed to the *Celsius scale*. Temperatures on this scale are specified as °C. For the Celsius scale, the temperature of the ice point is defined as 0°C with the ice point being defined as the temperature at which ice and water exist together at a pressure of 1.0132×10^5 Pa, the standard atmospheric pressure. The temperature of the steam point is defined as 100°C with the steam point being defined as the temperature at which distilled water boils at a pressure of 1.0132×10^5 Pa. The temperature at which water boils is very dependent on the pressure. The temperature interval of 100 degrees between the ice point and the steam point is termed the *fundamental interval*.

Consider the problem of interpolation with the Celsius scale. Suppose we have two different forms of thermometer, perhaps a mercury-in-glass thermometer and one constructed with water-in-glass. We can use the reference points of the ice point and the steam point to mark the 0°C and 100°C points on the stems of each thermometer. However, if we assume that the scale between these two fixed marks is equally divided into 100 degrees then when we use the thermometers to measure some intermediate temperature we can end up with different values. In fact if we had the mercury-in-glass thermometer indicating 4°C then the water-in-glass thermometer would indicate a less than zero temperature. This is because water has its maximum density at 4°C. Thus to be able to specify, without ambiguity, temperatures between the fixed points we need to specify which form of thermometer we propose to use.

If we use the expansion of liquids as the means of defining temperatures between fixed points we have the problem that we obtain different values depending on the liquid used. However, suppose we use gases instead of liquids. If a gas is kept at constant volume then when the temperature increases its pressure increases (or if the pressure is kept constant then when the temperature increases its volume increases). We might thus have a thermometer of the form shown in Figure 13.1.

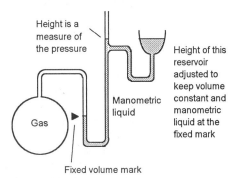

Figure 13.1 *Constant volume gas thermometer*

With the gas thermometer we can arbitrarily define the relationships between the pressure of the fixed volume and temperature such that we have:

$$\frac{T_s}{T_i} = \frac{P_s}{P_i}$$

where P_s is the pressure at the steam point, a temperature designated as T_s, and P_i is the pressure at the ice point, a temperature designated as T_i. If we use this relationship with gases such as nitrogen, oxygen, hydrogen, argon and helium, we obtain virtually the same value of P_s/P_i ratio regardless of which gas we have in the thermometer bulb. We thus have a thermometer which seems to give results almost independent of which substance we use. The value of the ratio is slightly different for the different gases and does depend to some extent on the amount of gas we have in the thermometer bulb. However, if we measure the ratio for a number of gases at different pressures and extrapolate back to zero pressure (Figure 13.2), i.e. to low density conditions, we find that all the gases give exactly the same limiting value for the ratio, namely 1.36609 ± 0.00004.

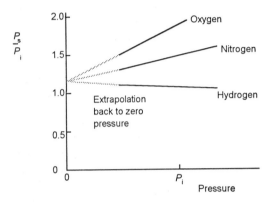

Figure 13.2 *Constant volume gas thermometer measurements*

Thus we can define a temperature scale by specifying that the temperatures at the ice point and at the steam point are related by:

$$\frac{T_s}{T_i} = \lim_{P_i \to 0} \frac{P_s}{P_i} = 1.36609$$

If we also specify that the interval between the two temperatures is to be 100 degrees then a scale has been specified. Thus:

$$\frac{T_i + 100}{T_i} = 1.36609$$

and hence, on this scale, the temperature of the ice point is 273.15 and that of the steam point 373.15. The zero value for the temperature is said to occur at absolute zero. For any other temperature, we measure the pressure p and then use:

$$\frac{T}{T_i} = \lim_{P_i \to 0} \frac{P}{P_i}$$

Hence:

$$T = T_i \lim_{P_i \to 0} \frac{P}{P_i} = 273.15 \lim_{P_i \to 0} \frac{P}{P_i}$$

T is zero when the pressure P is zero. This defines the absolute zero of temperature.

Because each degree is the same size as the Celsius degree, the interval between the ice point and the steam point being divided into 100 degrees, we can quote the temperature values given as being in degrees Celsius though we have different values for the ice point and the steam point. We have a temperature scale with the ice point having the value 273.15°C and the steam point 373.15°C. Room temperature of 20°C on the Celsius scale is thus 293.15°C on this scale. Temperatures on this scale are generally said to be on the *gas thermometer scale* and are said to be absolute temperatures.

The unfortunate point about a temperature scale realised by means of a constant volume gas thermometer is that such a thermometer is a very bulky item and is slow and cumbersome to use. Consequently for everyday use a more practical scale, the *International Practical Temperature Scale*, has been adopted which gives results as close as possible to the gas thermometer scale. See Section 13.1.3 for details.

Lord Kelvin in 1848 proposed the theoretical basis for a temperature scale, referred to as the *Kelvin thermodynamic scale*, which is independent of any material property and based on the Carnot cycle. This involved consideration of a perfectly reversible heat engine. The gas thermometer scale does, however, give results which are identical to this thermodynamic scale. Temperatures on this scale are referred to as having the unit K, the degree symbol not being used. Thus a temperature of t °C on the Celsius scale is a temperature of $t + 273.15$ on the Kelvin scale, i.e. 273.15 K.

The *Rankine scale* is the thermodynamic equivalent based on the use of degrees Fahrenheit rather than degrees Celsius for the size of the degree. On the Rankine scale the ice point is 491.67°R. To convert temperatures from Fahrenheit to Rankine add 459.67.

Table 13.1 shows a comparison of the various temperature scales for the absolute zero, ice point and steam point temperatures.

Table 13.1 *Comparison of temperature scales*

Temperature	Kelvin K	Celsius °C	Fahrenheit °F	Rankine °R
Absolute zero	0	−273.15	−523.67	0
Ice point	273.15	0	32	523.67
Steam point	373.15	100	212	671.67

13.1.3 International Practical Temperature Scale

In 1927 the Seventh General Congress on Weights and Measures adopted the International Practical Temperature Scale as a more convenient way of specifying temperatures in such a way that the results are virtually identical with those realised by the gas thermometer scale. This scale is based on the specification of a large number of fixed points and the interpolation procedures that have to be used for temperatures between fixed points.

The fixed points are highly reproducible points corresponding to the melting, boiling or triple points of pure substances under specified conditions. The interpolation methods are standard instruments with specified output versus temperature relationships obtained by calibration at fixed points. The numbers assigned to the fixed points are such that there is exactly 100 degrees between the freezing point and the boiling point of water. Thus a change of one degree on this scale is equal to a change of 1°C on the older Celsius scale. Note that the ice point is not used as one of the fixed points. This is because there is difficulty in accurately reproducing this temperature, i.e. the temperature of a well-mixed mixture of ice and water saturated with air at one atmosphere pressure. In its place the triple point of water is used. The triple point is the temperature at which ice, liquid water and water vapour coexist. This temperature is given the value equivalent to 0.01°C.

Standard instruments used for interpolation include the platinum resistance thermometer, special thermocouples for higher temperatures and, for even higher temperatures, pyrometers based on the Planck law of radiation.

13.2 Expansion type instruments

Instruments for the measurement of temperature based on the expansion property can be grouped as:

1   *Solid expansion*
    This includes devices such as metal rods which, on expansing, can be used to activate a switch and bimetallic strips which change their curvature when the temperature changes.

2   *Liquid expansion*
    This includes liquid-in-glass and liquid-in-metal thermometers.

3   *Gas expansion*
    For general use this includes the gas-in-metal thermometer, the pressure of a fixed volume of gas changing when the temperature changes.

4   *Change of state*
    The vapour pressure thermometer depends on the saturated vapour pressure of a liquid changing with temperature and thus the pressure can be used as a measure of temperature.

13.2.1 Bimetallic strips

The bimetallic strip consists of two different metal strips of the same length bonded together (Figure 13.3). Because the metals have different coefficients of expansion, a temperature change results in the curvature of the strip changing, the metal with the larger coefficient of expansion being on

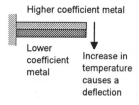

Higher coefficient metal

Lower coefficient metal

Increase in temperature causes a deflection

Figure 13.3 *Bimetal strip*

the outside of the curve. This curvature allows the higher coefficient metal to expand more than the lower coefficient metal. The amount by which the strip curves depends on the two metals used, the length of the composite strip, and the change in temperature. The movement may be used to open or close electrical contacts, as in the simple thermostat that was used (now generally based on the use of a junction diode) for many domestic heating systems. It can also be used to give a robust thermometer (Figure 13.4). Because the longer the length of the bimetallic strip the greater the movement, bimetallic strip thermometers usually have the strip in the form of a helix. Movement of the free end of the helix is then used to directly move a pointer across a scale. Such a thermometer is robust, cheap, can be used within the range of about −30°C to 600°C, has an accuracy of about ±1%, is direct reading but fairly slow reacting to changes in temperature.

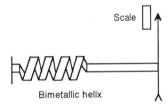

Figure 13.4 *Bimetallic thermometer*

13.2.2 Liquids in glass or metal

The *liquid-in-glass thermometer* consists of a liquid contained in a sealed container which is in the form of a bulb with capillary tube attached. With mercury used as the liquid, since the coefficient of thermal expansion of mercury is about eight times greater than that of the glass the mercury expands more than the glass when the temperature increases. Thus the mercury expands up the capillary tube and the position of the mercury meniscus can be used as a measure of temperature. Mercury boils at 357°C at atmospheric pressure and so to extend the range of a mercury-in-glass thermometer to higher temperatures, the top end of the capillary tube is enlarged to form a small bulb which is then filled with nitrogen or carbon dioxide at a pressure which is sufficient to stop the mercury boiling until a much higher temperature. With mercury as the liquid the range over which such a thermometer can be used is −35°C to +600°C, with alcohol −80°C to +70°C, with toluene −80°C to +100°C, with pentane −200°C to +30°C, with creosote −5°C to +200°C. Such thermometers are direct reading, fragile, capable of an accuracy of about ±1% under standardised conditions, fairly slow reacting to changes in temperature and cheap.

Thermometers are calibrated for use partially immersed up to some particular mark on the stem, totally immersed when the thermometer is immersed to the level of the liquid in the thermometer stem or completely immersed when the entire thermometer is immersed. If a thermometer is not immersed to the extent at which it was calibrated, then errors occur.

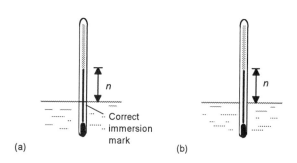

(a) (b)

Figure 13.5 *Corrections with: (a) partial immersion, (b) complete immersion thermometers*

When a thermometer calibrated for partial immersion is used to the correct immersion but in surrounding air which is at a different temperature t_{act} to that used for the calibration t_{cal} (Figure 13.5(a)), then the correction on the Celsius scale is:

$$\text{correction} = 0.00016n(t_{cal} - t_{act})$$

where n is the number of degrees of the emergent mercury column exposed. For a thermometer calibrated for total immersion or complete immersion the correction occurring when the thermometer is only partially immersed (Figure 13.5(b)) is:

$$\text{correction} = 0.00016n(t_{th} - t_{sur})$$

where t_{th} is the reading given by the thermometer and t_{sur} the temperature of the air surrounding the thermometer at the midpoint of the exposed mercury column.

Liquid-in-metal thermometers consist of a metal bulb which is connected to a Bourdon tube pressure gauge by a capillary tube. The bulb and the entire capillary tube are filled with a liquid (Figure 13.6). When the temperature increases the liquid endeavours to expand and as a consequence there is an increase in pressure which is registered by the pressure gauge. A variety of liquids are used with such thermometers, the general range covered being about −90°C to +650°C. With mercury as the liquid the range is −39°C to +650°C, with alcohol −46°C to +150°C, with xylene −40°C to +400°C, with ether +20°C to +90°C. Accuracy is about ±1%.

A source of error with this type of thermometer is the liquid in the connecting capillary tube, the temperature of this having an effect on the resulting pressure. The error is reduced by making the volume of this small, hence the use of capillary tubing. Another way is to have a second capillary

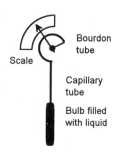

Figure 13.6 *Liquid-in-metal thermometer*

tube alongside the main capillary tube but terminating just before the bulb (Figure 13.7). It is connected to a second Bourdon tube and the display pointer is driven by the difference in movement between the two Bourdon tubes. Another method of correcting for the liquid in the capillary tube is to use a bimetallic strip. The strip is connected to the end of the Bourdon tube in such a way that it causes it to be displaced by an amount which depends on the ambient temperature and compensates for the liquid in the capillary tube.

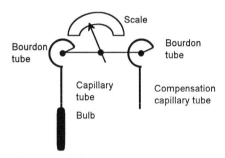

Figure 13.7 *Compensated liquid-in-metal thermometer*

Other sources of error with this type of thermometer are head errors, ambient pressure errors and immersion errors. Head errors occur if the height of the thermometer bulb changes with respect to the Bourdon tube. This is due to the height of the liquid in the thermometer exerting a pressure. The Bourdon gauge measures the gauge pressure and thus changes in the ambient pressure will affect its readings. The thermometer bulb needs to be fully immersed if correct readings are to be obtained.

13.2.3 Gas thermometers

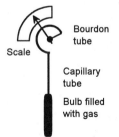

Figure 13.8 *Gas-in-metal thermometer*

The industrial form of gas thermometer consists of a thermometer bulb connected to a Bourdon gauge and filled with a gas such as nitrogen (Figure 13.8). When the temperature rises the gas pressure increases and is indicated by the Bourdon gauge. The bulb of the thermometer tends to be fairly large, about 50 to 100 cm^3. The thermometer is robust, has a range of about −100°C to +650°C, is direct reading, can be used to give a display at a distance from where the temperature is being measured, and has an accuracy of about ±0.5% of the full-scale reading. The sources of error are as outlined above for the liquid-in-metal thermometer (see Section 13.2.2). Because the errors due to the gas in the capillary tubing tend to be small with gas-filled thermometers, compensation methods are not generally used.

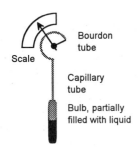

Figure 13.9 *Vapour pressure thermometer*

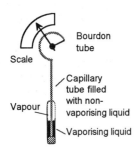

Figure 13.10 *Vapour pressure thermometer*

13.2.4 Vapour pressure thermometers

The vapour pressure thermometer (Figure 13.9) consists of a thermometer bulb connected to a Bourdon gauge and partially filled with liquid, the space above the liquid containing vapour from the liquid. When the temperature is increased, and the amount of liquid that evaporates increases, the vapour pressure increases. The vapour pressure is indicated by the Bourdon gauge and is a measure of the temperature. The relationship between the vapour pressure and the temperature is, however, not linear.

With methyl chloride as the liquid the range is about 0°C to +50°C, with sulphur dioxide +30°C to +120°C, with di-ethyl ether +60°C to 160°C, ethyl alcohol +30°C to +180°C, with water +120°C to +220°C, with toluene +150°C to +250°C. The instrument is robust, direct reading, can be used for displays at a distance from the thermometer bulb, has a non-linear scale and an accuracy of about ±1%. The non-linear scale means that divisions increase in size as the temperature increases. In some instruments a system of levers is used to give a linear scale over a limited range of temperatures. Vapour pressure thermometers are widely used, being cheaper than liquid- or gas-filled instruments.

Problems can arise as a result of the liquid distilling into and out of the capillary and Bourdon tubes as a result of temperature differences between them and the thermometer bulb. One method that has been used to overcome this problem is to fill the capillary tube and Bourdon tube with a non-vaporising liquid. This liquid is then just used to communicate the vapour pressure from the vapour above the vaporising liquid to the Bourdon tube (Figure 13.10).

13.3 Resistance thermometers

Resistance thermometers are based on the electrical resistance of an element changing when the temperature changes. Essentially there are three forms of such thermometer:

1 *Metal resistance thermometers*
These depend on the resistance of a coil of metal wire changing with temperature.

2 *Thermistors*
Thermistors are mixtures of metal oxides and semiconductors. They have resistances which change substantially when temperature changes occur.

3 *Semiconductors*
Junction diodes have forward voltages which depend on temperature and can be used as a measure of temperature. Such diodes are often incorporated in integrated circuits to give a temperature sensor complete with signal conditioning.

13.3.1 Metal resistance thermometers

The resistance of metals generally increases with temperature, the change in resistance being reasonably proportional to the temperature change (see Section 5.2.2). The resistance thermometer consists of a coil of wire connected as one arm of a Wheatstone bridge (see Section 7.1.2 for a discussion of how temperature compensation for the effects of temperature on the leads to the resistance coil can be eliminated). The coil can consist of the resistance wire wound over a ceramic-coated tube or a film of metal deposited on ceramic, the assembly then being further coated with ceramic and mounted in a protecting tube. Because of the poor thermal contact between the coil and the medium outside the tube for which the temperature is being measured, the response time is fairly slow, often of the order of a few seconds. The metals mainly used for the resistance coil are platinum, nickel and copper.

Platinum has a closely linear relationship between resistance and temperature, gives good repeatability, has long-term stability, can give an accuracy of ±0.5% or better, has a temperature range of about −200°C to +850°C, is relatively inert and can be used in a wide range of environments without deterioration. It is more expensive than the other metals but is, however, the most widely used. Nickel and copper are cheaper but have less stability, are more prone to interaction with the environment and cannot be used over such large ranges of temperature. Nickel has a range of about −80°C to +300°C and copper −200°C to +250°C.

13.3.2 Thermistors

Thermistors give much larger resistance changes per degree than metal wire elements (see Section 5.2.2). However, the resistance variation with temperature is non-linear. Their small size, often just a small bead, means a small thermal capacity and hence a rapid response to temperature changes. The temperature range over which they can be used will depend on the thermistor concerned, ranges generally being within about −100°C to +300°C. Over a small range the accuracy can be 0.1°C or better. However, their characteristics tend to drift with time. A potential divider circuit (see Section 7.1) or a Wheatstone bridge might be used for signal conditioning, there being generally no need for compensation for lead resistance since the resistance of the leads is negligible compared with that of the thermistor. Special circuits are sometimes used to linearise the output of thermistors.

13.3.3 Semiconductors

Junction diodes have forward voltages which depend on temperature and can be used as a measure of temperature (see Section 6.2.1). Such diodes are often incorporated in integrated circuits to give a temperature sensor complete with signal conditioning (see Section 6.2.1). Junction semiconductor devices have good linearity, good sensitivity and need comparatively simple external circuitry. They are, however, limited to temperatures below about 200°C because the junction is destroyed at higher temperatures.

13.4 Thermocouples

Thermocouples (see Section 5.6) have very small thermal capacity and so respond rapidly to changes in temperature. The base metal thermocouples, E, J, K and T, are relatively cheap, with accuracies of about ±1 to 3%, but deteriorate with age. The noble metal thermocouples, R and S, are more expensive, with accuracies of the order of ±1% or better, and are more stable with a long life. Standard tables are available which give the e.m.f.s of commonly used thermocouples as a function of temperature when one junction is at 0°C. An alternative to having one junction at 0°C is to leave it at the ambient temperature and use a compensation circuit which will give a potential difference which just compensates for the junction not being at 0°C (see Section 7.1.4).

Thermocouples can be used with galvanometer indicator circuits. When cold junction compensation circuits are used, the measurement circuit usually has a specific resistance. Thus in such a case a ballast resistance is included with the galvanometer and adjusted to give the required resistance value for the circuit (Figure 13.11).

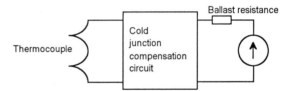

Figure 13.11 *Galvanometer circuit*

An alternative which avoids the need for a ballast resistance to adjust the resistance of the measurement circuit is a *potentiometer circuit*. With such a circuit the output from the thermocouple is opposed by an equal and opposite potential difference from the potentiometer so that they cancel out and give no current. Figure 13.12 shows the basic circuit. Because the thermoelectric e.m.f. is small compared with that of the supply voltage, in order to give a small enough potential drop per unit length of the potentiometer track a resistor is included in series with the potentiometer wire in order that the supply voltage is effectively dropped across a very long potentiometer.

With electronic instruments for indicating the output of thermocouples, the arrangement might use an operational amplifier and be of the form shown in Figure 7.17, with the addition of a circuit for the cold junction compensation.

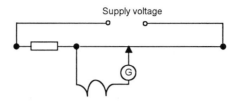

Figure 13.12 *Potentiometer circuit*

The output from a thermocouple, or from a number of thermocouples, might be processed by a computer. The arrangement could then be that the thermocouples are connected via a signal conditioning unit to a suitable digital acquisition card and into a computer. As an example of a signal conditioning unit for use in such a situation there is the National Instruments analogue multiplexer with temperature sensor AMUX-64T. This has a multiplexer which can take 64 single-ended or 32 differential inputs and also incorporates a temperature sensor for thermocouple cold junction compensation. The output from the multiplexer might then be used to provide analogue inputs to the National Instruments MIO digital acquisition board. The board controls the scanning by the multiplexer and provides suitable size digital outputs for the computer.

Instead of using tables to compute the temperature, computer software can be used to calculate temperatures using a suitable equation. In general, the temperature T is related to the e.m.f. e by an equation of the form:

$$T = a_0 + a_1 e + a_2 e^2 + a_3 e^3 + a_4 e^4 + \dots$$

where a_0, a_1, a_2, a_3, a_4, etc. are polynomial coefficients. For example, for a type E thermocouple the coefficients might be used up to the ninth in order to give an accuracy of $\pm 0.5°C$ in the temperature range $-100°C$ to $+1000°C$, the coefficients being:

a_0	$+0.104\ 967\ 248$
a_1	$+1.718\ 945\ 82 \times 10^4$
a_2	$-2.826\ 390\ 850 \times 10^5$
a_3	$+1.269\ 533\ 95 \times 10^7$
a_4	$-4.487\ 030\ 846 \times 10^8$
a_5	$+1.108\ 66 \times 10^{10}$
a_6	$-1.768\ 07 \times 10^{11}$
a_7	$+1.718\ 42 \times 10^{12}$
a_8	$-9.192\ 78 \times 10^{12}$
a_9	$+2.061\ 32 \times 10^{13}$

For a type J thermocouple in the range $0°C$ to $760°C$ with an accuracy of $\pm 0.1°C$, the coefficients up to the fifth are required:

a_0	$-0.048\ 868\ 252$
a_1	$+1.987\ 314\ 503 \times 10^4$
a_2	$-2.186\ 145\ 353 \times 10^5$
a_3	$+1.156\ 919\ 978 \times 10^7$
a_4	$-2.649\ 175\ 314 \times 10^8$
a_5	$+2.018\ 441\ 314 \times 10^9$

Often several thermocouples might be connected in series or parallel. When thermocouples are connected in series (Figure 13.13(a)) with all the hot junctions at one temperature and all the cold junctions at another temperature, the output is the sum of the e.m.f.s due to each and so there is an increase in the sensitivity compared with that given by just one thermocouple. Such an arrangement is called a *thermopile*. The output from a thermopile may be high enough to enable a cheaper instrument to be used for the voltage measurement than otherwise would be the case with a single thermocouple. The parallel connection of a number of thermocouples (Figure 13.13(b)) generates the same e.m.f. as a single thermocouple if all the hot junctions are at one temperature and all the cold junctions at another. If the junctions are not all at the same temperatures then the voltage indicated is the average of the individual voltages if the thermocouples are linear over the temperature range being measured. An alternative way of obtaining the average is to connect the thermocouples in series and then just divide the result by the number of hot junctions.

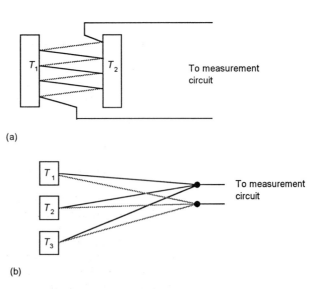

(a)

(b)

Figure 13.13 *(a) Series connection of thermocouples, (b) parallel connection of thermocouples*

13.5 Pyrometers

The amount of thermal energy emitted by a body by radiation and the wavelengths of that radiation depend on the temperature of the body. These two functions of temperature provide the basis of temperature measurement involving radiation. Instruments used for measuring temperature by means of the radiation emitted are termed *pyrometers*.

When discussing the radiation emitted by an object the term *black body* is used. An ideal block body is one that at all temperatures will absorb all the radiation falling on it without reflecting any back. The *absorption power* is the fraction of the incident radiation that is absorbed. An ideal black body has thus an absorption power of 1. Non-black bodies will have an absorption power less than 1. A black body is also a perfect radiator. It will radiate more radiation than a non-black body. The term *emissivity* is used for a surface and is the ratio of the radiation emitted from that surface at a particular temperature compared with that emitted by an ideal black body at the same temperature. Table 13.2 shows approximate values for the emissivities of some surfaces. The emissivity of a surface depends on the surface shape and texture, its temperature, and the wavelength considered.

Table 13.2 *Emissivity values*

Material	Surface	Temperature °C	Emissivity
Aluminium	oxidised	600	0.2
Brass	oxidised	600	0.6
Cast iron	oxidised	600	0.8
Cast iron	strongly oxidised	250	0.95
Copper	oxidised	200	0.6
Fire clay		1000	0.6

The total power P of the radiation emitted per unit surface area at a particular temperature from a black body is proportional to the fourth power of the temperature T on the kelvin scale:

$$P = \sigma T^4$$

This is known as the *Stefan–Boltzmann law* and the constant of proportionality σ as the Stefan–Boltzmann constant, its value being $5.670\ 32 \times 10^{-8}\ \text{W m}^{-2}\ \text{K}^{-4}$.

A black body at a particular temperature emits radiation at all wavelengths. However, the distribution of the energy is not uniform across the wavelengths. For example, a body at a temperature of about 900 K glows with a dull red colour. At a temperature of about 1200 K it glows bright red, at 1400 K it is orange and at temperatures above about 1700 K it is white hot. The colour changes because of the relative amounts of radiation emitted at the various wavelengths. Figure 13.14 shows how the power P_λ emitted per unit surface area at a particular wavelength varies with the wavelength λ from a black body at a number of temperatures.

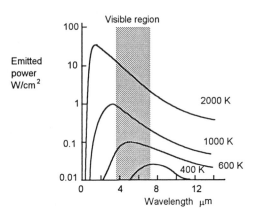

Figure 13.14 *Power distribution with wavelength*

The distribution is described by *Planck's radiation law*, the power P_λ at a particular wavelength λ being given by:

$$P_\lambda = \frac{c_1}{\lambda^5 (e^{c_2/\lambda T} - 1)}$$

where c_1 and c_2 are constants with the values 3.75×10^{-16} W m^2 and 1.44×10^{-2} m K. As the graph shows, the wavelength of the radiation at the maximum intensity depends on the temperature, getting shorter as the temperature increases. Differentiation of the above equation can be used to determine the maximum. The result is known as *Wien's displacement law*:

$$\lambda_{max} T = \text{a constant}$$

The constant has the value 2898 μm K. Note that the area under the line in Figure 13.14, i.e. the integral of the P_λ equation over all wavelengths, for a particular temperature is the total power emitted and is proportional to T^4, i.e. the Stefan–Boltzmann law.

The techniques involving the radiation from a body that can be used for the measurement of temperature include:

1 *Optical pyrometer*
   This is based on comparing the brightness of the light emitted by the hot body with that from a known standard.

2 *Total radiation pyrometer*
   This involves the measurement of the total amount of radiation emitted by the hot body. These utilise a resistance element or a thermopile.

3 *Photon radiation pyrometers*
   These are based on the use of photoelectric elements to detect the incident photon flux.

13.5.1 Optical pyrometer

The *optical pyrometer,* known generally as the *disappearing filament pyrometer,* involves just the visible part of the radiation emitted by a hot object. The radiation is focused onto a filament so that the radiation and the filament can both be viewed in focus through an eyepiece (Figure 13.15). The filament is heated by an electrical current until the filament and the hot object seem to be the same colour, the filament image then disappearing into the background of the hot object. The filament current is then a measure of the temperature. A red filter between the eyepiece and the filament is generally used to make the matching of the colours of the filament and the hot object easier. Another red filter may be introduced between the hot object and the filament with the effect of making the object seem less hot and so extending the range of the instrument.

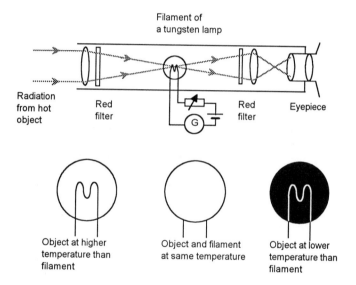

Figure 13.15 *Disappearing filament pyrometer*

The disappearing filament pyrometer has a range of about 600°C to 3000°C, an accuracy of about ±0.5% of the reading and involves no physical contact with the hot object. It can thus be used for moving or distant objects.

Corrections have to be made for the emissivity of the hot object. When the brightness of the two objects is matched we have the same radiation power from the hot object and from the filament. Thus:

$$P_\lambda = \frac{ec_1}{\lambda^5(e^{c_2/\lambda T} - 1)} = \frac{c_1}{\lambda^5(e^{c_2/\lambda T_f} - 1)}$$

where T is the temperature of the hot object and T_f that of the filament. We will assume that the red filter is being used and so the wavelength is about 0.63 μm, then e is the emissivity at that wavelength. Since the exponential terms in the brackets are greater than 1 we can write:

$$\frac{e}{e^{c_2/\lambda T}} = \frac{1}{e^{c_2/\lambda T_f}}$$

Hence:

$$\frac{1}{T} - \frac{1}{T_f} = \frac{\lambda \ln e}{c_2}$$

The temperature T of the hot object is thus only the same as the temperature T_f of the filament if the emissivity is 1. The above equation can be used to determine the temperature T when the emissivity is not 1.

13.5.2 Total radiation pyrometer

The *total radiation pyrometer* involves the radiation from the hot object being focused onto a radiation detector. Figure 13.16 shows the basic form of an instrument which uses a mirror to focus the radiation onto the detector. Some forms use a lens to focus the radiation. The detector is typically a thermopile with often up to 20 or 30 thermocouple junctions, a resistance element or a thermistor. The detector is said to be *broad band* since it detects radiation over a wide band of frequencies and so the output is the summation of the power emitted at every wavelength. It represents the area under the power-distribution graph in Figure 13.14 for a particular temperature and thus is proportional to the fourth power of the temperature (the Stefan–Boltzmann law).

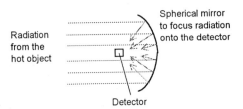

Figure 13.16 *Total radiation pyrometer*

Consider radiation from a source at a temperature T. This radiation heats the pyrometer detector. The temperature of the detector increases until the rate at which the radiation heats it is balanced by the rate at which it loses heat to the surroundings. As a rough approximation, the rate at which the detector at temperature T_d loses heat to the surroundings at temperature T_s is proportional to $(T_d - T_s)$. Thus:

$$K(T_d - T_s) = \sigma T^4$$

If the detector is a thermopile with the cold junction at the ambient temperature then the voltage output V from it is likely to be reasonably proportional to $(T_d - T_s)$. Thus the output V is proportional to T^4.

The above discussion has assumed that the hot body for which the temperature is being measured is a black body. For a body with an emissivity e, the energy emitted per second per unit area by such a body at a temperature T will be $e\sigma T^4$. The equivalent black body will be at a temperature T_b where:

$$e\sigma T^4 = \sigma T_b^4$$

Thus:

$$T = \frac{T_b}{e^{1/4}}$$

If the observed temperature is, on the assumption it is a black body, at 1200 K, and in fact it has an emissivity of 0.60, then the actual temperature is $1200/0.60^{1/4} = 1363$ K.

The accuracy of broad band total radiation pyrometers is typically about ±0.5% and ranges are available within the region 0°C to 3000°C. The time constant for the instrument varies from about 0.1 s when the detector is just one thermocouple or small bead thermistor to a few seconds with a thermopile involving many thermocouples. Some instruments use a rotating mechanical chopper to chop the radiation before it impinges on the detector. The aim is to obtain an alternating output from the detector, since amplification is easier with an alternating voltage. It is thus of particular benefit when the level of radiation is low. However, choppers can only be used with detectors which have a very small time constant and thus tend to be mainly used with small bead thermistor detectors.

13.5.3 Photon radiation pyrometers

Photon detectors are semiconductor devices in which the incident photons cause electrons to be excited from the valence band to the conduction band. The energy of a photon is $hf = hc/\lambda$, where h is Planck's constant, c the velocity of light, f the frequency and λ the wavelength of the photons. These detectors only respond to photons whose energy is equal to the energy gap between the bands. Thus such detectors only respond to photons of a particular wavelength and have a narrow wavelength response with the peak response being at the wavelength hc/E, where E is the size of the energy gap between the valence and conduction bands. Photodiodes and photoconductive cells are examples of such devices. Thus a radiation pyrometer employing a photon detector will only be monitoring the energy in a narrow band of wavelengths. Often a narrow pass-band filter is used to further restrict the wavelengths being detected.

A photon detector generates a voltage that is proportional to the photon flux density impinging on the detector, i.e. the number of photons per unit area. The power P_λ of the energy radiated per unit area from a hot surface is:

$$P_\lambda = \frac{c_1}{\lambda^5 (e^{c_2/\lambda T} - 1)}$$

Each photon has an energy of $hf = hc/\lambda$, where f is the frequency, c the velocity of light and h Planck's constant. Hence the number of photons N_λ emitted per unit area per second at a wavelength λ is:

$$N_\lambda = \frac{P_\lambda}{(hc/\lambda)} = \frac{c_1}{hc\lambda^4 (e^{c_2/\lambda T} - 1)}$$

Since $c_1 = 2\pi c^2 h$, then:

$$N_\lambda = \frac{2\pi c}{\lambda^4 (e^{c_2/\lambda T} - 1)}$$

To obtain the number of photons in a wavelength band of width $\Delta\lambda$, distributed about the wavelength λ of the radiation, we need to integrate the above expression between $\lambda + \Delta\lambda/2$ and $\lambda - \Delta\lambda/2$. The result is that the number of photons emitted per second is proportional to T^3.

Suppose we have the photon radiation pyrometer with a lens used to focus the radiation from the hot surface onto the detector (Figure 13.17). Consider the emitted radiation from an area A_s, say a circular patch of diameter d_s, of the hot surface to be collected by the lens and focused onto a photon detector of area A_d, say a circular area of diameter d_d. From similar triangles:

$$\frac{d_s}{u} = \frac{d_f}{v}$$

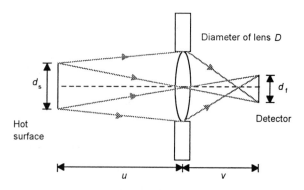

Figure 13.17 *Photon-detector system*

So the area A_s of the surface scanned is:

$$A_s = \tfrac{1}{4}\pi d_s^2 = \tfrac{1}{4}\pi\left(\frac{ud_f}{v}\right)^2 = \left(\frac{u}{v}\right)^2 A_d$$

But for a thin lens:

$$\frac{1}{u} + \frac{1}{v} = \frac{1}{F}$$

where F is the focal length of the lens. Thus, eliminating v:

$$A_s = \left(\frac{u-F}{F}\right)^2 A_d$$

Since u is generally much greater than f we can write:

$$A_s = \left(\frac{u}{F}\right)^2 A_d$$

The radiation from the hot surface can be considered to be emitted in all directions but only a segment of it hits the lens and so passes through to the detector. The solid angle subtended by an element of area which is a radial distance from a point source is surface area/distance^2. Thus for the pyrometer, the radiation which passes through the lens is in the solid angle and subtended by the lens diameter (Figure 13.18). This is is $\tfrac{1}{4}\pi D^2/u$. If we sum this for each element of area in the source area A_s then the fraction of the radiation passing through the lens is:

$$\text{fraction of radiation passing through lens} = \frac{D^2}{4u^2}A_s$$

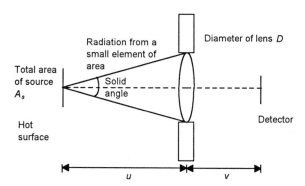

Figure 13.18 *Solid angle of radiation incident at lens*

Hence, if we assume that the system is focused so that all the radiation that passes through the lens is incident on the detector:

$$\text{fraction of the radiation incident on the detector} = \frac{D^2}{4u^2}A_s$$

$$= \frac{D^2}{4u^2}\left(\frac{u}{F}\right)^2 A_d$$

$$= \frac{D^2}{4F^2}A_d$$

If N_λ is the number of photons emitted per second for a particular narrow wavelength band then the number detected is:

$$\text{number of photons detected/s} = \frac{D^2}{4F^2}A_d N_\lambda$$

Since N_λ is proportional to T^3, then:

$$\text{number of photons detected/s} \propto \frac{D^2}{F^2}A_d T^3$$

The voltage output from the detector is proportional to the number of photons detected and thus:

$$\text{output voltage} \propto \frac{D^2}{F^2}A_d T^3$$

For a non-black body we can include the emissivity e to give:

$$\text{output voltage} \propto e\frac{D^2}{F^2}A_d T^3$$

As long as the radiation is focused onto the detector, the output voltage is independent of the distance of the hot surface from the instrument. Without the lens, or a spherical mirror, to focus the radiation onto the detector the output voltage would depend on the distance of the hot surface from the instrument since the angle subtended at the detector by the hot surface would depend on its distance away.

13.6 Dynamic characteristics of thermal sensors

Consider a temperature sensor, e.g. a resistance element or a thermocouple, at a temperature T_0 placed in surroundings at a constant temperature T_s (Figure 13.19). The system is a first order system and its temperature T variation with time t is described by the differential equation:

$$mc\frac{dT}{dt} = hA(T_s - T)$$

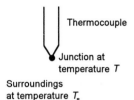

Figure 13.19 *Sensor*

where m is the mass of the sensor, c its specific heat capacity, A the sensor heat transfer area and h the heat transfer coefficient. See Section 3.2 for the derivation of the equation and its solution. The solution of the equation is:

$$T = T_s + (T_0 - T_s)\, e^{-t/\tau}$$

where $\tau = mc/hA$ and is termed the *time constant*. The time taken for the sensor to begin to approach the temperature of the surroundings depends on the value of the time constant. The larger the time constant the longer the time taken. Small values of the time constant mean for the sensor a small mass and a low specific heat capacity. Thus a thermistor in the form of a small bead would have a smaller time constant, and hence faster response time, than a larger more bulky resistance coil.

Suppose we put the sensor, initially at temperature T_0, in surroundings where the temperature starting at T_0 is increasing at a constant rate, i.e. a ramp function. We then have $T_s = at + T_0$, where a is a constant:

$$mc\frac{\mathrm{d}T}{\mathrm{d}t} = hAat - hAT$$

The solution to this equation (see Section 3.2.2) is:

$$T = a(\tau\, e^{-t/\tau} + t - \tau) + T_0$$

The measurement error at any instant, i.e. the difference between the temperature T indicated and that of the surroundings T_s which is being measured, is:

$$\text{error} = T - T_s = T - at - T_0 = at\, e^{-t/\tau} - a\tau$$

The first time varies with time and eventually will die away. The second time is a constant and so is an error that will always be present. The thermometer will always lag behind the actual temperature (see Figure 3.5). The size of this steady-state error depends on the time constant; the smaller the time constant the smaller the error. Thus for a sensor to be used for a fast-changing temperature, a small time constant is necessary.

We can analyse such situations by using the transfer function for the sensor. If we take the Laplace transform of the initial differential equation (see Section 3.4), we have:

$$mcsT(s) = hAT_s(s) - hAT(s)$$

and can describe the relationship between the input and the output for the sensor as:

$$\text{transfer function} = \frac{T(s)}{T_s(s)} = \frac{hA}{mcs + hA} = \frac{1}{\tau s + 1}$$

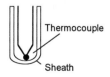

Figure 13.20 *Sensor*

Temperature sensors are often enclosed in a protective sheath, often termed a protective well (Figure 13.20) and thus, in such situations, account has to be taken of the heat having to flow through the sheath and heat it up. For the sensor we have:

$$m_s c_s \frac{dT_s}{dt} = h_{sw} A_s (T_w - T_s)$$

where m_s is the mass of the sensor, c_s its specific heat, h_{sw} the heat transfer coefficient between the sensor and the sheath, A_s the sensor heat transfer area, T_s the temperature of the sensor and T_w the temperature of the sheath. We can write this differential equation as:

$$\tau_1 \frac{dT_s}{dt} = T_w - T_s$$

where $\tau_1 = m_s c_s / h_{sw} A_s$. For the sheath we have a heat flow into it from the surroundings and a heat flow out to the sensor. Thus:

$$m_w c_w \frac{dT_w}{dt} = -h_{sw} A_s (T_w - T_s) + h_{wf} A_w (T_f - T_w)$$

where m_w is the mass of the sheath, c_w its specific heat, h_{wf} the heat transfer coefficient between the sheath and the surrounding fluid, A_w the sheath heat transfer area and T_f the temperature of the surrounding fluid. We can write this differential equation as:

$$\tau_2 \frac{dT_w}{dt} = -\delta(T_w - T_s) + (T_f - T_w)$$

where $\tau_2 = m_w c_w / h_{wf} A_w$ and $\delta = h_{sw} A_s / h_{wf} A_w$.

The Laplace transforms of the two differential equations are:

$$\tau_1 s T_s(s) = T_w(s) - T(s)$$

and:

$$\tau_2 s T_w(s) = -\delta T_w(s) + \delta T_s(s) + T_f(s) - T_w(s)$$

Eliminating $T_w(s)$ between these equations gives:

$$\frac{T_s(s)}{T_f(s)} = \frac{1}{\tau_1 \tau_2 s^2 + (\tau_1 + \tau_2 + \delta \tau_1)s + 1}$$

The transfer function of the sensor with a sheath is thus second order. If $\delta \tau_1$, i.e. $m_s c_s / h_{wf} A_w$ and termed the coupling term, is small compared with τ_1 and τ_2 then the equation approximates to:

$$\frac{T_s(s)}{T_f(s)} = \frac{1}{\tau_1 \tau_2 s^2 + (\tau_1 + \tau_2)s + 1} = \frac{1}{\tau_1 s + 1} \frac{1}{\tau_2 s + 1}$$

13.7 Radiation errors

Problems can occur in the measurements of the temperature of gases in that a thermometer placed in a gas can receive energy radiated from the containing vessel's walls, radiate energy itself and also respond to the temperature of the gas in contact with it. If the walls of the container are at a lower temperature than the gas, then while the thermometer is being heated by contact with the gas there is a net flow of radiation from the thermometer to the container walls. The result is that the thermometer will register a lower temperature than that of the gas.

Consider the case where energy is transferred to and from the sensor by convection from the immediate gas environment and to and from the sensor by radiation to the container walls. If we assume that no energy is lost or gained by thermal conduction through the supports for the thermometer, then at the steady-state condition we have:

convective heat transfer to the thermometer
= radiative energy transfer from the thermometer

The convective heat transfer can be considered to be $hA_s(T_g - T_s)$, where h is the heat transfer coefficient, A_s the surface area of the sensor and T_s its temperature. T_g is the temperature of the gas. Assuming that the sensor is a small body completely enclosed by the larger one of the container walls, the net amount of radiation from the sensor is given by the Stefan–Boltzmann law as $A_s \sigma e(T_s^4 - T_w^4)$, where e is the emissivity of the sensor and T_w the temperature of the surrounding walls. Thus:

$$hA_s(T_g - T_s) = A_s \sigma e(T_s^4 - T_w^4)$$

The error due to the radiation is thus:

$$\text{error} = T_s - T_g = \frac{\sigma e}{h}\left(T_w^4 - T_s^4\right)$$

The temperature error can thus be made small by insulating the enclosure wall so that it is at virtually the same temperature as the gas and hence the sensor. A sensor with a shiny surface, so that it has a low emissivity, will also reduce the error. Radiation shields can also be used to reduce the error. A radiation shield is an opaque surface interposed between the sensor and the container walls so as to reduce the radiation interchange between the two, i.e. it stops the sensor 'seeing' the walls. In principle the shield attains an equilibrium temperature closer to the gas temperature than the walls.

Problems

1 A constant volume gas thermometer contains nitrogen. If the pressure is 5 MPa at the triple point of water, what will be the pressure at +200°C?

2 A liquid-in-metal thermometer has a bulb with a volume ten times that of the capillary plus Bourdon tube. No compensating device is used to compensate for the liquid in the capillary tube and Bourdon tube. What

is the error resulting from the ambient temperature changing by 15°C from that at which the thermometer was calibrated?

3   An iron–constantan thermocouple has a cold junction maintained at 0°C and its output e.m.f. is measured using a potentiometer. Using appropriate tables, determine the temperature if the potentiometer indicates an e.m.f. of 10.777 mV.

4   If the thermocouple referred to in problem 3 did not have the cold junction at 0°C but at 20°C, what would the e.m.f. have been for the same temperature hot junction?

5   A Wheatstone bridge is to be used with a resistance thermometer. The bridge has equal arms of resistance 25 Ω and is balanced at 0°C. The accuracy of each of the resistors is ±0.1%. What will be the accuracy with which the temperature can be determined if the resistance of the thermometer element is determined by balancing the bridge? The temperature coefficient of resistance for the thermometer element is 0.0039 K^{-1}.

6   A resistance thermometer element has a temperature coefficient of resistance of 0.004 K^{-1} and is used as one arm of a Wheatstone bridge, the other bridge resistors all being 100 Ω and the bridge supply being 10 V. At 4.7°C the bridge is balanced. What will be (a) the resistance of the element at 25.0°C, (b) the voltage output change from the bridge per degree change in temperature?

7   A platinum resistance thermometer element has a coil of resistance 100 Ω. What will be the change in resistance per degree change in temperature if the platinum has a temperature coefficient of resistance of 0.003 92 K^{-1}?

8   A thermistor is used in the circuit shown in Figure 13.21. The thermistor has a resistance of 10 kΩ at a temperature of 300 K and its variation of resistance with temperature is described by the equation:

$$R = A\ e^{\beta/T}$$

where R is the resistance at temperature T, A a constant and β another constant which has the value 3000 K. Determine the output from the circuit at a temperature of 305 K.

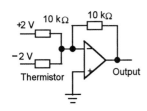

Figure 13.21 *Problem 8*

9   A thermopile consists of ten junction pairs in series. How will the output from the thermopile compare with that from a single thermocouple?

10 When a total radiation pyrometer is used for the measurement of the temperature of a hot object, the emissivity of the object is 0.2 ± 0.05.

What will be the percentage accuracy of the result due to this uncertainty in the value of the emissivity?

11 What will be the error in the temperature of 1100°C given by a total radiation pyrometer if the emissivity which had been assumed to be 0.8 was in fact 0.7?

12 Suggest temperature measurement systems that could be used to:
(a) Measure the temperature in an oven at about 150°C with an accuracy of ±2°C and display the temperature on a dial on the fascia of the oven.
(b) Measure the temperature of molten iron in a furnace and give the result on a scale situated some distance from the furnace. The temperature may be assumed not to be changing rapidly.
(c) Switch an alarm on when the temperature in a liquid rises above 50°C.

Appendix A: Distributions

This appendix is a brief consideration of distributions commonly encountered in dealing with errors and is intended to provide some background to the discussion of measurement errors in Chapter 2. For more details the reader is referred to texts dealing with experimental methods in science or engineering or statistics texts, e.g. *Experimental Methods* by W. Bolton (Newnes, Butterworth-Heinemann 1996), *Data Analysis for Physical Science Students* by L. Lyons (Cambridge University Press 1991), *Introduction to Probability and Statistics for Engineers and Scientists* by S.M. Ross (Wiley 1987).

What is meant by a distribution?

Consider some experiment in which repeated measurements are made of some quantity, e.g. the time taken for 100 oscillations of a simple pendulum. Suppose we take 20 readings and have the following results:

> 20.1, 20.3, 20.8, 20.5, 21.0, 20.8, 20.3, 20.4, 20.7, 20.6,
> 20.5, 20.7, 20.5, 20.1, 20.6, 20.4, 20.7, 20.5, 20.6, 20.3

We can display this data pictorially by means of a *histogram*. Thus, if we divide the data range into a number of convenient, equally sized, segments of, in this case, 0.2 we have:

> values between 20.0 and 20.2 (>20.0 and ≤20.2) come up twice
> values between 20.2 and 20.4 (>20.2 and ≤20.4) come up five times
> values between 20.4 and 20.6 (>20.4 and ≤20.6) come up seven times
> values between 20.6 and 20.8 (>20.6 and ≤20.8) come up five times
> values between 20.8 and 21.0 (>20.8 and ≤21.0) come up once

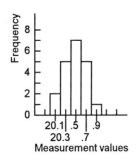

Figure A.1 *Histogram of the data*

The term *frequency* is used for the number of times a measurement occurs within a segment and the histogram represents the *frequency distribution*, showing how the data values group together. Figure A.1 shows the histogram for the above data. The horizontal axis gives the measurement values for the midpoints of each segment.

Using the frequency, it is difficult to compare a histogram obtained with the 20 values with one obtained with, say, 100 values. For this reason, the *relative frequency* tends to be used. The relative frequency is the fraction of the total number of readings in a segment. Thus, for the data given above we have:

> values between 20.0 and 20.2, relative frequency 2/20 = 0.1
> values between 20.2 and 20.4, relative frequency 5/20 = 0.25

values between 20.4 and 20.6, relative frequency 7/20 = 0.35
values between 20.6 and 20.8, relative frequency 5/20 = 0.25
values between 20.8 and 21.0, relative frequency 1/20 = 0.05

The relative frequency always has a value less than 1 and the sum of all the relative frequencies is 1, since this is the relative frequency for all the readings. The area of a rectangular segment in the histogram is made equal to relative frequency for that segment and hence the total area of all the strips is 1. Thus we can write for the relative frequency of a segment of width Δx and ordinate y:

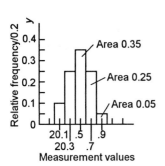

Figure A.2 *Histogram with relative frequency*

$$\text{relative frequency} = y\, \Delta x$$

Figure A.2 shows the above results with the vertical axis as the relative frequency per unit segment width, in this case the relative frequency per 0.20. The first segment has an ordinate y of (0.1/0.2) and thus an area of (0.1/0.2) × 0.20 = 0.1. A distribution plotted using relative frequencies per unit segment is said to be *normalised*.

The higher the relative frequency of a segment, the greater the probability that if we take a single measurement at random from the entire set that it will lie in that segment. The *probability* of an event occurring is the frequency with which it occurs as a fraction of the total number of possible outcomes:

$$\text{probability} = \frac{\text{number of ways an event can occur}}{\text{total number of ways possible}}$$

A coin can land either heads or tails uppermost. There are thus two possible outcomes of tossing a coin. The probability of a single toss of a coin landing heads uppermost, i.e. just one way out of the possible two, is 1 in 2 or 0.5. The relative frequency may thus be considered to be the probability. Thus, for the segment 20.6 to 20.8 there is a relative frequency of 0.25. This means the probability of a single measurement having a value between 20.6 and 20.8 is 0.25, i.e. 1 in 4. Of the total 20 values, 1 in 4, i.e. 5, lie between 20.6 and 20.8.

The histogram shown in Figure A.2 has a jagged appearance. This is because it represents only a few values. If we had taken a very large number of readings then we could have divided the range into smaller segments and still had an appreciable number of values in each segment. The result of plotting the histogram would now be to give one with a much smoother appearance. As the number of readings used increases, not only does the histogram become smoother, but it settles down into a constant shape. For example, the normalised histograms for 10 and 20 readings will be jagged and could vary quite significantly in their general shape. However, the normalised histograms for 500 and 1000 readings are likely to be smooth with virtually no differences in their shape. Thus, with such large numbers, we might have a distribution of the form shown in Figure A.3, a smooth curve having been drawn through the tops of the infinitely small segments of the histogram. The y axis is generally termed the *frequency function*,

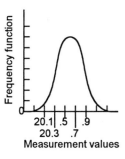

Figure A.3 *Distribution with large numbers of readings*

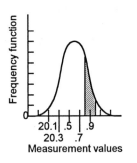

Figure A.4 *Probability of a single measurement of 20.9*

being the relative frequency per unit segment when the segments are infinitesimally small. This distribution represents the *limiting frequency distribution* we would obtain with an infinite number of values.

If we consider a segment of a limiting frequency distribution then the relative frequency of that segment, i.e. its area, gives the *probability* that a single measurement taken at random from the distribution will lie in that segment. Consider the probability, with a very large number of readings, of obtaining a value between 20.8 and 21.0 with the distribution shown in Figure A.4. If we take a segment 20.8 to 21.0 then the area of that segment is the relative frequency. Suppose the result in this case is 0.30. The probability of taking a single measurement and finding it in that interval is 0.30, i.e. on average 30 times in every 100 values taken.

Mean of a distribution

The mean value $\bar{x}$ of a set of readings can be obtained in a number of ways, depending on the form with which the data is presented:

1. For a list of discrete readings, sum all the readings and divide by the number N of readings, i.e.:

$$\bar{x} = \frac{x_1 + x_2 + x_3 + \ldots + x_j}{N} = \frac{\Sigma x_j}{N}$$

2. For a distribution of discrete readings, if we have n_1 readings with value x_1, n_2 readings with value x_2, n_3 readings with value x_3, etc., then the above equation for the mean becomes:

$$\bar{x} = \frac{n_1 x_1 + n_2 x_2 + n_3 x_3 + \ldots + n_j x_j}{N}$$

But n_1/N is the relative frequency of value x_1, n_2/N is the relative frequency of value x_2, etc. Thus, to obtain the mean, multiply each reading by its relative frequency y and sum over all of the values.

$$\bar{x} = \sum_{j=1}^{n_j} y_j x_j$$

3. For readings presented as a histogram plotted with relative frequencies (Figure A.5), the above relationship translates into: the mean is given by multiplying the mid-ordinate of each measurement value segment by the relative frequency for that segment and summing over all the possible values. Thus if we consider segment j in the histogram then it has a mid-ordinate measurement value of x_j and a relative frequency of $y_j \Delta x$. Thus the mean is given by:

$$\bar{x} = \sum_{j=1}^{n_j} y_j x_j \Delta x$$

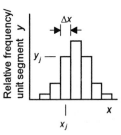

Figure A.5 *Histogram*

4 For readings presented as a continuous distribution curve, we can consider that we have a histogram with very large numbers of very thin segments. Thus if y, a function of x, represents relative frequency values and x the measurement values, the rule given above for histograms translates into:

$$\bar{x} = \int_{-\infty}^{\infty} xy \, dx \text{ or } \int_{-\infty}^{\infty} xf(x) \, dx$$

For a symmetrical distribution, as in Figure A.4, the mean will be the value with the greatest frequency.

With a very large number of readings, the mean value is taken as being the *true value* about which the random fluctuations occur.

Standard deviation

Any single reading x in a distribution (Figure A.6) will deviate from the mean of that distribution by some error e, i.e.:

$$e = x - \bar{x}$$

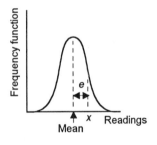

Figure A.6 *Error e*

With a particular measurement we might obtain a series of readings which is widely scattered around the mean while another has readings closely grouped round the mean. The measurement with the large random fluctuations will give a distribution which is a broad peaked distribution curve while that with smaller random fluctuations will give a narrow peaked curve. Figure A.7 shows the type of curves that might occur. The broad peaked distribution curve is said to be for a more imprecise set of readings than the narrow peaked curve. A measurement is said to be *precise* when it is determined from readings in which the random errors are small. This is irrespective of whether or not systematic errors are present. An accurate measurement is when it is determined from readings in which the random and systematic errors are small.

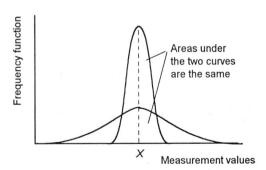

Figure A.7 *Distributions with different precisions*

A measure of the precision can thus be obtained by some measure of the 'width' of the distribution curves. The average error cannot be used, since for every positive value of an error there will be a negative error and so the sum of all the errors will be zero. The measure used is the standard deviation. The *standard deviation* σ is the root-mean-square value of e for all the measurements in the distribution, i.e. σ^2 is the mean value of e^2. The quantity σ^2 is known as the *variance* of the distribution. Thus, for a number of discrete values, $x_1, x_2, x_3, \ldots$, etc., we can write for the mean value of the sum of the squares of their deviations from the mean of the set of results:

$$\text{sum of squares of deviation} = \frac{(x_1 - \bar{x})^2 + (x_2 - \bar{x})^2 + (x_3 - \bar{x})^2 + \ldots}{N}$$

Hence the mean of the square root of this sum of the squares of the deviations, i.e. the standard deviation, is:

$$\sigma = \sqrt{\frac{\left((x_1 - \bar{x})^2 + (x_2 - \bar{x})^2 + (x_3 - \bar{x})^2 + \ldots\right)}{N}}$$

However, we need to distinguish between the standard deviation of a sample s and the standard deviation σ of the entire population of readings that are possible and from which we have only considered a sample (many statistics textbooks adopt the convention of using Greek letters when referring to the entire population and Roman for samples). When we are dealing with a sample we need to write:

$$s = \sqrt{\frac{\left((x_1 - \bar{x}_s)^2 + (x_2 - \bar{x}_s)^2 + (x_3 - \bar{x}_s)^2 + \ldots\right)}{N - 1}}$$

with $\bar{x}_s$ being the mean value of the sample rather than the true value $\bar{x}$. The reason for the $N - 1$ rather than N is that in calculating the standard deviation we are using the sample mean rather than the true mean value that would have been used if we had used the entire population. The root-mean-square of the deviations of the readings in a sample around the sample mean is less than around any other figure. Hence, if the true mean of the entire population were known, the estimate of the standard deviation of the sample data about it would be greater than that about the sample mean. Therefore, by using the sample mean, an underestimate of the population standard deviation is given. This bias can be corrected by using one less than the number of observations in the sample in order to give the sample mean.

Normal distribution

A particular form of distribution, known as the *normal distribution* or *Gaussian distribution*, works well as a model for measurements when there are random errors. This form of distribution has a characteristic bell shape

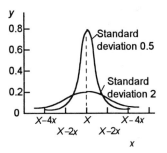

Figure A.8 *Typical forms of normal distribution*

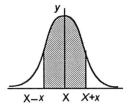

Figure A.9 *Area between −x and +x of X*

(Figure A.8). It is symmetric about its mean value X, having its maximum value at that point. It tends rapidly to zero as x increases or decreases from the mean. It can be completely described in terms of its mean X and its standard deviation σ. The following equation describes how the values are distributed about the mean.

$$y = f(x) = \frac{1}{\sigma\sqrt{2\pi}}\,e^{-(x-X)^2/2\sigma^2}$$

The fraction of the total number of measurements that lies between $-x$ and $+x$ from X is the fraction of the total area under the curve that lies between those ordinates (Figure A.9). We can obtain areas under the curve by integration.

To save the labour of carrying out the integration, the results have been calculated and are available in tables. As the form of the graph depends on the value of the standard deviation, as illustrated in Figure A.8, the area depends on the value of the standard deviation σ. In order not to have to give tables of the areas for different values of x for each value of σ, the distribution is considered in terms of the value of $(x - X)/\sigma$, this commonly being designated by the symbol z, and areas tabulated against this quantity. Table A.1 shows examples of the type of data given in such tables:

Table A.1 *Areas under normal curve*

$(x - X)/\sigma$	Area from X	$(x - X)/\sigma$	Area from X
0	0.0000	1.6	0.4452
0.2	0.0793	1.8	0.4641
0.4	0.1555	2.0	0.4772
0.6	0.2257	2.2	0.4861
0.8	0.2881	2.4	0.4918
1.0	0.3413	2.6	0.4953
1.2	0.3849	2.8	0.4974
1.4	0.4192	3.0	0.4987

When $x - X = 1\sigma$, then $(x - X)/\sigma = 1.0$ and the area between the ordinate at the mean and the ordinate at 1σ, as a fraction of the total area, is 0.3413. The total area within $\pm 1\sigma$ of the mean is thus the fraction 0.6816. Expressed as a percentage, the area is 68.16%. This means that the chance of a reading being within $\pm 1\sigma$ of the mean is 68.16%, i.e. roughly two-thirds of the readings (Figure A.10).

When $x - X = 2\sigma$, then $(x - X)/\sigma = 2.0$ and the area between the ordinate at the mean and the ordinate at 1σ, as a fraction of the total area, is 0.4772. The total area within $\pm 2\sigma$ of the mean is thus the fraction 0.9544.

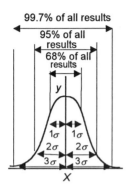

Figure A.10 *Percentages of results in bands of 1, 2 and 3 standard deviations of the mean*

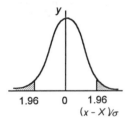

Figure A.11 *Area for which there is 5% significance*

Expressed as a percentage, the area is 95.44%. This means that the chance of a reading being within $\pm 2\sigma$ of the mean is 95.44%.

When $x - X = 3\sigma$, then $(x - X)/\sigma = 3.0$ and the area between the ordinate at the mean and the ordinate at 3σ, as a fraction of the total area, is 0.4987. The total area within $\pm 1\sigma$ of the mean is thus the fraction 0.9974. Expressed as a percentage, the area is 99.74%. This means that the chance of a reading being within $\pm 3\sigma$ of the mean is 99.74%. Thus, virtually all the readings will lie within $\pm 3\sigma$ of the mean.

Significance and confidence levels

With the normal distribution, 95% of the readings fall within 1.96 standard deviations of the mean. Thus, if we obtained a value which differed by more than 1.96 standard deviations from the mean, it could be said to be significantly different from most of the reading values since it would not be one of the 95% and so expected to occur often. In statistical terms we say that a reading which was not within 1.96 standard deviations of the mean is *significant at the 5% level* (Figure A.11). For a reading to be *significant at the 1% level* its deviation from the mean must be at least 2.576 standard deviations.

Consider a situation in which we have a box in which we think there are five red and five black counters. Suppose we now draw, at random, five counters from the box. If these are all red then we might suspect that the box did not contain five red and five black counters, considering our sample to be a significant result which was highly improbable if there had been five red and five black counters in the box. How small must the probability of this 'improbable' event be for us to doubt the initial hypothesis of five red and five black when it actually occurs? Traditionally the probability, i.e. significance, levels are taken as:

5% and less is deemed significant
1% and less is deemed highly significant
0.1% and less is deemed very highly significant

Thus if an event occurs which has a 5% or less significance level, then the event is regarded as significant. For example, if the value of the acceleration due to gravity in a particular place is expected to be 9.813 m/s² and the mean result of a number of experiments is found to be 9.892 m/s², then if this mean has a 5% significance level we begin to doubt that the expected value of 9.813 m/s² is valid. If the significance level is 1% or less then the result is deemed highly significant and we now have serious doubts about it being 9.813 m/s² and regard it as highly improbable.

Since 95% of the readings in a normal distribution lie within 1.96 standard deviations of the mean X, we can be 95% confident that a reading taken at random will occur within these limits. We thus define what is termed a *95% confidence interval* as:

$$X - 1.96\sigma \leq x \geq X + 1.96\sigma$$

This means that if we consider 95% of the readings, we will be able to describe them as having means that fall within the interval $X \pm 1.96\sigma$.

In a similar way we can define a *99% confidence interval* as:

$$X - 2.576\sigma \leq x \geq X + 2.576\sigma$$

99% of the readings occur within 2.576 standard deviations of the mean.

t-distribution

It has often been assumed, in earlier examples in this chapter, that the standard deviation of the population is known. Often, however, only the standard deviation of a small sample is known and this has been used as though it were the standard deviation of the population. But this introduces some element of unreliability. Thus if we want to be sure of, say, the 95% confidence level then we need to broaden the interval to allow for it having been calculated from the standard deviation of a small sample. We do this by replacing the values of the areas for $(x - X)/\sigma$ taken from the normal distribution by larger values for a similar distribution, this being called the *Student's t-distribution*. It is called Student because W.S. Gossett, who studied the distribution, published papers about it under the pen name of 'Student'. If we use the sample standard deviation s then, for a sample size n:

$$\text{estimated standard error of the mean } \hat{\sigma} = \frac{s}{\sqrt{n}}$$

and we define t as:

$$t = \frac{x - X}{\hat{\sigma}} = \frac{x - X}{s/\sqrt{n}}$$

When n is more than about 25, the sample standard deviation is close enough to the population standard deviation for $\hat{\sigma}$ to be virtually the same as σ and thus t to be effectively the same as z, i.e. $(x - X)/\sigma$.

When using the t distribution, the 95% confidence interval is given by:

$$X - t_{0.025}\frac{s}{\sqrt{n}} \leq x \geq X + t_{0.025}\frac{s}{\sqrt{n}}$$

The 0.025 subscript to the t indicates that we are taking an ordinate which gives the area fraction 0.025, i.e. 2.5%, from the outer limit of the distribution. For the 99% confidence interval we have:

$$X - t_{0.005}\frac{s}{\sqrt{n}} \leq x \geq X + t_{0.005}\frac{s}{\sqrt{n}}$$

The 0.005 subscript to the *t* indicates that we are taking an ordinate which gives the area fraction 0.005, i.e. 0.5%, from the outer limit of the distribution.

Table A.2 gives *t* values. To use the table you need to know the *degree of freedom* of the data being used. The degree of freedom, usual symbol v, is the denominator used in calculating s, i.e. $n - 1$.

Table A.2 *t critical points*

v	$t_{0.25}$	$t_{0.10}$	$t_{0.05}$	$t_{0.025}$	$t_{0.010}$	$t_{0.005}$	$t_{0.0025}$	$t_{0.0010}$	$t_{0.0005}$
1	1.00	3.08	6.31	12.7	31.8	63.7	127	318	637
2	0.82	1.89	2.92	4.30	6.96	9.92	14.1	22.3	31.6
3	0.76	1.64	2.35	3.18	4.54	5.84	7.45	10.2	12.9
4	0.74	1.53	2.13	2.78	3.75	4.60	5.60	7.17	8.61
5	0.73	1.48	2.02	2.57	3.36	4.03	4.77	5.89	6.87
6	0.72	1.44	1.94	2.45	3.14	3.71	4.32	5.21	5.96
7	0.71	1.41	1.89	2.36	3.00	3.50	4.03	4.79	5.41
8	0.71	1.40	1.86	2.31	2.90	3.36	3.83	4.50	5.04
9	0.70	1.38	1.83	2.26	2.82	3.25	3.69	4.30	4.78
10	0.70	1.37	1.81	2.23	2.76	3.17	3.58	4.14	4.59
11	0.70	1.36	1.80	2.20	2.72	3.11	3.50	4.02	4.44
12	0.70	1.36	1.78	2.18	2.68	3.05	3.43	3.93	4.32
13	0.69	1.35	1.77	2.16	2.65	3.01	3.37	3.85	4.22
14	0.69	1.35	1.76	2.14	2.62	2.98	3.33	3.79	4.14
15	0.69	1.34	1.75	2.13	2.60	2.95	3.29	3.73	4.07
16	0.69	1.34	1.75	2.12	2.58	2.92	3.25	3.69	4.01
17	0.69	1.33	1.74	2.11	2.57	2.90	3.22	3.65	3.97
18	0.69	1.33	1.73	2.10	2.55	2.88	3.20	3.61	3.92
19	0.69	1.33	1.73	2.09	2.54	2.86	3.17	3.58	3.88
20	0.68	1.33	1.72	2.09	2.53	2.85	3.15	3.55	3.85
25	0.68	1.32	1.71	2.06	2.49	2.79	3.08	3.45	3.73
30	0.68	1.31	1.70	2.04	2.46	2.75	3.03	3.39	3.65

Appendix B: Solving differential equations

This appendix is a brief consideration of the solution of first and second order differential equations. It does not aim to cover all the methods that might be used but just those commonly used. For a more detailed consideration, the reader is referred to such texts as *Ordinary Differential Equations* by W. Bolton (Longman 1994), *Introduction to Ordinary Differential Equations* by R.C. McCann (Harcourt Brace Jovanovich 1982), *Introduction to Ordinary Differential Equations* by S.L. Ross (Wiley 1989) or to sections in books such as *Engineering Mathematics* by A. Croft, R. Davison and M. Hargreaves (Addison-Wesley 1992) or *Foundation Mathematics for Engineers* by J. Berry and P. Wainwright (Macmillan 1991).

Differential equations

The term *differential equation* is used for any equation that contains a derivative or a number of derivatives. The *order* of a differential equation is equal to the order of the highest derivative that appears in it. dy/dx is termed the first derivative or the derivative of order 1, d^2y/dx^2 is termed the second derivative or the derivative of order 2 and thus, in general, d^ny/dx^n is termed the *n*th derivative or the derivative of order *n*.

The term *linear differential equation* is used for one where the dependent variable and all its derivatives only occur to the first power, i.e. we have no terms such as $(dy/dx)^2$.

Solving first order equations

If a differential equation can be put into the form:

$$g(y) \frac{dy}{dx} = f(x)$$

where $g(y)$ is some function of y and $f(x)$ is some function of x, then the solution can be obtained by *separation of the variables* and integrating:

$$\int g(y)\, dy = \int f(x)\, dx$$

For example, with the differential equation:

$$a_1 \frac{dx}{dt} + a_0 x = b_0 y$$

where y is constant and not varying with time t, we can rearrange the equation to give:

$$\frac{dx}{dt} = \frac{b_0 y - a_0 x}{a_1}$$

and then separate the variables to give:

$$\int \frac{1}{(b_0/a_0)y - x} \, dx = \int \frac{a_0}{a_1} dt$$

Then the solution is:

$$-\ln\left[\left(\frac{b_0}{a_0}\right)y - x\right] = \frac{a_0}{a_1} t + C$$

where C is a constant. If we have $x = x_0$ at time $t = 0$, and let $(b_0/a_0) = G$, then:

$$-\ln(Gy - x_0) = C$$

If we let τ be (a_1/a_0) then we can write the equation as:

$$-\ln[Gy - x] = (t/\tau) - \ln[Gy - x_0]$$

and so:

$$x = Gy - (Gy - x_0) \, e^{-t/\tau}$$

There is another way of solving the above differential equation which makes use of the fact that it is a linear differential equation. With a linear differential equation the solution can be obtained as the sum of two parts, one called the *particular integral* and one the *complementary function*. Thus if we consider the above differential equation, namely:

$$a_1 \frac{dx}{dt} + a_0 x = b_0 y$$

and let $x = u + v$, then:

$$a_1 \frac{d(u + v)}{dt} + a_0(u + v) = b_0 y$$

Rearranging this gives:

$$\left(a_1 \frac{du}{dt} + a_0 u\right) + \left(a_1 \frac{dv}{dt} + a_0 v\right) = b_0 y$$

If we let:

$$a_1\frac{du}{dt} + a_0u = b_0y$$

then we must have:

$$a_1\frac{dv}{dt} + a_0v = 0$$

This last equation describes a situation where there is no input to the system and its solution gives the transient part of the output. The equation with the y term describes the situation when there is an input and for this reason is often termed the forced response.

We can solve the transient differential equation by using the separation of variables method. However, because all such differential equations yield solutions of the same form, we can jump to just trying a solution with an arbitrary constant. Thus trying $u = A\ e^{st}$, where A and s are constants, the differential equation becomes:

$$a_1As\ e^{st} + a_0A\ e^{st} = 0$$

Thus we must have $s = -(a_0/a_1)$ and so, if we let $\tau = (a_1/a_0)$, the solution of the transient equation is:

$$u = A\ e^{-t/\tau}$$

To obtain the solution of the forced response differential equation we can also try a solution. The form of solution to try depends on the form of the input signal. For a step input we try $v = k$, where k is a constant. For a ramp signal where $y = ct$ we would try $v = kt$. Thus for a step signal, the differential equation becomes:

$$0 + a_0k = b_0y$$

and so $k = (b_0/a_0)y$. If we let $(b_0/a_0) = G$, then the solution for the forced differential equation is:

$$v = Gy$$

The full solution is thus:

$$x = u + v = A\ e^{-t/\tau} + Gy$$

We can determine the value of the constant A by using the fact that $x = x_0$ at $t = 0$. Thus $x_0 = A + Gy$ and so, as before:

$$x = Gy - (Gy - x_0)\ e^{-t/\tau}$$

Solving second order equations

To solve a linear second order differential equation we can employ the same method of particular integral and complementary function that was used with the first order equation. Consider the differential equation:

$$a_2 \frac{d^2x}{dt^2} + a_1 \frac{dx}{dt} + a_0 x = b_0 y$$

Let $x = u + v$. Then we can obtain, after substituting for x:

$$\left(a_2 \frac{d^2u}{dt^2} + a_1 \frac{du}{dt} + a_0 u \right) + \left(a_2 \frac{d^2v}{dt^2} + a_1 \frac{dv}{dt} + a_0 v \right) = b_0 y$$

Then, for the transient element we have:

$$a_2 \frac{d^2u}{dt^2} + a_1 \frac{du}{dt} + a_0 u = 0$$

and for the forced element:

$$a_2 \frac{d^2v}{dt^2} + a_1 \frac{dv}{dt} + a_0 v = b_0 y$$

For the transient equation we can try a solution of the form $u = A\,e^{st}$, where A and s are constants. The differential equation then becomes:

$$a_2 A s^2\, e^{st} + a_1 A s\, e^{st} + a_0 A\, e^{st} = 0$$

which simplifies to:

$$a_2 s^2 + a_1 s + a_0 = 0$$

This equation is termed the *auxiliary equation*. The roots of the equation can be obtained by using the formula for the roots of a quadratic equation. Thus:

$$s = \frac{-a_1 \pm \sqrt{a_1^2 - 4a_2 a_0}}{2a_2}$$

If we let $(a_1/2a_2) = \zeta$ and $\omega_n = (a_0/a_2)$, then we have $(a_1/2a_2) = \zeta\omega_n$ and so we can write:

$$s = -\zeta\omega_n \pm \omega_n \sqrt{\zeta^2 - 1}$$

With ζ greater than 1, we have two real roots s_1 and s_2, where:

$$s_1 = -\zeta\omega_n + \omega_n \sqrt{\zeta^2 - 1}$$

$$s_2 = -\zeta\omega_n - \omega_n\sqrt{\zeta^2 - 1}$$

Thus the transient solution is of the form:

$$u = A\,e^{s_1 t} + B\,e^{s_2 t}$$

With $\zeta = 1$ there are two equal roots and we have $s_1 = s_2 = -\omega_n$. It might seem that for this condition the solution should be $u = A\,e^{st}$. However, such a solution is not capable of satisfying the initial conditions for a second order system and so the solution is written as:

$$u = (At + B)\,e^{-\omega_n t}$$

With ζ less than 1, there are two complex roots since we can write the equation for the roots as:

$$s = -\zeta\omega_n \pm \omega_n\sqrt{-1}\,\sqrt{1 - \zeta^2}$$

Both roots involve $\sqrt{(-1)}$. Writing j for $\sqrt{(-1)}$, we have:

$$s = -\zeta\omega_n \pm j\omega_n\sqrt{1 - \zeta^2}$$

If we let $\omega = \omega_n\sqrt{1 - \zeta^2}$ then:

$$s = -\zeta\omega_n \pm j\omega$$

The solution is then:

$$u = A\,e^{(-\zeta\omega_n + j\omega)t} + B\,e^{(-\zeta\omega_n - j\omega)t}$$

This can be rewritten as:

$$u = e^{-\zeta\omega_n t}(A\,e^{j\omega t} + B\,e^{-j\omega t})$$

Euler's formula (see mathematics texts on complex numbers, e.g. *Complex Numbers* by W. Bolton (Longman 1995)), gives:

$$e^{j\omega t} = \cos\omega t + j\sin\omega t$$

$$e^{-j\omega t} = \cos\omega t - j\sin\omega t$$

Hence:

$$u = e^{-\zeta\omega_n t}\left(A\cos\omega t + jA\sin\omega t + B\cos\omega t - jB\sin\omega t\right)$$

Grouping terms we can write:

$$u = e^{-\zeta\omega_n t}\left[(A + B)\cos\omega t + j(A - B)\sin\omega t\right]$$

and if we substitute new constants P and Q, where $(A + B) = P$ and $j(A - B) = Q$, then we have:

$$u = e^{-\zeta\omega_n t}\left(P\cos\omega t + Q\sin\omega t\right)$$

We can express this equation in alternative forms. Since:

$$\sin(\omega t + \phi) = \sin\omega t\cos\phi + \cos\omega t\sin$$

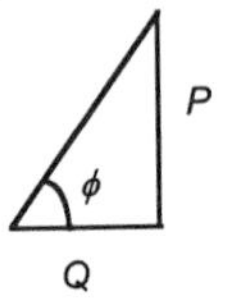

then if we let P and Q represent the sides of a right-angled triangle (Figure B.1) with angle ϕ such that $P/\sqrt{(P^2 + Q^2)} = \sin\phi$ and $Q/\sqrt{(P^2 + Q^2)} = \cos\phi$ we can write:

Figure B.1 *Right-angled triangle*

$$u = \sqrt{P^2 + Q^2}\ e^{-\zeta\omega_n t}\sin(\omega t + \phi) = C\,e^{-\zeta\omega_n t}\sin(\omega t + \phi)$$

Alternatively, using $\cos(\omega t - \theta) = \cos\omega t\cos\theta + \sin\omega t\sin\theta$, we can write:

$$u = \sqrt{P^2 + Q^2}\ e^{-\zeta\omega_n t}\cos(\omega t - \theta) = C\,e^{-\zeta\omega_n t}\cos(\omega t - \theta)$$

where C is a constant.

To solve the forced differential equation, if we have a unit step input at time $t = 0$ then we can try $v = k$. Substituting this in the differential equation:

$$a_2\frac{d^2v}{dt^2} + a_1\frac{dv}{dt} + a_0 v = b_0 y$$

gives:

$$0 + 0 + a_0 v = b_0$$

Thus $v = (b_0/a_0) = G$, where G is the steady-state gain.

The solutions of the differential equation are thus:

1 *Damping factor less than 1, under-damped*

$$x = e^{-\zeta\omega_n t}\left(P\cos\omega t + Q\sin\omega t\right) + G$$

2 *Damping factor equal to 1, critically damped*

$$x = (At + B)\,e^{-\omega_n t} + G$$

3 *Damping factor more than 1, over-damped*

$$x = A\,e^{s_1 t} + B\,e^{s_2 t} + G$$

The constants in the equations can be obtained by substituting the initial conditions. Because there are two constants involved in each equation, two sets of initial conditions are required. Thus we might have the initial conditions that $x = 0$ at $t = 0$ and also $dx/dt = 0$ at $t = 0$. Thus, for example, with $x = 0$ at $t = 0$ the under-damped solution gives:

$$0 = e^{-0}(P + 0) + G$$

Thus $P = -G$. For $dx/dt = 0$ at $t = 0$, we have to differentiate the output equation before applying this condition. Thus:

$$\frac{dx}{dt} = e^{-\zeta\omega_n t}\left(\omega P \sin\omega t - \omega Q \cos\omega t\right) - \zeta\omega_n\, e^{-\zeta\omega_n t}\left(P\cos\omega t + Q\sin\omega t\right)$$

Thus:

$$0 = 1(0 - \omega Q) - \zeta\omega_n(P + 0) = 1(0 - \omega Q) - \zeta\omega_n(-G + 0)$$

Hence $Q = \zeta G(\omega/\omega_n)$.

Appendix C: The Laplace transform

This appendix gives more details of the Laplace transform than appears in Chapter 3. For a more detailed discussion of the Laplace transform, and examples of its use, the reader is referred to *Laplace and z-Transforms* by W. Bolton (Longman 1994) or *An Introduction to the Laplace Transform and the z-Transform* by A.C. Grove (Prentice-Hall 1991).

The Lapace transform

A quantity which is a function of time is said to be in the *time domain* and represented by a function such as $f(t)$. For functions for which we have $t \geq 0$ the Laplace transform can be obtained by:

1   Multiplying the function by e^{-st}.

2   Then integrating with respect to time from zero to infinity.

s is a constant with the unit of 1/time. The result of the integration is termed the *Laplace transform* and the function is then said to be in the *s-domain*. Thus the Laplace transform of the function of time $f(t)$, which is written as $\mathcal{L}\{f(t)\}$, is given by:

$$\mathcal{L}\{f(t)\} = \int e^{-st} f(t)\, dt$$

The transform is *one-sided* in that values are only considered between 0 and $+\infty$, and not over the full range of time from $-\infty$ to $+\infty$. In the *s*-domain a function is usually written, since it is a function of s, as $F(s)$. A capital letter F is generally used for the Laplace transform and a lower-case letter f for the time-varying function $f(t)$. Thus:

$$\mathcal{L}\{f(t)\} = F(s)$$

For the inverse operation, when the function of time is obtained from the Laplace transform, we can write:

$$f(t) = \mathcal{L}^{-1}\{F(s)\}$$

This reads as: $f(t)$ is the inverse transform of the Laplace transform $F(s)$.

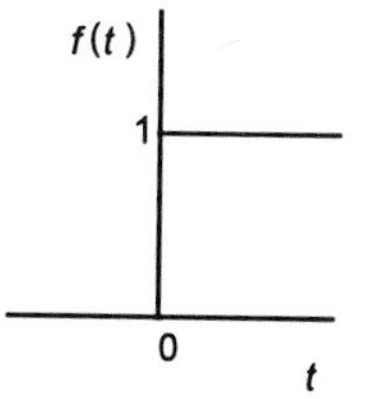

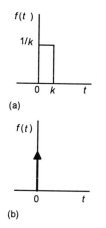

$f(t)$

1

0

t

Figure C.1 *Unit step*

$f(t)$

$1/k$

0 k t

(a)

$f(t)$

0 t

(b)

Figure C.2 *Impulse*

The Laplace transform from first principles

Consider a *unit step function* (Figure C.1). This has the constant value of 1 for all values of time greater than 0, i.e. $f(t) = 1$ for $t \geq 0$. The Laplace transform is then:

$$\mathcal{L}\{f(t)\} = F(s) = \int_0^\infty 1\, e^{-st}\, dt = -\frac{1}{s}[e^{-st}]_0 = \frac{1}{s}$$

Consider now the Laplace transform of the exponential function e^a, where a is a constant. The Laplace transform of $f(t) = e^a$ is then:

$$F(s) = \int_0^\infty e^{at}\, e^{-st}\, dt = \int_0^\infty e^{-(s-a)t}\, dt = -\frac{1}{s-a}\left[e^{-(s-a)t}\right] = \frac{1}{s-a}$$

In order to derive the transform for a *unit impulse* we first need to consider a rectangular pulse of size $1/k$ that occurs at time $t = 0$ and which has a pulse width of k, i.e. the area of the pulse is 1 (Figure C.2(a)). If we maintain this constant pulse area of 1 and then decrease the width of the pulse (i.e. reduce k), the height increases. Thus, in the limit as $k \to 0$ we end up with just a vertical line at $t = 0$ with the height of the graph going off to infinity. The result is a graph that is zero except at a single point where there is an infinite spike (Figure C.2(b)) and thus represents an impulse. The impulse is said to be a unit impulse because the area enclosed by it is 1.

The Laplace transform for the unit area rectangular pulse in Figure C.2(a) is given by:

$$F(s) = \int_0^\infty f(t)\, e^{-st}\, dt = \int_0^k \frac{1}{k}\, e^{-st}\, dt + \int_k^\infty 0\, e^{-st}\, dt$$

$$= \left[-\frac{1}{sk}\, e^{-st}\right] = -\frac{1}{sk}(e^{-sk} - 1)$$

To obtain the Laplace transform for the unit impulse we find the value of the above in the limit as $k \to 0$. Expanding the exponential term as a series gives:

$$e^{-sk} = 1 - sk + \frac{(-sk)^2}{2!} + \frac{(-sk)^3}{3!} + .$$

and so:

$$F(s) = 1 - \frac{sk}{2!} + \frac{(sk)^2}{3!} + ..$$

In the limit as $k \to 0$ the Laplace transform tends to the value 1. Hence the Laplace transform of the unit impulse is 1.

In determining the Laplace transforms of functions it is not generally necessary to evaluate integrals since tables are available that give the Laplace transforms of commonly occurring functions. Table C.1, at the end of this appendix, lists some of the more common time functions and their Laplace transforms.

Properties of Laplace transforms

The following are some of the basic properties of the Laplace transform.

Linearity property
If two separate time functions, e.g. $f(t)$ and $g(t)$, have Laplace transforms then the transform of the sum of the time functions is the sum of the two separate Laplace transforms:

$$\mathcal{L}\{f(t) + g(t)\} = \mathcal{L}f(t) + \mathcal{L}g(t)$$

Since mulitplying a function by a constant a is the same as adding that function a times then:

$$\mathcal{L}\{af(t)\} = a\mathcal{L}f(t)$$

s-Domain shifting property
This property is used to determine the Laplace transform of functions that have an exponential factor and is sometimes referred to as the *first shifting property*. If $F(s)$ is the Laplace transform of $\{f(t)\}$ then:

$$\mathcal{L}\{e^{at}f(t)\} = F(s-a)$$

Time domain shifting property
If a signal is delayed by a time T then its Laplace transform is multiplied by e^{-sT}. If $F(s)$ is the Laplace transform of $f(t)$ then:

$$\mathcal{L}\{f(t-T)u(t-T)\} = e^{-sT}F(s)$$

This delaying of a signal by a time T is referred to as the *second shift theorem*. The time domain shifting property can be applied to all Laplace transforms. Thus for an impulse which is delayed by a time T, i.e. it occurs at time $t = T$ and not $t = 0$, the Laplace transform 1 is multiplied by e^{-sT} to give $1e^{-sT}$ as the transform for the delayed function.

Periodic functions
For a function $f(t)$ which is a periodic function of period T, the Laplace transform of that function is:

$$\mathcal{L}f(t) = \frac{1}{1 - e^{-sT}} F_1(s)$$

where $F_1(s)$ is the Laplace transform of the function for the first period.

Initial value theorem
The *initial value theorem* can be stated as: if a function of time $f(t)$ has a Laplace transform $F(s)$ then in the limit as the time tends to zero the value of the function is given by:

$$\lim_{t\to 0} f(t) = \lim_{s\to\infty} sF(s)$$

Final value theorem

The *final value theorem* can be stated as: if a function of time $f(t)$ has a Laplace transform $F(s)$ then in the limit as the time tends to infinity the value of the function is given by

$$\lim_{t\to\infty} f(t) = \lim_{s\to 0} sF(s)$$

Derivatives and integrals

The Laplace transform of a derivative of a function $f(t)$ is given by:

$$\mathcal{L}\left\{\frac{d}{dt}f(t)\right\} = sF(s) - f(0)$$

where $f(0)$ is the value of the function when $t = 0$. For a second derivative:

$$\mathcal{L}\left\{\frac{d^2}{dt^2}f(t)\right\} = s^2F(s) - sf(0) - \frac{d}{dt}f(0)$$

where $df(0)/dt$ is the value of the first derivative at $t = 0$.

The Laplace transform of the integral of a function $f(t)$ which has a Laplace transform $F(s)$ is given by:

$$\mathcal{L}\left\{\int_0^t f(t)\,dt\right\} = \frac{1}{s}F(s)$$

The inverse transform

The inverse Laplace transformation is the conversion of a Laplace transform $F(s)$ into a function of time $f(t)$. This operation can be written as:

$$\mathcal{L}^{-1}\{F(s)\} = f(t)$$

The inverse operation can generally be carried out by using the standard forms given in Table C.1. The linearity property of Laplace transforms means that if we have a transform as the sum of two separate terms then we can take the inverse of each separately and the sum of the two inverse transforms is the required inverse transform:

$$\mathcal{L}^{-1}\{aF(s) + bG(s)\} = a\mathcal{L}^{-1}F(s) + b\mathcal{L}^{-1}G(s)$$

However, $F(s)$ is often a ratio of two polynomials and cannot be readily identified with a standard transform in Table C.1. It has then to be converted into simple fraction terms before the standard transforms can be used, this being said to be decomposing it into *partial fractions*. This technique can be used provided the degree of the numerator is less than the degree of the denominator, the degree of a polynomial being the highest power of s in the expression. When the degree of the numerator is equal to or higher than that of the denominator, the denominator must be divided into the numerator until the result is the sum of terms with the remainder fractional term having a numerator of lower degree than the denominator.

Thus when we have partial fractions of a form for which:

1 The denominator contains factors which are only of the form $(s + a)$, $(s + b)$, $(s + c)$, etc. and so the expression is of the form

$$\frac{f(s)}{(s+a)(s+b)(s+c)}$$

The partial fractions are of the form:

$$\frac{A}{(s+a)} + \frac{B}{(s+b)} + \frac{C}{(s+c)}$$

2 There are repeated $(s + a)$ factors in the denominator, i.e. the denominator contains powers of such a factor, and the expression is of the form:

$$\frac{f(s)}{(s+a)^n}$$

The partial fractions are of the form:

$$\frac{A}{(s+a)^1} + \frac{B}{(s+a)^2} + \frac{C}{(s+a)^3} + \cdots + \frac{N}{(s+a)^n}$$

3 The denominator contains quadratic factors and the quadratic does not factorise without imaginary terms and is of the form:

$$\frac{f(s)}{(as^2 + bs + c)(s + d)}$$

The partial fractions are of the form:

$$\frac{As+B}{as^2 + bs + c} + \frac{C}{s+d}$$

The values of the constants A, B, C, etc. in the partial fraction expressions above can be found by either making use of the fact that the equality between the expression and the partial fractions must be true for all values of s or that the coefficients of s^n in the expression must equal those of s^n in the partial fraction expansion. The use of the first method is illustrated by the following example where the partial fractions of:

$$\frac{s+4}{(s+1)(s+2)} \text{ are } \frac{A}{s+1} + \frac{B}{s+2}$$

Then, for the expressions to be equal, we must have:

$$\frac{s+4}{(s+1)(s+2)} = \frac{A(s+2)+B(s+1)}{(s+1)(s+2)}$$

and consequently:

$$s+4 = A(s+2)+B(s+1)$$

This must be true for all values of s. The procedure is to pick values of s that will enable some of the terms involving constants to become zero and so enable other constants to be determined. If we let $s = -2$ then we have:

$$-2+4 = A(-2+2)+B(-2+1)$$

and so $B = -2$. If we now let $s = -1$ then

$$-1+4 = A(-1+2)+B(-1+1)$$

and so $A = 3$. Thus

$$\frac{s+4}{(s+1)(s+2)} = \frac{3}{s+1} - \frac{1}{s+2}$$

The two fractions can then be evaluated using the standard forms given in Table C.1.

Table C.1 *Laplace transforms*

Time function $f(t)$	Laplace transform $F(s)$
1 Unit impulse	1
2 Unit impulse delayed by T	e^{-sT}
3 Unit step	$\dfrac{1}{s}$
4 Unit step delayed by T	$\dfrac{e^{-sT}}{s}$
5 t, a unit ramp	$\dfrac{1}{s^2}$
6 t^n, nth order ramp	$\dfrac{n!}{s^{n+1}}$
7 e^{-at}, exponential decay	$\dfrac{1}{s+a}$
8 $1-e^{-at}$, exponential growth	$\dfrac{a}{s(s+a)}$
9 $t\,e^{-at}$	$\dfrac{1}{(s+a)^2}$
10 $t^n\,e^{-at}$	$\dfrac{n!}{(s+a)^{n+1}}$
11 $t-\dfrac{1-e^{-at}}{a}$	$\dfrac{a}{s^2(s+a)}$
12 $e^{-at}-e^{-bt}$	$\dfrac{b-a}{(s+a)(s+b)}$
13 $(1-at)e^{-at}$	$\dfrac{s}{(s+a)^2}$
14 $1-\dfrac{b}{b-a}e^{-at}+\dfrac{a}{b-a}e^{-bt}$	$\dfrac{ab}{s(s+a)(s+b)}$
15 $\dfrac{e^{-at}}{(b-a)(c-a)}+\dfrac{e^{-bt}}{(c-a)(a-b)}+\dfrac{e^{-ct}}{(a-c)(b-c)}$	$\dfrac{1}{(s+a)(s+b)(s+c)}$
16 $\sin\omega t$, a sine wave	$\dfrac{\omega}{s^2+\omega^2}$
17 $\cos\omega t$, a cosine wave	$\dfrac{s}{s^2+\omega^2}$
18 $e^{-at}\sin\omega t$, a damped sine wave	$\dfrac{\omega}{(s+a)^2+\omega^2}$
19 $e^{-at}\cos\omega t$, a damped cosine wave	$\dfrac{s+a}{(s+a)^2+\omega^2}$

Time function f(t)	*Laplace transform F(s)*
20 $1 - \cos \omega t$	$\dfrac{\omega^2}{s(s^2 + \omega^2)}$
21 $t \cos \omega t$	$\dfrac{s^2 - \omega^2}{(s^2 + \omega^2)^2}$
22 $t \sin \omega t$	$\dfrac{2\omega s}{(s^2 + \omega^2)^2}$
23 $\sin(\omega t + \theta)$	$\dfrac{\omega \cos \theta + s \sin \theta}{s^2 + \omega^2}$
24 $\cos(\omega t + \theta)$	$\dfrac{s \cos \theta - \omega \sin \theta}{s^2 + \omega^2}$
25 $e^{-at} \sin(\omega t + \theta)$	$\dfrac{(s+a) \sin \theta + \omega \cos \theta}{(s+a)^2 + \omega^2}$
26 $e^{-at} \cos(\omega t + \theta)$	$\dfrac{(s+a) \cos \theta - \omega \sin \theta}{(s+a)^2 + \omega^2}$
27 $\dfrac{\omega}{\sqrt{1-\zeta^2}} e^{-\zeta\omega t} \sin \omega \sqrt{1-\zeta^2}\, t$	$\dfrac{\omega^2}{s^2 + 2\zeta\omega s + \omega^2}$
28 $1 - \dfrac{1}{\sqrt{1-\zeta^2}} e^{-\zeta\omega t} \sin\left(\omega \sqrt{1-\zeta^2}\, t + \phi\right),\ \cos\phi = \zeta$	$\dfrac{\omega^2}{s(s^2 + 2\zeta\omega s + \omega^2)}$
29 $\sinh \omega t$	$\dfrac{\omega}{s^2 - \omega^2}$
30 $\cosh \omega t$	$\dfrac{s}{s^2 - \omega^2}$
31 $e^{-at} \sinh \omega t$	$\dfrac{\omega}{(s+a)^2 - \omega^2}$
32 $e^{-at} \cosh \omega t$	$\dfrac{s+a}{(s+a)^2 - \omega^2}$
33 Half-wave rectified sine, period $T = 2\pi/\omega$	$\dfrac{\omega}{(s^2 + \omega^2)(1 - e^{-\pi s/\omega})}$
34 Full-wave rectified sine, period $T = 2\pi/\omega$	$\dfrac{\omega}{(s^2 + \omega^2)} \dfrac{(1 + e^{-\pi s/\omega})}{(1 - e^{-\pi s/\omega})}$
35 Rectangular pulses, period T, amplitude +1 to 0	$\dfrac{1}{s(1 + e^{-sT/2})}$

Note: $f(t) = 0$ for all negative values of t.

Appendix D: Fourier series

This appendix is a brief consideration of the Fourier series and how it can be used to represent both periodic and non-periodic waveforms. For a more detailed consideration the reader is referred to texts such as *Fourier Series* by W. Bolton (Longman 1995), *Fourier Series and Harmonic Analysis* by K.A. Stroud (Stanley Thornes 1984), *Lectures on Fourier Series* by L. Solymar (Oxford University Press 1988).

The Fourier series

The Fourier series is a way to represent a signal which varies with time in terms of sinusoidal signals. The basic sinusoidal signal has the equation:

$$y = A \sin \omega t$$

where y is the value of the function at a time t, the value at time $t = 0$ being zero. A is the amplitude. Such a function is periodic and repeats itself every cycle, i.e. 360° or 2π radians. ω is the angular frequency and is equal to $2\pi f$ with f being the frequency of the waveform. If the frequency of the waveform is doubled then the equation becomes:

$$y = A \sin 2\omega t$$

If the frequency is trebled:

$$y = A \sin 3\omega t$$

If the above functions had not started off with zero values at time $t = 0$ but with a phase angle ϕ then for the single, double and treble frequencies with different amplitudes and phase angles we would have:

$$y = A_1 \sin\left(\omega t + \phi_1\right)$$

$$y = A_2 \sin\left(2\omega t + \phi_2\right)$$

$$y = A_3 \sin\left(3\omega t + \phi_3\right)$$

In 1822, Jean Baptiste Fourier showed that a periodic function, whatever the waveform, can be built up from a series of sinusoidal waves of multiples of a basic frequency. Thus any periodic signal can be represented by a constant signal of size A_0, i.e. a d.c. term, plus sines of multiples of a basic frequency, each possibly having a different amplitude and phase angle.

$$y = A_0 + A_1 \sin\left(\omega t + \phi_1\right) + A_2 \sin\left(2\omega t + \phi_2\right)$$
$$+ A_3 \sin\left(3\omega t + \phi_3\right) + \dots + A_n \sin\left(n\omega t + \phi_n\right)$$

Thus, for example, a square waveform can approximately be built up with the addition of the fundamental waveform, the waveform with an amplitude of one-third that of the fundamental and three times the frequency, the waveform with an amplitude of one-fifth that of the fundamental and five times the frequency, etc., the following being the equation which represents the square waveform:

$$y = A_1 \sin \omega t + \tfrac{1}{3}A_1 \sin 3\omega t + \tfrac{1}{5}A_1 \sin 5\omega t + \tfrac{1}{7}A_1 \sin 7\omega t + \dots$$

Thus the Fourier series enables us to represent a periodic waveform by a spectrum of sinusoidal frequency terms.

Fourier coefficients

The Fourier series is concisely expressed as

$$y = A_0 + \sum_{n=1}^{\infty} A_n \sin\left(n\omega t + \phi_n\right)$$

Since $\sin (A + B) = \sin A \cos B + \cos A \sin B$, we can write:

$$A_n \sin\left(n\omega t + \phi_n\right) = A_n \sin \phi_n \cos n\omega t + A_n \cos \phi_n \sin n\omega t$$

If we represent the non-time varying terms $A_n \sin \phi_n$ and $A_n \cos \phi_n$ by constants a_n and b_n we can then write:

$$A_n \sin\left(n\omega t + \phi_n\right) = a_n \cos n\omega t + b_n \sin n\omega t$$

and the Fourier series equation can be written as:

$$y = \tfrac{1}{2}a_0 + a_1 \cos \omega t + a_2 \cos 2\omega t + \dots + a_n \cos n\omega t$$
$$+ b_1 \sin \omega t + b_2 \sin 2\omega t + \dots + b_n \sin n\omega t$$

with, for convenience, a_0 being taken as $2A_0$ (see later for the reason). Hence we can write the Fourier series equation as:

$$y = \tfrac{1}{2}a_0 + \sum_{n=1}^{\infty} a_n \cos n\omega t + \sum_{n=1}^{\infty} b_n \sin n\omega t$$

The a and b terms are called the *Fourier coefficients*.

Since we have $a_n = A_n \sin \phi_n$ and $b_n = A_n \cos \phi_n$ then:

$$\phi_n = \tan^{-1}\left(\frac{a_n}{b_n}\right)$$

and, since:

$$a_n^2 + b_n^2 = A_n^2 \sin^2 \phi_n + A_n^2 \cos^2 \phi_n = A_n^2$$

we have:

$$A_n = \sqrt{a_n^2 + b_n^2}$$

Establishing the Fourier coefficients

Now consider how we can establish the Fourier coefficients for a waveform. Suppose we have the Fourier series in the form:

$$y = \tfrac{1}{2}a_0 + a_1 \cos \omega t + a_2 \cos 2\omega t + ... + a_n \cos n\omega t$$
$$+ b_1 \sin \omega t + b_2 \sin 2\omega t + ... \, b_n \sin n\omega t$$

Integrate both sides of the equation over one period T of the fundamental. The integral for each cosine and sine term will be the area under the graph of that expression for one cycle and thus is zero. A consequence of this is that the only term which is not zero when we integrate the equation is the integral of the a_0 term. Thus, integrating from some arbitrary time t_0 over one period T gives:

$$\int_{t_0}^{t_0+T} y \, dt = \int_{t_0}^{t_0+T} \tfrac{1}{2}a_0 \, dt = \tfrac{1}{2}a_0 T$$

and so:

$$a_0 = \frac{2}{T} \int_{t_0}^{t_0+T} y \, dt$$

We can obtain the a_1 term by multiplying the equation by $\cos \omega t$ and then integrating over one period. Thus the equation becomes:

$$y \cos \omega t = \tfrac{1}{2}a_0 \cos \omega t + a_1 \cos \omega t \cos \omega t + a_2 \cos \omega t \cos 2\omega t$$
$$+ ... + b_1 \cos \omega t \sin \omega t + b_2 \cos \omega t \sin 2\omega t + ...$$

$$= \tfrac{1}{2}a_0 \cos \omega t + a_1 \cos^2 \omega t + a_2 \cos \omega t \cos 2\omega t$$
$$+ ... + b_1 \cos \omega t \sin \omega t + b_2 \cos \omega t \sin 2\omega t + ...$$

The integration over a period T of all the terms involving $\sin \omega t$ and $\cos \omega t$ will be zero. Thus we are only left with the $\cos^2 \omega t$ term and so:

$$\int_{t_0}^{t_0+T} y \cos \omega t \, dt = \int_{t_0}^{t_0+T} a_1 \cos^2 \omega t \, dt$$

Since $\cos^2 A = \frac{1}{2}(\cos 2A + 1)$ then:

$$\int_{t_0}^{t_0+T} y \cos \omega t \, dt = a_1 \int_{t_0}^{t_0+T} \frac{1}{2}(\cos 2\omega t + 1) \, dt$$

$$= a_1 \int_{t_0}^{t_0+T} \frac{1}{2} \cos 2\omega t \, dt + a_1 \int_{t_0}^{t_0+T} \frac{1}{2} \, dt$$

The integral of the $\cos 2\omega t$ term over a period will be zero. Thus:

$$\int_{t_0}^{t_0+T} y \cos \omega t \, dt = \frac{1}{2} a_1 T$$

and so:

$$a_1 = \frac{2}{T} \int_{t_0}^{t_0+T} y \cos \omega t \, dt$$

In general, multiplying the equation by $\cos n\omega t$ we obtain

$$a_n = \frac{2}{T} \int_{t_0}^{t_0+T} y \cos n\omega t \, dt$$

This equation gives for $n = 0$ the equation given earlier for a_0. This would not have been the case if the first term in the Fourier series had been written as a_0 instead of $a_0/2$. This simplification is thus the reason for the ½ being used with a_0 in the defining equation for the series.

In a similar way, multiplying the equation by $\sin \omega t$ and integrating over a period enables us to obtain the b coefficients. Thus:

$$y \sin \omega t = \frac{1}{2} a_0 \sin \omega t + a_1 \sin \omega t \cos \omega t + a_2 \sin \omega t \cos 2\omega t$$
$$+ \dots + b_1 \sin \omega t \sin \omega t + b_2 \sin \omega t \sin 2\omega t + \dots$$

$$= \frac{1}{2} a_0 \sin \omega t + a_1 \sin \omega t \cos \omega t + a_2 \sin \omega t \cos 2\omega t$$
$$+ \dots + b_1 \sin^2 \omega t + b_2 \sin \omega t \sin 2\omega t + \dots$$

The integration over a period T of all the terms involving $\sin \omega t$ and $\cos \omega t$ will be zero and so:

$$\int_{t_0}^{t_0+T} y \sin \omega t \, dt = \int_{t_0}^{t_0+T} b_1 \sin^2 \omega t \, dt$$

Since $\sin^2 A = \frac{1}{2}(1 - \cos 2A)$ then:

$$\int_{t_0}^{t_0+T} y \sin \omega t \, dt = \int_{t_0}^{t_0+T} b_1 \frac{1}{2}(1 - \cos 2\omega t) \, dt$$

$$= \int_{t_0}^{t_0+T} \frac{1}{2} b_1 \, dt - \int_{t_0}^{t_0+T} b_1 \cos 2\omega t \, dt$$

The integral of the cos $2\omega t$ term over a period will be zero and so:

$$\int_{t_0}^{t_0+T} y \sin \omega t \, dt = \tfrac{1}{2} b_1 T$$

and:

$$b_1 = \frac{2}{T} \int_{t_0}^{t_0+T} y \sin \omega t \, dt$$

In general, multiplying the equation by sin $n\omega t$ gives:

$$b_n = \frac{2}{T} \int_{t_0}^{t_0+T} y \sin n\omega t \, dt$$

Fourier series for common wave forms

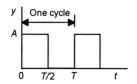

Figure D.1 *Rectangular waveform*

Consider the rectangular wave form shown in Figure D.1. It can be described as:

$y = A$ for $0 \leq t < T/2$
$y = 0$ for $T/2 \leq t < T$, period T

Now consider the determination of the coefficients. Since:

$$a_0 = \frac{2}{T} \int_0^T y \, dt$$

and this is the area under the graph of y against t for the period T, then since this area is $AT/2$, we have $a_0 = A$. To obtain a_n we use:

$$a_n = \frac{2}{T} \int_0^T y \cos n\omega t \, dt$$

Since y has the value A up to $T/2$ and is zero thereafter, we can write the above equation in two parts as:

$$a_n = \frac{2}{T} \int_0^{T/2} A \cos n\omega t \, dt + \frac{2}{T} \int_{T/2}^T 0 \cos n\omega t \, dt$$

The value of the second integral is 0 and so:

$$a_n = \frac{2}{T} \left[\frac{A}{n\omega} \sin n\omega t \right]_0^{T/2}$$

Since $\omega = 2\pi/T$ then we have, for the sine term, sin $2n\pi t/T$. Thus with $t = T/2$ we have sin $n\pi$ which is zero and since sin $0 = 0$, we have $a_n = 0$.
For the b_n terms we can use:

$$b_n = \frac{2}{T} \int_0^T y \sin n\omega t \, dt$$

Since we have $y = A$ from 0 to $T/2$ and then $y = 0$ for the remainder of the period this equation can be written in two parts as:

$$b_n = \frac{2}{T} \int_0^{T/2} A \sin n\omega t \, dt + \frac{2}{T} \int_{T/2}^T 0 \sin n\omega t \, dt$$

The value of the second integral is 0 and so:

$$b_n = \frac{2}{T} \left[-\frac{A}{n\omega} \cos n\omega t \right]_0^{T/2}$$

Since $\omega = 2\pi/T$ then $\cos n\omega t = \cos 2\pi t/T$ and:

$$b_n = \frac{A}{\pi n}(1 - \cos n\pi)$$

Hence:

$$b_1 = \frac{A}{\pi}(1 - \cos \pi) = \frac{2A}{\pi}$$

$$b_2 = \frac{A}{2\pi}(1 - \cos 2\pi) = 0$$

$$b_3 = \frac{A}{3\pi}(1 - \cos 3\pi) = \frac{2A}{3\pi}$$

etc.

Thus the Fourier series for the rectangular waveform can be written as:

$$y = A\left(\frac{1}{2} + \frac{2}{\pi} \sin \omega t + \frac{2}{3\pi} \sin 3\omega t + ... \right)$$

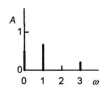

Figure D.2 *Rectangular waveform*

We can thus represent the rectangular waveform by a d.c. component of $0.5A$ plus a sinusoidal component of the basic frequency and amplitude $(2/\pi)A = 0.64A$ plus a sinusoidal component of three times the basic frequency and amplitude $(2/3\pi)A = 0.21A$, etc. Figure D.2 shows the representation of the rectangular waveform as a frequency spectrum.

As a further example, consider the sawtooth waveform shown in Figure D.3. It can be described by:

$$y = At/T \text{ for } 0 \le t < T, \text{ period } T$$

To determine a_0 we can use:

$$a_0 = \frac{2}{T} \int_0^T y \, dt$$

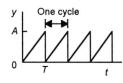

Figure D.3 *Sawtooth waveform*

The integral is the area under the graph of y against t between 0 and time T. This is $AT/2$ and so $a_0 = A$. To obtain a_n we can use:

$$a_n = \frac{2}{T} \int_0^T y \cos n\omega t \, dt$$

Since $\omega = 2\pi/T$ and $y = At/T$ then:

$$a_n = \frac{2}{T} \int_0^T \frac{At}{T} \cos \frac{2\pi nt}{T} \, dt$$

Using integration by parts gives:

$$a_n = \frac{2}{T} \left[\frac{At}{2\pi n} \sin \frac{2\pi nt}{T} + \frac{At}{4\pi^2 n^2} \cos \frac{2\pi nt}{T} \right]_0^T$$

$$= \frac{2A}{T} \left[\frac{T}{4\pi^2 n^2} - \frac{T}{4\pi^2 n^2} \right] = 0$$

The values of a_n are zero for all values other than a_0. The values of b_n can be found by:

$$b_n = \frac{2}{T} \int_0^T y \sin n\omega t \, dt = \frac{2}{T} \int_0^T \frac{At}{T} \sin \frac{2\pi nt}{T} \, dt$$

Integration by parts gives:

$$b_n = \frac{2}{T} \left[-\frac{At}{2\pi n} \cos \frac{2\pi nt}{T} + \frac{At}{4\pi^2 n^2} \sin \frac{2\pi nt}{T} \right]_0^T$$

$$= \frac{2A}{T} \left[-\frac{T}{2\pi n} \right]_0^T = -\frac{A}{\pi n}$$

The Fourier series for the sawtooth waveform is thus:

$$y = \frac{A}{2} - \frac{A}{\pi} \sin \omega t - \frac{A}{2\pi} \sin 2\omega - \frac{A}{3\pi} \sin 3\omega t - \dots$$

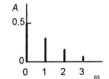

Figure D.4 shows the representation of the sawtooth waveform as a frequency spectrum. The minus signs indicate that those elements are out of phase with the others.

Figure D.4 *Sawtooth waveform*

Circuit analysis

When we have an electrical circuit with a signal which is described by a Fourier series then we can consider each of the Fourier components independently, find the effect of each such source and then determine the overall effect by summing the effects arising from each source.

To illustrate this, consider a circuit in which a capacitor is connected across a sawtooth voltage supply which has an output described by the equation:

$$v = \frac{V}{2} - \frac{V}{\pi} \sin \omega t - \frac{V}{2\pi} \sin 2\omega t - \frac{V}{3\pi} \sin 3\omega t$$

The current is given by $i = C \, dv/dt$. Thus the current due to the first term is zero. Due to the second term the current is $-(CV\omega/\pi) \cos \omega t$, due to the third term $-(2CV\omega/2\pi)$ cos $2\omega t$ and the third term $-(3CV\omega/3\pi)$ cos ωt. Thus the overall current in the circuit is described by:

$$i = -\frac{CV\omega}{\pi} \cos \omega t - \frac{CV\omega}{\pi} \cos 2\omega t - \frac{CV\omega}{\pi} \cos 3\omega t$$

The voltage signal had different amplitudes at each of the frequencies, the current has the same amplitudes. Thus the shape of the current waveform is different from that of the voltage waveform.

Non-periodic waveforms

The Fourier series has been considered for periodic waveforms. We can, however, use it to represent non-periodic waveforms. It may seem odd to consider that we can represent a non-periodic, finite-duration waveform by sinusoids which are periodic and infinite in duration. We can, however, regard the non-periodic waveform as being what a periodic signal would become when the periodic time approaches infinity and so effectively we can only ever see just one cycle of a periodic waveform. This approach leads into a method referred to as the *Fourier transform* where we end up with a non-repetitive waveform transformed into a spectrum of frequencies.

For more information, see texts such as *Fourier Series* by W. Bolton (Longman 1995) for an introduction to the topic or *A Student's Guide to Fourier Transforms* by J.F. James (Cambridge University Press 1995).

Appendix E: Sampled data systems

This appendix is a brief indication of some of the mathematics behind systems based on sampling of an analogue signal. It touches on the z-transform. For more details regarding the z-transform, the reader is referred to such texts as *Laplace and z-transforms* by W. Bolton (Longman 1994) and *An Introduction to the Laplace Transform and the z-transform* by A.C. Grove (Prentice-Hall 1991), and for illustrations of its use to texts on digital signal processing such as *Digital Signal Processing Design* by A. Bateman and W. Yates (Pitman 1988) and *Introductory Digital Signal Processing with Computer Applications* by P.A. Lynn and W. Fuerst (Wiley 1994).

Sampling

Suppose we take samples of some analogue waveform. We can consider the waveform to be represented by a series of sinusoidal waves of different frequencies, see Appendix D on the Fourier series. Suppose we have just a simple sinusoidal signal and use a sampling frequency which is twice the frequency f_a of the analogue signal (Figure E.1). If we take just the samples and reconstitute a waveform from these then we have a triangular wave form. According to the Fourier series, a triangular waveform has frequency components at f, $3f$, $5f$, etc., where f is the frequency of the triangular wave-form. Since this is here the same as that of the analogue signal f_a then the triangular waveform has components at frequencies of f_a, $3f_a$, $5f_a$, etc. Thus to recover the analogue signal at frequency f_a we can use a low-pass filter with a cut-off frequency between f_a and $3f_a$. The signal is thus recoverable.

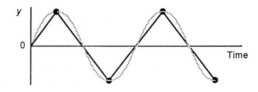

Figure E.1 *Sampling at frequency $2f_a$*

Figure E.2 *Sampling with a frequency less than* $2f_a$

Now suppose we sample at a frequency less than twice the analogue signal frequency (Figure E.2). If we now reconstitute the waveform from the samples then we end up with a triangular waveform which could fit a sinusoidal wave with a frequency which is different from f_a. It is no longer possible to recover the original signal. The higher frequency analogue waveform now has an *alias* in the lower frequency wave inferred from the sample. For example, we might have a 3 kHz sinusoidal signal and sample it at a frequency of 2 kHz and then end up with a 2 kHz alias signal.

A non-sinusoidal analogue signal can be represented by a spectrum of sinusoidal waves as a consequence of applying the Fourier series. Thus if we choose a sampling frequency which results in the sampling frequency not being more than twice every signal frequency, some of the frequency components will produce aliases and so the result will be a distorted waveform. To avoid such distortion *anti-aliasing filters* are used prior to the sampling to filter out those components which would give alias signals.

Mathematical representation of sampling

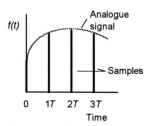

Figure E.3 *Sampled signal*

Consider a continuous analogue signal being sampled every T seconds (Figure E.3). The output is then a series of pulses at regular time intervals, the height of each pulse being a measure of the size of the analogue signal at the times concerned, i.e. at times of 0, $1T$, $2T$, $3T$, etc. The analogue signal is a function of time, i.e. $f(t)$, and the series of pulses resulting from the sampling are the values of that function at 0, $1T$, $2T$, $3T$, etc. Thus we can represent the series of pulses up to some time kT as:

$$f(0), f(1T), f(2T), f(3T), \dots f(kT)$$

Each of these pulses can be considered as a unit impulse multiplied by the value of $f(t)$ at the times concerned. Thus if we write a function $f^*(t)$ to describe the sequence of pulses resulting from the function $f(t)$ being sampled with a sampling period T, we can write:

$$f^*(t) = f(0)(\text{impulse at } 0) + f(1T)(\text{impulse at } 1T) + f(2T)(\text{impulse at } 2T)$$
$$+ f(3T)(\text{impulse at } 3T) + \dots + f(kT)(\text{impulse at } kT)$$

This is a sequence of impulses with the first at 0 time, the second delayed by a time T, the third delayed by a time $2T$, ... the kth delayed by a time kT.

z-transform

The Laplace transform of a unit impulse at time $t = 0$ is 1. The Laplace transform of an impulse at time T is e^{-Ts}, at time $2T$ is e^{-2Ts}, ... at time kT is e^{-kTs}. Thus the Laplace transform of $f^*(t)$ is:

$$F^*(s) = f(0)1 + f(1T)\,e^{-Ts} + f(2T)\,e^{-2Ts} + \ldots f(kT)\,e^{-kTs}$$

We can represent this as:

$$F^*(s) = \sum_{k=0} f(kT)\,e^{-kTs}$$

If we let $z = e^{Ts}$ then we can write the equation as:

$$F^*(s) = \sum_{k=0} f(kT)\,z$$

i.e.

$$F^*(s) = f(0)z^0 + f(1T)z^{-1} + f(2T)z^{-2} + \ldots f(kT)z^{-k}$$

But letting $z = e^{Ts}$ means $s = \frac{1}{T}\ln z$. $F^*(s)$ with this value of s is called the *z-transform* of $f^*(t)$ and denoted by $F(z)$. Thus:

$$F(z) = \sum_{k=0} f(kT)\,z$$

i.e.

$$F(z) = f(0)z^0 + f(1T)z^{-1} + f(2T)z^{-2} + \ldots f(kT)z^{-k}$$

Consider the z-transform for a unit step signal (Figure E.4). Such a signal has $f(t)$ equal to 1 for all values of time greater than 0. Thus:

$$F(z) = 1z^0 + 1z^{-1} + 1z^{-2} + \ldots 1z^{-k}$$

This is a geometric series with a sum of $1/(1 - z^{-1})$ and so:

$$F(z) = \frac{z}{z - 1}$$

Thus by transforming a signal into the z-domain we have an expression which is much easier to handle than a series. Tables exist to enable the z-transforms of other forms of signal to be looked up rather than having to sum the series. Table E.1 at the end of this appendix gives some transforms.

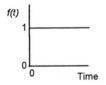

Figure E.4 *Unit step*

Digital transfer function

In the *s*-domain, the transfer function $G(s)$ is the Laplace transform of the output $Y(s)$ divided by the Laplace transform of the input $X(s)$. If we take the *z*-transforms of this equation then:

$$G(z) = \frac{Y(z)}{X(z)}$$

where $G(z)$ is termed the *digital transfer function*, $Y(z)$ is the *z*-transform of the output and $X(z)$ the *z*-transform of the input.

As an illustration, consider the electrical circuit shown in Figure E.5. If we apply Kirchhoff's law to the circuit then:

$$Ri + \frac{1}{C} \int i \, dt = v_{in}$$

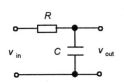

Figure E.5 *Electric circuit*

Taking the Laplace transform gives:

$$RI(s) + \frac{1}{sC}I(s) = V_{in}(s)$$

The output v_{out} is given by $v_{out} = (1/C) \int i \, dt$ and is:

$$V_{out}(s) = \frac{1}{sC}I(s)$$

Hence:

$$G(s) = \frac{V_{out}(s)}{V_{in}(s)} = \frac{1}{RCs+1} = \frac{(1/RC)}{s+(1/RC)}$$

This is of the form $a/(s + a)$ and, using tables, the *z*-transform of such an expression is $az/(z - e^{-aT})$, where T is the sampling period. Thus:

$$G(z) = \frac{(1/RC)z}{z - e^{-T/RC}}$$

Suppose we now have a unit step input to the circuit. Since the *z*-transform for a unit step is $1/(z - 1)$, then the sampled output is given by:

$$V_{out}(z) = G(z) \times V_{in}(z) = \frac{(1/RC)z}{z - e^{-T/RC}} \times \frac{1}{z - 1}$$

Suppose, for simplicity, we had $RC = 1$ and $T = 1$. Then:

$$V_{out}(z) = \frac{z}{(z - 0.37)(z - 1)} = \frac{z}{z^2 - 1.37z + 0.37}$$

We can use long division to write this in terms for which we can easily determine the inverse transformation. Thus:

$$\begin{array}{r} z^{-1} + 1.37z^{-2} + 0.51z^{-3} + \dots \\ z^2 - 1.37z + 0.37 \overline{) z} \\ \underline{z^2 - 1.37 + 0.37z^{-1}} \\ 1.37 - 0.37z^{-1} \\ \underline{1.37 - 1.88z^{-1} + 0.51z^{-2}} \\ 0.51z^{-2} - 0.51z^{-2} \end{array}$$

Thus:

$$Y(z) = z^{-1} + 1.37z^{-2} + 0.51z^{-3} + \dots$$

This is a transform of a sequence of impulses. Thus the output is a pulse of size 1 at $1T$, size 1.37 at $2T$, 0.51 at $3T$, etc.

Digital filters

There are two basic forms of digital filter, the *finite impulse response* (FIR) filter and the *infinite impulse response* (IIR) filter. The filters can be considered to basically act by simple arithmetic being performed on their input which is in the form of a series of impulses.

Figure E.6 shows a block diagram representation of the action of the finite impulse filter. For a particular impulse it multiplies its value by some constant, then adds to it the previous impulse multiplied by some constant, and adds to it the previous previous impulse multiplied by some constant. The z^{-1} term in a block indicates a time delay of one sampling period. The output depends on carrying out arithmetic on a finite number of impulses, hence the term finite impulse response. The pulse transfer function for such a filter is:

$$G(z) = a_0 + a_1 z^{-1} + a_2 z^{-1}$$

where a_0, a_1 and a_2 are the weighting factors used to multiply the impulses before they are summed. Practical filters may require many tens of such coefficients.

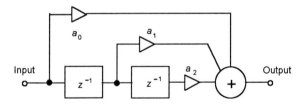

Figure E.6 *Finite impulse response filter*

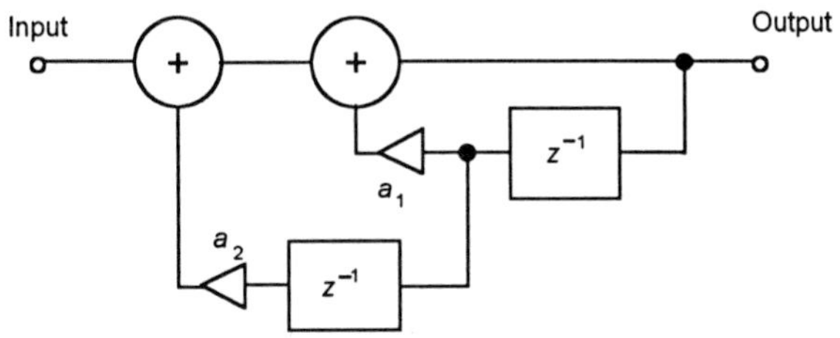

Figure E.7 *Infinite impulse response filter*

Figure E.7 shows a block diagram of a simple form of infinite impulse response filter. Such a filter has feedback paths and has the pulse transfer function of:

$$G(z) = \frac{1}{1 - a_1 z^{-1} - a_2 z^{-2}}$$

It is termed infinite impulse because theoretically the feedback means that the response to an impulse could go on for ever.

The above represents a very simplified consideration of digital filters. For more information the reader is referred to texts such as *Digital Signal Processing Design* by A. Bateman and W. Yates (Pitman 1988) and *Introductory Digital Signal Processing with Computer Applications* by P.A. Lynn and W. Fuerst (Wiley 1994).

Table E.1 *z-transforms*

	$f(t)$	$F(s)$	$F(z)$
1	Unit impulse at $t = 0$	1	1
2	Unit step at $t = 0$	$\dfrac{1}{s}$	$\dfrac{z}{z-1}$
3	Unit ramp t at $t = 0$	$\dfrac{1}{s^2}$	$\dfrac{Tz}{(z-1)^2}$
4	t^2	$\dfrac{2}{s^3}$	$\dfrac{T^2z(z+1)}{(z-1)^3}$
5	t^3	$\dfrac{6}{s^4}$	$\dfrac{T^3z(z^2+4z+1)}{(z-1)^4}$
6	e^{-at}	$\dfrac{1}{s+a}$	$\dfrac{z}{z-e^{-aT}}$
7	$1 - e^{-at}$	$\dfrac{a}{s(s+a)}$	$\dfrac{z(1-e^{-aT})}{(z-1)(z-e^{-aT})}$
8	$t\,e^{-at}$	$\dfrac{1}{(s+a)^2}$	$\dfrac{Tz\,e^{-aT}}{(z-e^{-aT})^2}$
9	$(1-at)\,e^{-at}$	$\dfrac{s}{(s+a)^2}$	$\dfrac{z[z-e^{-aT}(1+aT)]}{(z-e^{-aT})^2}$
10	$e^{-at} - e^{-bt}$	$\dfrac{b-a}{(s+a)(s+b)}$	$\dfrac{z(e^{-aT}-e^{-bT})}{(z-e^{-aT})(z-e^{-bT})}$
11	$\sin\omega t$	$\dfrac{\omega}{s^2+\omega^2}$	$\dfrac{z\sin\omega T}{z^2-2z\cos\omega T+1}$
12	$\cos\omega t$	$\dfrac{s}{s^2+\omega^2}$	$\dfrac{z(z-\cos\omega T)}{z^2-2z\cos\omega T+1}$
13	$e^{-at}\sin\omega t$	$\dfrac{\omega}{(s+a)^2+\omega^2}$	$\dfrac{z\,e^{-aT}\sin\omega T}{z^2-2z\,e^{-aT}\cos\omega T+e^{-2aT}}$
14	$e^{-at}\cos\omega t$	$\dfrac{s+a}{(s+a)^2+\omega^2}$	$\dfrac{z(z-e^{-aT}\cos\omega T)}{z^2-2z\,e^{-aT}\cos\omega T+e^{-2aT}}$
15	$\sinh\omega t$	$\dfrac{\omega}{s^2-\omega^2}$	$\dfrac{z\sinh\omega T}{z^2-2z\cosh\omega T+1}$
16	$\cosh\omega t$	$\dfrac{s}{s^2-\omega^2}$	$\dfrac{z(z-\cosh\omega T)}{z^2-2z\cosh\omega T+1}$

Note: T is the sampling period.

Appendix F: General reading

References have been included in the text to books of specific relevance to particular sections and chapters. The following are some general texts that are of relevance:

1   Barney G.C. *Intelligent Instrumentation* (Prentice-Hall 1988)
    Discussion of microprocessor applications in measurement and control.

2   Bentley J.P. *Principles of Measurement Systems* (Longman 1995)
    Comprehensive coverage of measurement and instrumentation systems.

3   Bolton W. *Instrumentation and Measurement Pocket Book* (Butterworth-Heinemann Newnes 1991, 1996)
    Concise details of a wide range of instrumentation.

4   Doeblin E.W. *Measurement Systems* (McGraw-Hill 1990)
    Detailed and comprehensive coverage of instrumentation systems.

5   Figliola R.S. and Beasley D.E. *Theory and Design for Mechanical Measurements* (Wiley 1995)
    Comprehensive coverage of instrumentation systems.

6   Ed. Noltingk B.E. *Instrumentation Reference Book* (Butterworth-Heinemann 1995)
    A very thick book comprising articles on all the major aspects of instrumentation.

Answers

Chapter 1   1   See section 1.1.1
2   See section 1.1.2
3   See section 1.3.1
4   (a) 1.0 ± 0.02 V, (b) 4.0 ± 0.2 V
5   3.0 cm
6   -3.9%
7   -0.89%
8   1000 ± 2.8 kPa
9   0.5 kN
10   228 s
11   ± 0.05 kN
12   33.0 mV
13   $1200 \pm 12°C$
14   See Section 1.6.1

Chapter 2   1 $\pm 0.25°C$
2   ± 0.05 units
3   (a) 50.2 s, 1.5 s, (b) 3.13 mm, 0.04 mm, (c) 50.2 cm^3, 2.2 cm^3
4   (a) 800 kN, (b) ± 0.07 kN
5   (a) $51.3\ \Omega$, (b) $\pm 0.07\ \Omega$
6   (a) 39.0 kV, (b) ± 0.11 kV
7   18.45 to $20.51\ \Omega$
8   19.33 to 20.27 s
9   $150 \pm 10\ \Omega$
10   $33.3 \pm 7.2\ \Omega$
11   $1.77 \pm 0.09 \times 10^6$ mm^2
12   787 ± 24 kg/m^3
13   ± 0.6 g
14   60 ± 2 km
15   $2.0 \pm 0.4 \times 10^5$ mm^3
16   0.400 ± 0.046 kN/mm^2
17   $\Delta s = \frac{1}{2}\,c \sin A . \Delta b + \frac{1}{2}\,b \sin A . \Delta c + \frac{1}{2}\,bc \cos A . \Delta a$
18   $\Delta N = e^{-\lambda t}\,\Delta N_0 + N_0 \lambda\,e^{-\lambda t}\,\Delta t$
19   $\Delta \eta = \frac{2}{9}g\left(\rho_s - \rho_1\right)\left(\frac{2r}{v}\Delta r + \frac{r^2}{v^2}\Delta v\right)$
20   (a) $\pm 0.60\%$, (b) $\pm 0.35\%$
21   (a) ± 0.0003 g, (b) $\pm 0.000\,22$ g
22   (a) $\pm 1.4\%$, (b) $\pm 0.8\%$
23   $y = 0.66x + 29.15$
24   $e = 6.0W + 5.6$
25   $m = 0.500\theta + 44.86$
26   $d = 9.884W + 0.062$

27 $y = 0.857x + 0$

28 $y = 1.538x + 0.640$

29 $E = 0.307W + 1.567$

30 $y = 3.071\ e^{0.5056x}$

Chapter 3

1 (a) $x = 1 - e^{-t/2}$, (b) $x = 8\ e^{-t/2} + 4t - 8$

2 $x = 5(1 - e^{-t/\tau})$

3 (a) 36.8%, (b) 13.5%

4 12 s

5 (a) $2\dfrac{dx}{dt} + x = 45$, (b) $x = 45 - 25\ e^{-t/2}$

6 $x = 1.5t - 0.75 + 0.75\ e^{-2t}$

7 0.44

8 4.0 s

9 (a) 5 rad/s, (b) 1.25

10 316 rad/s, 6.3 N s/m

11 (a) $\dfrac{1}{As + \rho g/R}$, (b) $\dfrac{1}{LCs^2 + RCs + 1}$

12 Over-damped

13 $0.167 - 0.5\ e^{-2t} + 0.333\ e^{-3t}$

14 (a) Over-damped, (b) under-damped, (c) critically damped

15 $10\ e^{-4t} - 10\ e^{-3t}$

16 $\dfrac{5}{1 + j2\omega}$

17 Ratio $= \dfrac{1}{\sqrt{1 + 25\omega^2}}$, $\phi = -\tan^{-1} 5\omega$

18 $2.6 \sin(3t - 0.21)$, $-12°$ phase angle

19 0 to 0.1 rad/s

20 204°C, 236°C

21 12.0 s

22 $\dfrac{8 \times 10^{-2}}{(1 + 10s)(1 + s)(2 \times 10^{-5}s^2 + 0.01s + 1)}$

Chapter 4

1 0.8

2 250 h, 0.004/h

3 0.998

4 0.08/year

5 (a) 0.002, (b) 0.02, (c) 0.18

6 (a) 0.63, (b) 0.86

7 5

8 160×10^{-8}/hour

9 0.56

10 0.388

11 0.020 25/1000 h

12 0.93

13 0.9999

14 0.40

Chapter 5

1 0.46 V
2 454 Ω
3 117 Ω
4 +1.54°C
5 0.13 Ω
6 0.050 Ω
7 −29.7 pF/mm
8 55.3 pF
9 5.2×10^{-5} Wb
10 $\dfrac{1.26 \times 10^{-5}}{0.12 + 200d}$
11 4.6 V
12 12.227 mV
13 −0.89%
14 0.22 mm
15 313 mV/V
16 Less than or equal to 0.14
17 36.66 kΩ
18 $0.02\dfrac{0.001s}{0.001s + 1}\dfrac{4 \times 10^{10}}{s^2 + 4000s + 4 \times 10^{10}}$

Chapter 6

1 9 s
2 Thermistor non-linear, much higher sensitivity
3 See Section 6.5.3
4 See Section 6.7
5 Metal (tin) oxide sensor
6 Silicon p-n or p-i-n photodiodes

Chapter 7

1 As in the question
2 As in the question
3 0.059 V
4 5.25×10^{-5} V
5 5 %
6 See Section 7.1.4, $R_2 = 19.9\ \Omega$
7 50
8 25 kΩ
9 10 kΩ
10 490 kΩ
11 100 kΩ
12 2.33 kΩ
13 1.59 s
14 0.796 ms
15 2.442×10^{-4}
16 9
17 12 μs
18 1, 2, 4, 8
19 19.6 mV
20 ±0.0122%

Chapter 8
   1   244 µV

   2   About 200 to 400 gives the range and a lower limit greater than the quantisation error.

   3   Loading error less than the quantisation error.

   4   (a) 40 kHz, (b) 8 kHz

   5   See Section 8.1.1

   6   See Figure 8.21 and Section 8.5.3

   7   See Section 8.5.3

   8   (a) (i) none, (ii) yes; (b) (i) yes, (ii) none.

   9   (a) Output when input to X400 and no input to X401. (b) Output when there is no input to X400 or an input to X401. (c) Output when there is an input to X400. However, if the input ceases the output will continue. This is termed a latching program. (d) When there is no input to X400 there is an output from Y430 but not from Y431. When there is an input to X400, Y430 is switched off and Y431 on.

  10   A system similar to that shown in Figure 8.38 but with perhaps a diaphragm pressure gauge monitoring the pressure at the base of the container and the output being to a valve.

Chapter 9
   1   NBA/K_s

   2   (a) 0.22 rad/V, (b) 2.0, (c) 3.6 Hz

   3   Damping factor becomes 0.4, voltage sensitivity decreased by a factor of 0.7.

   4   60 Ω

   5   19 kHz

   6   (a) Frequency modulated, (b) direct mode.

   7   See Section 9.5.2, more data per length of track.

   8   See Section 9.5

Chapter 10
   1   0.09%

   2   0.38 mV

   3   0.108 V/N

   4   1.57 mV

   5   15.0 mm

   6   0.94 Pa

   7   $+630 \times 10^{-6}$ at 39.5°, -430×10^{-6} at 129.5°

   8   $+190 \times 10^{-6}$ at 14.5°, -360×10^{-6} at 104.5°

   9   For example: (a) force balance (see Section 10.1.2), (b) load cell (see Section 10.1.3), (c) strain gauged shaft with transmitted radio signal (see Section 10.2.3), (d) diaphragm gauge (see section 10.3.2), (e) piezoelectric diaphragm gauge (see Section 10.3.2), (f) McLeod or ionisation gauge (see Section 10.3.4), (g) strain gauge rosette (see Section 10.4.1).

Chapter 11   1   0.1%, 10%
2   4.8 V
3   15
4   128
5   For successive numbers Gray gives only one change of digit and so reduces chance of error.
6   9.17 rev/s, 57.6 rad/s
7   D.C. tachogenerator coupled to driveshaft via gearing.
8   0.40g
9   (a) 0.92, (b) 1.0, error of −8% with 30 Hz but negligible with 1 kHz.
10   3.6 Hz
11   Integrator operational amplifier circuit with RC = 1, then inverting operational amplifier with gain of 245.
12   0.051 sin ($2\pi t$ − 50°) + 0.058 sin ($8\pi t$ − 120°)
13   5.9×10^{-4} Ω/g

Chapter 12   1   0.049 m/s
2   0.018 m³/s
3   0.092 m/s
4   0.033 m³/s
5   0.18 sin 200t V, 31.8 Hz
6   As given in the problem.
7   2.45 dm³/s
8   0.0029 m³/s
9   36.1 m/s
10   1.4 m/s
11   0.21 m³/s
12   0.096 m³/s

Chapter 13   1   8.66 MPa
2   +1.5°C
3   See Table 5.2, 200°C
4   9.758 mV
5   ±0.44°C
6   (a) 108.12 Ω, (b) 9.97 mV/°C
7   0.392 Ω/K
8   3.59 V
9   ×10
10   6.25%
11   46.6°C
12   For example: (a) liquid or gas in metal thermometer, (b) total radiation pyrometer, (c) resistance element with Wheatstone bridge and amplified output.

Index